Random Demands, Travel Times, and Water Quality in Deadends

The mission of the Awwa Research Foundation (AwwaRF) is to advance the science of water to improve the quality of life. Funded primarily through annual subscription payments from over 1,000 utilities, consulting firms, and manufacturers in North America and abroad, AwwaRF sponsors research on all aspects of drinking water, including supply and resources, treatment, monitoring and analysis, distribution, management, and health effects.

From its headquarters in Denver, Colorado, the AwwaRF staff directs and supports the efforts of over 700 volunteers, who are the heart of the research program. These volunteers, serving on various boards and committees, use their expertise to select and monitor research studies to benefit the entire drinking water community.

Research findings are disseminated through a number of technology transfer activities, including research reports, conferences, videotape summaries, and periodicals.

Random Demands, Travel Times, and Water Quality in Deadends

Prepared by:
Steven G. Buchberger, **Jason T. Carter**, **Yeongho Lee**, and **Trent G. Schade**
Department of Civil and Environmental Engineering
University of Cincinnati
Cincinnati, OH 45221-0071

Jointly sponsored by:
Awwa Research Foundation
6666 West Quincy Avenue, Denver, CO 80235-3098
and

National Science Foundation
4201 Wilson Boulevard
Arlington, VA 22230

Published by:

Awwa Research Foundation
Advancing the Science of Water®

IWA Publishing

DISCLAIMER

This study was funded by the Awwa Research Foundation (AwwaRF). AwwaRF assumes no responsibility for the content of the research study reported in this publication or for the opinions or statements of fact expressed in the report. The mention of trade names for commercial products does not represent or imply the approval or endorsement of AwwaRF. This report is presented solely for informational purposes.

Published by IWA Publishing, Alliance House, 12 Caxton Street, London SW1H 0QS, UK

Telephone: +44 (0) 20 7654 5500; Fax: +44 (0) 20 7654 5555; Email: publications@iwap.co.uk;
Web: **www.iwapublishing.com**

AwwaRF report number 90963F. Originally published by AwwaRF for its subscribers in 2003.
IWA Publishing version published 2004.

ISBN 978-1-84339-869-1

CONTENTS

* Chapter 8 ends on p. 257 and the References begin on p. 451. There are no pages missing. The Appendices on CD begin on p. 258 and end on p. 450.

TABLES

TABLES (cont'd)

TABLES (cont'd)

TABLES (cont'd)

FIGURES

FIGURES (cont'd)

FIGURES (cont'd)

FIGURES (cont'd)

FIGURES (cont'd)

FOREWORD

The Awwa Research Foundation is a nonprofit corporation that is dedicated to the implementation of a research effort to help utilities respond to regulatory requirements and traditional high priority concerns of the industry. The research agenda is developed through a process of consultations with subscribers and drinking water professionals. Under the umbrella of a Strategic Research Plan, the Research Advisory Council prioritizes the suggested projects based upon current and future needs, applicability, and past work; the recommendations are forwarded to the Board of Trustees for final selection. The foundation also sponsors research projects through the unsolicited proposal process; the Collaborative Research, Research Applications and Tailored Collaboration programs; and various joint research efforts with organizations such as the U.S. Environmental Protection Agency, the U.S. Bureau of Reclamation, and the Association of California Water Agencies.

This publication is a result of one of these sponsored studies, and it is hoped that its findings will be applied in communities throughout the world. The following report serves not only as a means of communicating the results of the water industry's centralized research program but also as a tool to enlist the further support of the nonmember utilities and individuals. Projects are managed closely from their inception to the final report by the foundation's staff and large cadre of volunteers who willingly contribute their time and expertise. The foundation serves a planning and management function and awards contracts to other institutions such as water utilities, universities, and engineering firms. The funding for this research effort comes primarily from the Subscription Program, through which water utilities subscribe to the research program and make an annual payment proportionate to the volume of water they deliver and consultants and manufacturers subscribe based on the annual billings. The program offers a cost-effective and fair method for funding research in the public interest.

A broad spectrum of water supply issues is addressed by the foundation's research agenda: resources, treatment and operations, distribution and storage, water quality and analysis, toxicology, economics, and management. The ultimate purpose of the coordinated effort is to assist water suppliers to provide the highest possible quality of water economically and reliably. The true benefits are realized when the results are implemented at the utility level. The foundation's trustees are pleased to offer this publication as a contribution toward that end.

Edmund G. Archuleta, P.E..
Chair, Board of Trustees
Awwa Research Foundation

James F. Manwaring, P.E.
Executive Director
Awwa Research Foundation

ACKNOWLEDGMENTS

The authors extend sincere appreciation to the Utilities Department of the City of Milford, Ohio for granting us access to many components of their water distribution system, including: production wells, treatment plant, storage tanks, booster pumps, and the dead-end loop. In particular, we express our sincere thanks to City of Milford employees, David Dahlheimer, Pamela Lee, and Ernest Stickler for their courteous cooperation and prompt help in troubleshooting field problems, securing system data and coordinating this study.

We are indebted to the 21 families residing in the study area who allowed us to install and operate flow monitoring devices inside their homes. We are especially grateful to Reverend Thomas and his family who, for over two years, granted weekly access to the water quality monitoring unit installed in the basement of their home at the apex of the dead-end loop.

We greatly appreciate the kind and thoughtful input from Dr. Lewis A. Rossman and Dr. Eugene W. Rice of the National Risk Management Research Laboratory at the US Environmental Protection Agency (USEPA) in Cincinnati, Ohio. Dr. Rossman and Dr. Rice helped conceive important water quality elements of the field study and provided timely advice and valuable insights while serving on the graduate committees for Trent Schade and Jason Carter, respectively. We also wish to recognize the expert technical assistance provided by Janet Blannon and Clifford Johnson who helped analyze hundreds of grab samples at the USEPA Laboratory in Cincinnati. The authors also gratefully acknowledge helpful review comments from Dr. Thomas M. Walski of Haestad Methods, Inc.

Finally, we appreciate the steady encouragement and patience of Dr. Martin Allen, the AwwaRF project officer. We also appreciate the helpful suggestions from members of the Project Advisory Committee (PAC) – Preston Luitweiler of the Philadelphia Suburban Water Company, Van Speight of Houston Public Works and Engineering, Joseph Vencill of the Chicago Water Department.

This project was supported with funds from the Awwa Research Foundation (Project 294), an NYI award from the National Science Foundation (BCS-9257608) and a University Research Council grant from the University of Cincinnati.

ACKNOWLEDGEMENTS

The authors extend sincere appreciation to the Utilities Department of the City of Milford, Ohio, for granting us access to many components of their water distribution system, including production wells, treatment plant, storage tanks, booster pumps, and the dead-end loop. In particular we express our sincere thanks to City of Milford employees, David Dahlheimer, Pamela Lee, and Ernest Stickler for their courteous cooperation and prompt help in troubleshooting field problems, securing system data and coordinating this study.

We are indebted to the 21 families residing in the study area who allowed us to install and operate flow monitoring devices inside their homes. We are especially grateful to Reverend Thomas and his family who, for over two years, granted weekly access to the water quality monitoring unit installed in the basement of their home at the apex of the dead-end loop.

We greatly appreciate the kind and thoughtful input from Dr. Lewis A. Rossman and Dr. Eugene W. Rice of the National Risk Management Research Laboratory of the US Environmental Protection Agency (USEPA) in Cincinnati, Ohio. Dr. Rossman and Dr. Rice helped conceive important water quality elements of the field study and provided timely advice and valuable insights while serving on the graduate committees for Trent Schade and Jason Carter, respectively. We also wish to recognize the expert technical assistance provided by Jane Brannon and Clifford Johnson who helped facilitate [illegible] samples at the USEPA laboratories in Cincinnati. The authors [illegible]

[illegible]

EXECUTIVE SUMMARY

BACKGROUND INFORMATION

Reliable water quality predictions for municipal pipe networks require accurate hydraulic models. This is a major challenge because pipe flows change continuously over time and across space in response to random demands imposed by many consumers dispersed throughout the service area. Issues of fluctuating flows and uncertain quality are especially acute in dead-end zones where water demands are (relatively) sparse, travel times are long and disinfectant residual is low.

Existing water quality models assume ideal plug flow with constant demand patterns. Dead-end mains, however, experience intermittent unsteady bursts of surging water punctuated by random periods of stagnation. It is probably not a coincidence that most network models have had limited success predicting water quality in peripheral zones of the distribution system.

In view of increasingly stringent regulations promulgated under the Safe Drinking Water Act and heightened concerns about the vulnerability of water distribution systems to bio-chemical terrorism, the need for accurate water quality modeling under realistic random flow conditions will grow. As a consequence, future generations of network water quality models will adopt spatial and temporal resolutions having much finer scales than have been customarily used in the past. It is inevitable that such ultra-fine resolution will require a more detailed description of flow conditions throughout the network, especially along dead-end mains and other gateways to consumers.

PROJECT OBJECTIVES

The broad goal of this study is to improve our ability to understand, model and predict the flow and quality of drinking water in municipal distribution systems. This study establishes a long-term continuous record of water consumption and water quality in a residential neighborhood served by a dead-end pipe in a small community near Cincinnati, Ohio. Using this record of water demands and quality, specific objectives of this project are to:

- Characterize statistical properties and patterns of water demand and water quality.
- Test the hypothesis that residential water use is a time dependent Poisson process.
- Derive travel time distributions for water delivered to consumers on the deadend loop.
- Determine if surrogate quality parameters are useful indicators of system integrity.
- Check for significant correlations between system water quality and travel times.
- Evaluate the usefulness of real-time monitoring of distribution system water quality.

STUDY SITE

The project site was located in Milford, Ohio. The city draws water from three shallow wells in an unconfined sand and gravel aquifer near the Little Miami River. Water treatment includes aeration, softening, settling, recarbonation, filtration, fluoridation and disinfection. Finished water with an average free chlorine concentration of 0.5 mg/L and a pH of 9.0 is delivered to about 6,000 people. The study site is a closed dead-end loop that traverses 1,800 feet with a 6-inch cast iron pipe, supplying water to 54 residents at 21 homes in the heart of the distribution system.

EXPERIMENTAL APPROACH

Water Demands

Instrument packs to monitor residential water consumption were installed at secure indoor locations on the service line in each home on the loop. Residential water demands were recorded around-the-clock during a 22-month period from February 1996 through November 1997. Water demands were downloaded nightly via dedicated modems linked to a computer at the University of Cincinnati. Nearly 365,000 residential water demands from 21 homes spanning 214 consecutive days of use (April through October 1997) were converted to **S**ingle **E**quivalent **R**ectangular **P**ulses (SERPs). The SERPs preserved the start time, duration, volume, and average intensity of the original water demand. Each SERP was carefully screened, categorized as an indoor or outdoor use, archived in text files and then stored on compact disc (57 megabytes). The average rate of recovery for water demands approached 92 percent (i.e., data were not measured about 8 percent of the time). This collection of field data represents one of the most extensive high resolution records (1-second interval) of continuous residential water use assembled. It is available for researchers, practitioners, consultants, regulators and others who may be investigating demands on water distribution systems.

Water Quality

Water quality at the entrance and the apex of the dead-end loop was sampled three ways. Continuous monitoring units (CMUs) took measurements of eight physico-chemical parameters at 10 minute intervals during the period March 8 to November 18, 1997 (256 days, corresponding roughly to the period of archived residential water demands). Weekly grab samples taken from the dead-end loop and two other locations in the distribution system were analyzed for HPC, TC, AOC and TOC over the period January 3 to November 18, 1997 (320 days). Two modified Robbins devices, one at the entrance and the other at the apex of the loop, provided coupon samples to quantify biofilm growth from July 13 to November 18, 1997 (130 days). In total, over 245,000 water quality measurements were obtained at the study site from January through November 1997.

RESULTS AND CONCLUSIONS: WATER DEMANDS

Single Equivalent Rectangular Pulses (SERPs)

SERP statistics for indoor water use remain remarkably consistent from month to month at any given household. This implies that indoor SERPs can be adequately characterized with demand monitoring programs that last only a few weeks (rather than a few months). The log normal probability distribution provides a reasonably good description of indoor SERP intensity, duration and volume. Weighted mean values rather than simple arithmetic averages (as tabulated by Mayer *et al* in the 1999 REUW study) should be used when reporting water use statistics. Across the entire dead-end loop, the mean volume of an indoor pulse was 1.71 gallons with a mean duration of 0.77 minutes and a mean intensity of 2.22 gpm. Per capita consumption at the study site was about 55 gpcd, consistent with national norms for indoor residential water use without leaks.

Flow Regimes and Stagnation

Network models assume that *quasi-steady continuous turbulent flow* exists throughout the water distribution system. However, in dead-ends zones, the predominant flow regime is likely to be *random intermittent laminar flow*. A good estimate of the probability of stagnation in any branching pipe supplying a neighborhood of N homes is given by: $P_0 = \exp[-N\overline{\rho}]$, where $\overline{\rho}$ is the average utilization factor for a typical single family home in the service zone. The utilization factor is a measure of how often a home uses water. A typical residence has $\overline{\rho} \approx$ 4 to 5% (i.e., water is used for about one hour each day). Across the dead-end loop study site with N=21 active homes in 1997, stagnant conditions existed about 40 percent of the time, $P_0 = \exp[-21\times0.045] = 39\%$.

Hydraulic Modeling

Current network models that perform extended period simulations are driven by average *total-time* demands applied at the neighborhood scale. To improve water quality predictions, it may be necessary for hydraulic models to adopt event-driven *busy-time* demands at the household scale. This degree of resolution is the only way to mimic the abrupt random shifts among flow regimes (i.e., stagnant, laminar, transition, turbulent) that occur in peripheral zones of the water distribution system. Furthermore, in service zones where laminar flow (and hence Taylor mixing) is common, it may be prudent for water quality models to incorporate dispersive transport along with advective transport.

Poisson Rectangular Pulse (PRP) Hypothesis

A compelling body of diverse evidence supports the hypothesis that residential water use behaves as a time dependent Poisson Rectangular Pulse (PRP) process. Water demand data from the study site show excellent agreement between the observed and predicted statistical characteristics of busy servers and flow rates. A busy server is one using water. The PRP model requires only three parameters, namely: (i) α–the average demand at a *busy* server, (ii) β^2–the variance of the demand at a *busy* server, and (iii) $\rho(t)$–the time dependent utilization factor for a typical single family residence. Values of α and β^2 are quite stable at a given home and can be obtained directly from measured SERPs. The utilization factor displays a characteristic diurnal pattern (i.e., the "rho profile") that reflects the time-dependence of the PRP model. The versatility of the PRP model was demonstrated with several novel applications that demonstrated how to: (i) estimate peak demands in branching pipes, (ii) identify leaks in pipe networks, (iii) simulate residential water demands for network models, and (iv) describe the joint occurrence of indoor and outdoor water demands. Applying concepts pioneered by 20th century communication utilities (dealing with demands on telephone exchanges), the PRP model switches from "phones to faucets" to provide a rigorous, robust framework for describing residential water demands on pipe networks. The probabilistic PRP approach complements existing deterministic network models which have become indispensable tools for solving contemporary problems on design and operation of municipal water distribution systems.

Travel Times

The theoretical mean travel time for an entire dead-end neighborhood is given by the ratio of total pipe volume to average water demand (T=V/Q). However, the mean travel time to individual households in dead-end zones is a strong function of location. Consumers near the entrance have mean travel times that are much less than the neighborhood average; conversely, consumers near the terminal end of a stem or near the apex of a loop have mean travel times that are much longer than the neighborhood average. Travel times to all 21 homes at the study site were estimated with a computer program that tracked the age of small parcels of water moving through the 6-inch supply line to the demand nodes. Based on residential demands from Winter 1996, the overall flow-weighted mean travel time in the neighborhood was about 30 hours. On an individual household basis, the mean travel time ranged from 8 hours for homes near the entrance to almost 96 hours for homes near the apex. The coefficient of variation for travel times stayed relatively constant at about 0.30. This implies that the mean and the standard deviation of travel times grew at nearly the same rate.

RESULTS AND CONCLUSIONS: WATER QUALITY

During the course of this study, the City of Milford consistently produced and delivered a high quality finished water to all residences along the dead-end loop.

Continuous Monitoring Units

Real time monitoring of system water quality as part of a SCADA program is feasible for certain physico-chemical parameters. Over a 9-month period, four probes (pH, conductivity, water temperature and free chlorine) provided reliable readings with data recovery rates approaching 85 percent. The DO probes seemed to function properly, but both expired early in the monitoring exercise. Three other probes (turbidity, nitrate, and ORP) proved unreliable during data collection. Consequently, data were insufficient to test whether or not ORP could be used as a viable surrogate for the concentration of disinfectant residual at the study site. pH was conserved between the entrance and the apex of the loop. Not surprisingly, the mean chlorine concentration dropped from 0.27 mg/L to 0.18 mg/L, a 33 percent loss across the loop. The rate of chlorine loss increased with water temperature. Significant heat exchange occurred at the dead-end loop. During winter, the pipe network lost heat to the ground and hence water temperatures dropped. Conversely, during summer, the network gained heat and temperatures rose. No definitive reproducible correlation between travel time and chlorine residual could be established for *individual water parcels* moving through the loop. Owing to uncertainties in water quality inputs, to questions about residential water demands, and to changing flow regimes at the loop, the precise correspondence between water parcels at the entrance and water parcels at the apex remained elusive.

Grab Samples

A total of 359 bulk fluid grab samples were taken from January to November 1997 at four locations in Milford's water distribution system. All tests for total coliform were negative. HPC values (in low nutrient R2A media) fluctuated dramatically with location and season and were consistently higher in the dead-end region than at the water treatment plant. By contrast, AOC readings (<1 mg-C/L) and TOC (0.4 to 1.5 mg/L) were stable and in the range expected for distribution systems fed by a groundwater source. Spearman rank correlations showed significant associations between several physico-chemical water quality parameters and bulk fluid HPC levels at the apex but not at the entrance to the loop. This finding underscores the recognition that multiple confounding factors affect the viability of bacteria within a municipal water distribution system.

Modified Robbins Devices

Two MRDs provided weekly biofilm samples from the entrance and the apex of the dead-end loop during the period July 13 to November 18, 1997. Samples were collected in triplicate and plated on R2A, PCA and TSA-SB nutrient media. Within five weeks, the biofilm reached a pseudo steady condition inside the MRD with counts approaching 10^6 CFU/cm^2 on CPVC plugs plated in R2A. In contrast, counts obtained from ductile iron plugs plated in R2A did not exceed 10^3 CFU/cm^2 . The biofilm growth potential observed for the low nutrient R2A media was similar at the entrance and the apex of the dead-end loop. However, the biofilm growth potential measured for the high nutrient agars (PCA and TSA-SB) showed a significant drop between the entrance and the apex of the loop. TSA-SB plate counts were not correlated with R2A plate counts. Hence, TSA-SB plate counts could complement counts from conventional R2A media in this distribution system.

CONTINUOUS WATER QUALITY MONITORING

Design and operation of a water quality monitoring program should be tailored to the water utility and must reflect source quality, treatment train, pipe network and applicable regulations. If the program intends to track water quality in real time, then a key element will be continuous monitoring unit(s) with dependable water quality probes. Reliable probes for pH, temperature and conductivity are available; durable monitors for chlorine are available. However, ion specific electrodes are susceptible to membrane fouling and calibration drift when deployed in the field. Staff dedicated to the continuous monitoring program are needed to maintain instrument calibration, perform periodic tune-ups and troubleshoot minor glitches.

The precise placement of a CMU in the field depends on many factors including: ease of access, availability of power, site safety and security, history of consumer complaints, proximity to storage tanks, frequency of stagnation, zones of multi-source blending, and allowance for wastewater drainage (if the CMU requires a bleed line). Hydraulic modeling is advisable to identify service areas in the distribution system that may warrant further attention (e.g., zones with frequent flow reversals, long travel times, unusual demand patterns, etc).

Sampling frequency should reflect operational objectives. If the intent is to quickly detect a water quality excursion, then the sampling frequency should be high. However, if the intent is to characterize finished water at a few locations in the network (say for regulatory reporting), then the sampling frequency can be much lower. Water quality observations can have extremely high autocorrelations. This implies that consecutive measurements contain redundant information. Consequently, the equivalent number of statistically independent observations may be relatively small. Grab samples have the advantage of yielding independent water quality measurements. However, infrequent grab samples may not accurately characterize water quality variability in the network nor identify real-time excursions occurring in the distribution system.

The CMUs will generate large sets of water quality data in a short period of time. Exploratory data analysis and hypothesis testing of these field data can be performed automatically with commercially available statistical software packages. To aid data analysis and interpretation, the statistical program should offer convenient graphical options and report generation capabilities. Like any new initiative, a start-up period will be needed to iron out wrinkles. Once operational, however, real time water quality data from a CMU can provide many benefits to the water utility, including:

- ❖ Dynamic snapshot on the status of finished water at strategic points in the network.
- ❖ Data to estimate travel times and dispersion rates in pipe networks.
- ❖ Data to assist in calibration of hydraulic and water quality models.
- ❖ Early detection and ample warning of water quality excursions in the network.
- ❖ Detection of trends and problems which require maintenance intervention.

CHAPTER 1
INTRODUCTION

PROBLEM STATEMENT

A major challenge facing the water industry is to better understand and predict how the quality of treated water changes within the distribution system. Modeling water quality between the points of treatment and consumption in a municipal network is confounded by many variables. In a typical distribution system, complicating factors can include: multi-source blending, storage tank mixing, uncertain hydraulic conditions, complex reaction kinetics, and a host of biological, physical and chemical interactions among constituents in the bulk fluid and along the pipe wall.

Reliable water quality predictions require accurate hydraulic models. There is growing awareness that flow conditions in a municipal pipe network are highly variable. Flow rates change continuously over time and across space in response to random demands imposed by many consumers dispersed throughout the service area. In essence, the water distribution system is forced by a nonstationary stochastic field.

From the standpoint of *water supply reliability*, variability in the rate of flow generally is not a serious concern. Although velocities and pressures may fluctuate over time throughout the network, operating conditions usually satisfy performance standards (e.g., the water supply is generally adequate to meet fire flows, even during peak demands). However, from the perspective of *water quality reliability*, the inherent uncertainty in the rate (and direction) of flow is an important consideration. Water quality modeling requires knowledge of the entire travel history of each water parcel moving through the pipe network. Transit times, travel paths and, hence, water quality are inextricably linked to the stochastic flow patterns of the hydraulic network.

The intertwined issues of fluctuating flows and uncertain quality are especially acute in dead-end regions of the pipe network. Here water demands from individual consumers measurably influence the pattern of flow. Dead-end mains experience intermittent bursts of surging water punctuated by random periods of stagnation. Long travel times through dead-end mains often lead to low disinfectant residual and an increased risk of microbial regrowth in the water distribution system.

Existing water quality models are based on ideal plug flow conditions resulting from a fixed demand pattern. This is in sharp contrast to the unpredictable transient hydraulic conditions that characterize dead-end mains. It is probably not a coincidence, therefore, that most network models have had limited success predicting the quality of water in peripheral zones of a municipal drinking water distribution system.

In view of increasingly stringent regulations promulgated under the Safe Drinking Water Act (SDWA) and heightened concerns about the vulnerability of water distribution systems to threats from bio-chemical terrorism, the need for accurate water quality modeling under realistic random flow conditions will grow. As a consequence, future generations of network water quality models will adopt spatial and temporal resolutions having much finer scales than have been customarily used in the past.

Already there is movement in this direction. Coupled with recent advances in geographic information systems (GIS), the "all pipes" network model is emerging as the new industry standard. It is inevitable that such ultra-fine resolution will require a more detailed description of flow conditions throughout the network, especially along dead-end mains. This, in turn, can be achieved only if there is a more accurate picture of residential water use than is now available.

PROJECT OBJECTIVES

The broad goal of this study is to improve our ability to understand, model and predict the flow and quality of drinking water in municipal distribution systems. This study establishes a long-term continuous record of water consumption and water quality along a residential dead-end loop in a municipal distribution system operated by a small community near Cincinnati, Ohio. Using this record of water demands and quality, specific objectives of this project are to:

Objective 1: Characterize statistical properties and patterns of water demand and water quality.

Objective 2: Test the hypothesis that residential water use is a time dependent Poisson process.

Objective 3: Derive travel time distributions for water delivered to consumers on the deadend loop.

Objective 4: Determine if surrogate quality parameters are useful indicators of system status.

Objective 5: Evaluate the usefulness of real-time monitoring of distribution system water quality.

Objective 6: Check for significant correlations between system water quality and travel times.

CHAPTER 2
BACKGROUND INFORMATION

Field measurements, laboratory tests, and data analyses were conducted on water demands and water quality parameters sampled over a 22-month period along a dead-end main in a residential zone of a municipal distribution system. Demand data were needed to test the hypothesis that the random intermittent behavior of residential indoor water use follows a time dependent Poisson Rectangular Pulse (PRP) process. The water quality data were needed to explore associations between water travel time and water quality measurements. This chapter begins with a brief overview of the links between residential water demands and the PRP water use hypothesis and then ends with a short discussion of the water quality constituents monitored in the distribution system.

MUNICIPAL WATER USE

Residential Demands

Residential demand is the largest component of municipal water use (Flack, 1982) often exceeding 75 percent of total system consumption. Owing to different demand characteristics, residential water use is usually segregated into domestic (indoor) and landscape (outdoor) demands. Indoor water use corresponds closely to winter residential demands and includes water used for drinking, cooking, bathing, washing, cleaning and disposal of wastes. In the classic study by Linaweaver et al., (1966) the duration of indoor water use was found to average less than three minutes and seldom exceed 25 minutes.

A rough breakdown of average residential indoor water use according to various household fixtures and appliances is given in Table 2.1. Most indoor water use occurs in the bathroom where toilets, showers and baths account for about 60 percent of the average demand. Neglecting leaks, water use at a typical single family residence is intermittent with inflows occurring only a small percentage of the time. Outdoor water use, driven primarily by landscape irrigation, exhibits a strong seasonal pattern with highly variable peak demands (Litke and Kauffman, 1993).

Table 2.1. Typical Residential Indoor Water Use [a]

Fixture/Server Category	Percent of Total	Daily Vol (gal/day)	Demand Rate	Daily Count (uses/day)	Flow Duration
Toilet	28	80	2 – 6 gal/use	20	32 min
Laundry	22	65	25 gal/use	2.5	21 min
Shower	20	60	2 – 4 gpm	3	20 min
Faucets	12	35	up to 4 gpm	40	17 min
Baths	10	30	30 gal/use	1	5 min
Dishwasher	3	8	8 gal/use	1	5 min
Leaks	5	12	0.008 gpm	continuous	24 hr
Total	100	290 [b]	0.20 gpm	68 [c]	100 min [d]

(a) adopted from Maddaus (1987a).
(b) based on single-family residence of four with 72.5 gallons per capita-day average water use.
(c) excludes water leak category.
(d) average *daily* utilization factor, ρ=(100 min per day)/(1440 min per day)≈0.07.
(e) average demand at busy server, α=(278 gal/day)/(100 min/day) = 2.78 gpm (excludes leaks).

Poisson Rectangular Pulse (PRP) Premise

Residential water use can be analyzed with methods from queuing theory first developed by Erlang (1917-18) in his pioneering study of telephone demands for the communications industry. In the context of municipal distribution systems, the basic premise is that residential water demands at single family homes behave as a time dependent Poisson Rectangular Pulse (PRP) process. In the jargon of queueing theory, home occupants are customers while water fixtures and appliances are servers[(1)]. When occupants draw water from the distribution network, the home is considered "busy"; otherwise the home is considered "idle".

The PRP model for residential water use has a very interesting and useful property, namely, it is *scalable*. The PRP concept can be applied to a water fixture, to an individual home, to small and

(1) Historical note: In a National Bureau of Standards report, R.B. Hunter (1940) proposed a related statistical method for estimating demands on residential plumbing systems. Hunter's scheme used combinatorial methods and the binomial probability law. His approach, however, requires the use of arbitrary and empirical loading factors, called "fixture units", which render the NBS method cumbersome to apply (for further details see Chapter 4 in AWWA M22, 1975).

large residential neighborhoods, or to the entire distribution system. It is important, therefore, to keep in mind the "point of reference" (i.e., fixture, residence, neighborhood, network) when applying the PRP concept to a real distribution system. This leads to the notion of defining a hierarchy for the pipe network. The hierarchy is analogous to various ordering systems used to classify streams in a drainage network[(2)]. The hierarchy offers a convenient way to represent the point of reference in a branching network.

Irrespective of scale or hierarchy, the PRP model requires only three parameters (ρ, α, β^2) to characterize residential indoor water use. When applied to a typical individual home, these model parameters can be interpreted in the following way: (1) ρ is a dimensionless "utilization factor" representing the percentage of time the home uses water; (2) α is the average rate of water use at a *busy* fixture in the home and (3) β^2 is the variance of the rate of water use at a *busy* fixture in the home. For example, data from Table 2.1 suggest $\rho \approx 0.07$ and $\alpha \approx 2.78$ gpm at a typical single family residence of four [see notes (d) and (e) beneath the Table]. The utilization factor from Table 2.1 represents an *average* daily value; ρ will vary during the course of a day and this variability is the primary source of time dependence in the proposed PRP model.

Under the PRP hypothesis, the frequency of residential water use is assumed to follow a Poisson arrival process with a time dependent rate parameter. When a water use occurs, it is represented as a single rectangular pulse of random duration and random but steady intensity as illustrated in Figure 2.1. Buchberger and Wells (1996) found that over 80 percent of indoor residential water demands occur as single pulses. They demonstrated that more complex demand patterns are easily converted to an equivalent single pulse. By virtue of the Poisson assumption, it is unlikely that more than one pulse will start at the same instant. Owing to the finite duration of each water pulse, however, it is possible that two or more pulses with different starting times will overlap for a limited period. When this occurs, the total water use at the residence is the sum of the joint intensities from the coincident pulses.

[(2)] See any geomorphology text for a discussion of Strahler's stream ordering system. More details on this idea are presented later in this chapter.

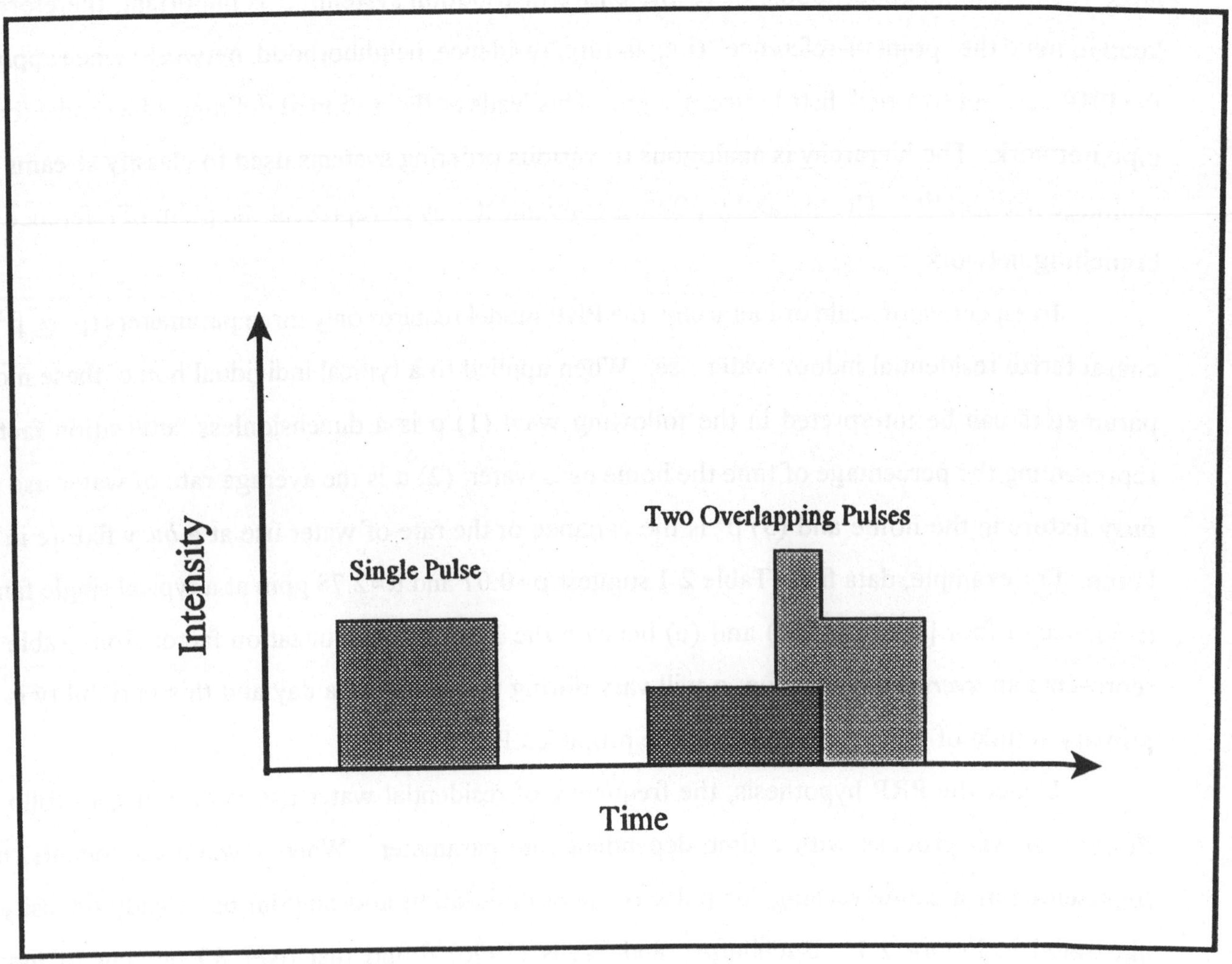

Figure 2.1 Definition sketch for Poisson rectangular pulse concept.

It can be shown that the PRP premise leads to explicit expressions for the theoretical moments (i.e., mean and variance) and probability distributions of busy fixtures, busy homes, flow rates, travel times and disinfectant concentrations in dead-end pipes of municipal water distribution systems (Buchberger and Wu 1995, Wu 1996, Buchberger and Schade 1997, Buchberger *et al.*, 1998, Buchberger and Lee 1999). Owing to its simplicity and versatility, the PRP approach offers a promising new way to model the temporal and spatial variability of the flow regime and the water quality in dead-end zones of the distribution system.

A suitable data base, however, is not yet available to test the PRP hypothesis. Apart from the field work of Orndorff (1966), Anderson and Watson (1967), and more recently by Buchberger and Wells (1996) and the REUW study (Mayer *et al*, 1999), there are few reported attempts to monitor the intensity, duration and frequency of residential water use on a continuous basis. A key objective of this study, therefore, is to collect around-the-clock measurements of domestic water use and then check various statistical properties of the observed data against theoretical predictions from the PRP hypothesis for residential water demands.

Dead-End Mains

Dead-end mains are pipes that have only one connection to the looped portion of the water distribution system. As shown in Figure 2.2, there are two basic types of dead-end mains: (1) stems and (2) loops. Water moves through a dead-end stem in one direction only, from the entrance to the point of withdrawal, at rates dictated by local downstream use. Only under extraordinary circumstances (e.g., upstream main break or upstream fire demand) is it possible for significant amounts of water to flow "backwards" to the entrance of a dead-end main.

A dead-end loop can be viewed as a dead-end stem that closes on itself. In contrast to a dead-end stem, flow in a dead-end loop is bi-directional and, therefore, slightly more complicated. The important feature, however, is that the dead-end loop is appended only to a dead-end stem and does not connect at any other point to the main portion of the water supply network.

Dead-end mains are surprisingly common. A survey of 14 randomly selected service zones in Cincinnati's water distribution system (see Table 2.2) revealed an average of 54 deadends per zone accounting for about 23 percent of the total pipe length in the district. The average length of a dead-end stem in the Cincinnati survey was about 800 feet. Dead-end mains are most prevalent in the peripheral suburban regions of the service zone; hence, they reach a disproportionately high percentage of the residential consumer base. Owing to chronic water quality problems with taste, odor, color and turbidity, many utilities periodically purge dead-end zones by flushing out fire hydrants (*Antoun* et al 1999).

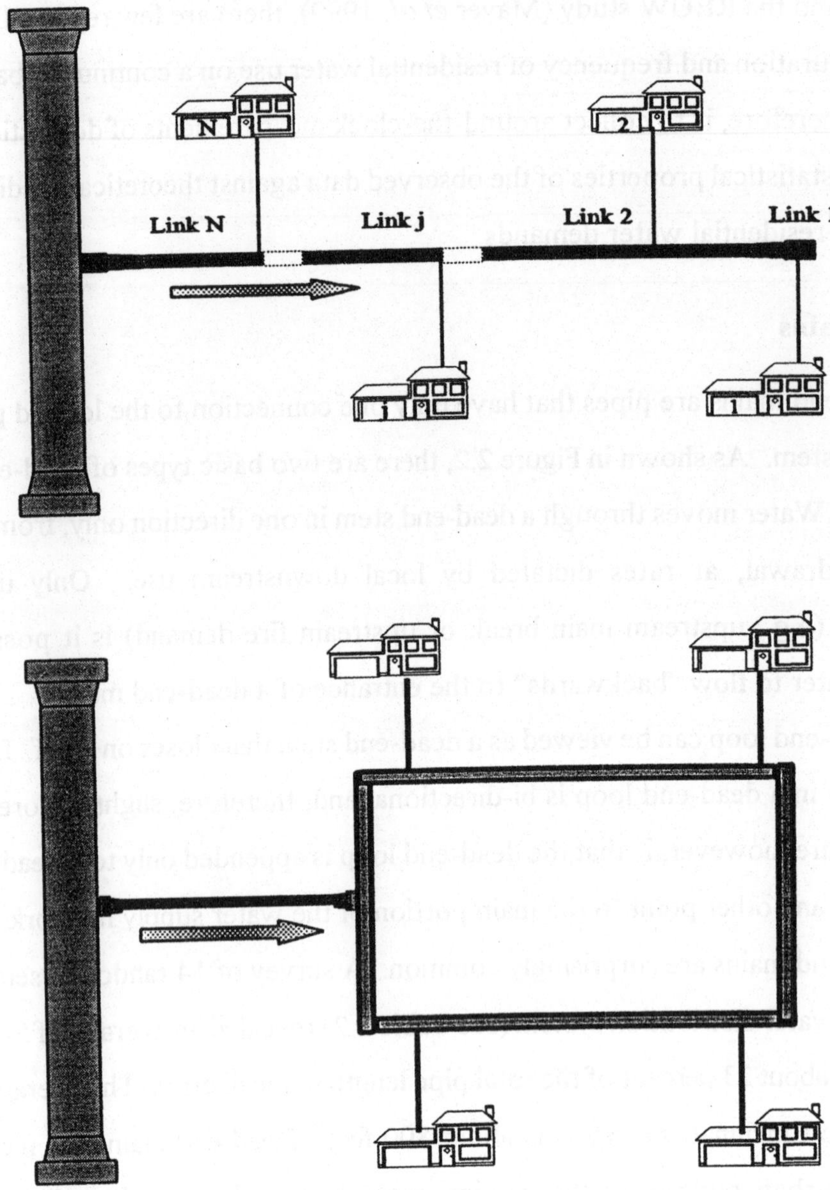

Figure 2.2 Definition sketch for dead-end stem (top) and dead-end loop (bottom).

Table 2.2

Dead-end Mains in a Random Survey of 14 Service Districts of Cincinnati's Water Distribution System.

Water District Number	Total Pipe Length (feet)	Number of Dead-end Stems	Total Dead-end Length (feet)	Average Deadend Length (feet)	Dead-end Mains as a Percent of District
3	218,600	94	69,200	736	31.7
5	279,800	98	94,000	959	33.6
21	86,100	23	15,300	665	17.8
31	96,100	1	530	530	0.6
33	138,900	32	22,200	694	16.0
48	284,100	92	77,600	843	27.3
52	181,100	58	48,000	828	26.5
57	132,500	32	23,200	725	17.5
66	311,000	144	105,600	733	34.0
71	175,300	59	38,000	644	21.7
78	87,700	15	4,200	280	4.8
93	139,400	28	21,600	771	15.5
94	452,000	56	53,300	952	11.8
95	53,900	23	22,700	987	42.1
Totals	2,636,500	755	595,600		
Average of 14 Districts	188,321	54	42,543	789	22.6

PRPs on Dead-end Mains

Figure 2.3 is an idealized sketch of residential water use on a dead-end mainline. It illustrates many of the concepts introduced in this study. On the left hand side of Figure 2.3, a dead-end stem branches from the mainline to serve three homes (labeled A, B, C). Acting independently, each home draws pulses of water from the dead-end main through their respective service line. On the right hand side are four demand hydrographs showing water use plotted against time (say, gpm versus minute for a one-hour period). The top three hydrographs correspond to flow through the service lines (i.e., "Point-of-Entry") to homes A, B and C. The different colors in each demand hydrograph indicate water drawn by different fixtures (i.e., "Point-of-Use") inside the home. These fixtures typically include toilets, showers, baths, dishwashers, clothes washers, and faucets.

Indoor residential water demands are intermittent and occasionally coincident. Overlapping red and yellow pulses indicate a joint water use at home A. Coincident water use is also evident at homes B and C. The string of digits above each demand hydrograph shows the number of busy fixtures in each home. Most of the time there is no demand for water, so there are no busy fixtures. When a demand occurs, usually only one fixture is involved. Brief periods of coincident water use show two busy fixtures in the same home. The horizontal black bars beneath the hydrograph identify busy periods in the home, that is, times when water is flowing through the service line. The intervals between the busy times correspond to periods when the water is stagnant.

The bottom hydrograph represents the total flow into the dead-end main resulting from the summation of instantaneous demands at homes A, B, and C. The number of busy fixtures and busy homes are identified along with the busy periods. Demands, fixtures and homes are cumulative along the deadend. A stagnant main implies that all fixtures and all homes are idle. Hence, the probability of zero busy homes must be the same as the probability of zero busy fixtures along the dead-end main. Note, however, that the number of busy fixtures can exceed the number of busy homes.

This example demonstrates that simple PRP sequences in residential service lines can generate complex intermittent random demands along the dead-end supply main. Complicated flow patterns often confront network modelers and confound water quality models. These flows cannot be analyzed with conventional deterministic methods. However, key statistical properties of random flow patterns can be readily described with the PRP concept. Subsequent chapters will demonstrate how to exploit these properties to improve the performance of network water quality models.

Table 2.4 Applicability of Results from Poisson Rectangular Pulse Model to Municipal Water Distribution Systems.

Feature of Distribution System	Hierarchy Level	Apply to Node or Link	Network Geometry	Flow Direction
busy fixtures	H-4	node	loops and branches	bi-directional
busy homes	H-4	node	loops and branches	bi-directional
residential demands	H-4	node	loops and branches	bi-directional
flow rate	H-3	link	branching pipes only	one way only
travel time	H-3	link	branching pipes only	one way only
disinfectant residual	H-3	link	branching pipes only	one way only

(a) The PRP model provides the probability distribution or the statistical moments (or both) for the system features listed in column 1.

(b) The PRP model can provide information about busy servers, busy homes and residential water demands for pipe nodes anywhere in the network.

(c) The PRP model can provide information for pipe links only when the direction of flow through the link is known at all times (e.g., branching pipes).

(d) A given hierarchy level subsumes all downstream levels (e.g., H-4 includes H-1 through H-3).

MUNICIPAL WATER QUALITY

Several water quality parameters were monitored at various locations in the municipal distribution system. A brief description of the physical, chemical and biological parameters is given here.

Physical Parameters

Temperature–Temperature is a widely used parameter in water quality studies. As temperature increases, dissolved oxygen decreases, and chemical and biochemical reaction rates increase, including disinfection byproducts formation, chlorine decay, pipe corrosion and microbiological growth in distribution systems (APHA et al, 1994; Tchobanoglous and Schroeder, 1985). It is not feasible to control temperature in distribution systems.

Turbidity–Turbidity measures water clarity based on light scattering due to the presence of fine suspended matter such as silt, clay, algae, bacteria and other microorganisms. Water treatment processes focus on removal of fine particulates because this material may harbor pathogens or toxic substances, can consume chlorine and reduce disinfection efficiency (APHA et al, 1994; Montgomery, 1985). To ensure that microorganisms are adequately removed, water utilities are advised to maintain turbidity below 0.1 NTU at the effluent of sand filters (Montgomery, 1985).

Conductivity–Conductivity is the capacity of a solution to conduct an electric current. The conductivity of potable water usually ranges from 50 to 1500 μS/cm. Conductivity measurements account for temperature, concentration, mobility and valence of dissolved inorganic ions such as: Ca^{2+}, Mg^{2+}, Na^{+}, K^{+}, HCO_3^{-}, SO_4^{2-} and Cl^{-} (APHA et al, 1994; De Zuane, 1990).

Oxidation reduction potential (ORP)–ORP is a electrometric measurement of redox reactions in water through potentiometric determination of electron activity among chemical constituents in water. ORP measurement is straightforward, but it is sensitive to pH and temperature variation, irreversible reactions, electrode poisoning, the presence of multiple redox couples, small exchange currents and inert redox couples (APHA et al, 1994; Montgomery, 1985). ORP may serve as a surrogate parameter for residual chlorine in the distribution systems since is provides the overall oxidation intensity of hypochlorous acid and hypochlorite ion. However, the reaction of the probe is slow and not durable in the field (Kim, 1993; Stumm and Morgan, 1981).

Chemical Parameters

pH–pH is the master variable in water chemistry. Although pH does not have a direct effect on human health, many important bio-chemical reactions that occur in water distribution systems (e.g., disinfection, precipitation, biofilm formation, and corrosion control) depend on pH. The National Secondary Drinking Water Standards set the range for pH between 6.5 and 8.5 for aesthetic reasons, such as impaired water taste or discolored water (APHA *et al*, 1994).

Dissolved Oxygen (DO)–DO is the most widely used water quality parameter because oxygen maintains aquatic life and the aesthetic quality of water. DO is not toxic, so no standard has ever been issued for drinking water (APHA *et al*, 1994; De Zuane, 1990). In distribution systems, the rate of pipe corrosion increases with DO levels since oxygen accepts electrons at the cathode (Montgomery, 1985; Tchobanoglous and Schroeder, 1985).

Nitrate–Nitrate is the most highly oxidized form of nitrogen (+5) and is very soluble. In high concentrations it can cause methemoglobinemia in babies. A major source of nitrate is nitrogen fertilizer. Conventional water treatment processes can not remove nitrate (APHA *et al*, 1994; Montgomery, 1985; De Zuane, 1990). The MCL of nitrate is 10 mg/L as N.

Free chlorine–Chlorine is used to deactivate potential pathogens in water distribution systems. Use of chlorine can intensify the taste and odor of phenol and organic compounds and generate potential chlorinated carcinogens—trihalomethanes and haloacetic acids. In water, "free chlorine" appears in two forms—hypochlorous acid and hypochlorite ion—through hydrolysis. The relative proportion of each species of free chlorine forms depends on pH and temperature. A detectable free chlorine residual of at least 0.2 mg/L is required in municipal water distribution systems fed by groundwater in Ohio (Ohio Administrative Code, 2002).

Total organic carbon (TOC) and assimilable organic carbon (AOC)–TOC represents the fraction of total carbon (TC) that is organically bound. Principal sources of TOC in water are living and decaying vegetation. TOC is closely related to occurrence of coliform bacteria in distribution systems. (LeChevallier *et al*, 1991, 1993, 1996; APHA *et al*, 1994; Montgomery, 1985). Not all organic compounds are susceptible to rapid microbial decomposition. The fraction of organic compounds that provides energy and carbon for bacterial growth is called labile dissolved organic carbon (AOC). Assuming that carbon is the limiting element, a bioassay is often used as an indirect method for quantifying AOC.

Biological Parameters

Biofilm–Biofilm refers to various microorganisms and the gelatinous matrix of extracellular organic polymers (exopolysaccharide chains, or EPS) that attach to the inside wall of water mains. Owing to limited nutrient availability, biofilms in drinking water systems form sporadically but can be found at some time in all distribution systems (van der Wende *et al.*, 1989). Biofilms are most likely to establish in areas where the pipe wall is interrupted (e.g. valves, elbows, unions, and hydrants). Biofilms have also formed on straight pipes provided that flow conditions do not exceed the velocity necessary to shear microbial layers from the pipe wall (Donlan and Pipes, 1988; Block, 1992; LeChevallier *et al.*, 1987).

Total Coliform–Coliforms are a group of gram-negative bacteria that express similar characteristics to pathogenic bacteria. *Escherichia coli* is a member of the coliform group. Coliforms were chosen as a pathogen surrogate because they are relatively easy to measure in addition to exhibiting a similar resistence to disinfection as many pathogenic microorganisms (Montgomery *et al.*, 1985). The USEPA requires all drinking water utilities to monitor for total coliforms at multiple points in the distribution system (total number of sample sites is dependent on population served). The test is used to detect inadequate treatment, compromised pipeline, or fecal contamination. Each of these events constitute a risk to the public health.

Heterotrophic Bacteria–Heterotrophic bacteria refers to a class of single-cell bacteria that require organic carbon for both respiration and cell synthesis. Heterotrophic bacteria are monitored in drinking water distribution systems to represent the general productivity of the finished water. The heterotrophic count provides a general indication of the degree of disinfection and can detect changes in water quality both in storage and distribution (Carter *et al.*, 2000). Heterotrophic bacteria can be measured in lieu of chlorine monitoring in the distribution system. Sites with an HPC count under 500 CFU/mL are considered equal to a measurable chlorine residual (Federal Register, 1989).

CHAPTER 3
METHODS AND MATERIALS

STUDY SITE

The project was located in Milford, Ohio, about 20 miles east of Cincinnati (see Figure 3.1). Milford draws its water from three shallow wells placed in an unconfined sand and gravel aquifer near the Little Miami River. Water treatment includes aeration, softening, settling, recarbonation, filtration, fluoridation and disinfection. Finished water has an average free chlorine concentration of 0.5 mg/L with pH of about 9.0. The treatment plant serves a population of about 6,000 people with more than 1,500 residential water accounts. In 1994, the average city-wide water use was 0.65 MGD with a maximum daily demand approaching 1.1 MGD.

The oldest sections of Milford's water distribution system date back to 1903. At the time of the study, the distribution system comprised nearly 25 miles of pipeline with about 35 percent classified as dead-end mains. Pipe diameters range from 4 to 10 inches and are composed of cast iron, ductile iron and PVC material. Copper service lines extend from the mainlines to homes and other consumers. The water distribution system includes two storage tanks (0.64 MG Wallace Street tank and 0.15 MG Edgecomb tank) which "float" on the network.

The study site is a dead-end loop in Milford's water distribution system (Figure 3.2). The loop traverses nearly 1,800 feet with 6 inch cast iron pipe and supplies water to 21 homes ranging from vintage 1960 ranch styles to new construction. The dead-end loop is fed by a short (~60 foot) one-way stub that branches from an 8 inch cast iron mainline. Water that enters the dead-end loop remains there until extracted by a consumer at one of the 21 homes connected to the supply loop. The mainline receives water directly from the Milford water treatment plant and from the Wallace Street storage tank. The loop is hydraulically isolated from the Edgecomb tank by flow control valves. A booster pump station is located on the 8 inch mainline about 120 feet downstream from the entrance to the dead-end loop (Figure 3.2). The booster station allowed direct access to the mainline for water quality samples.

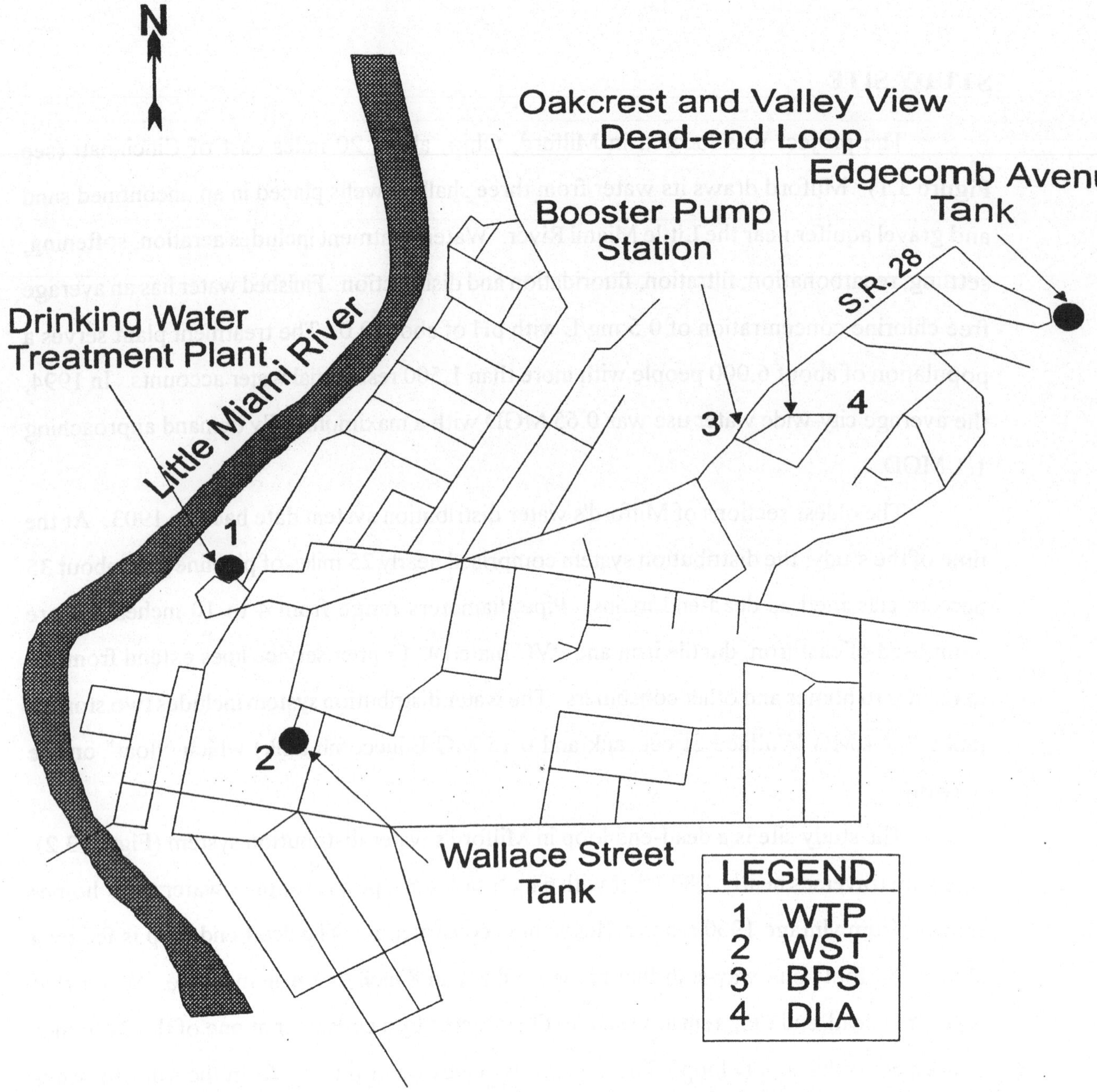

(Approximate scale: 1 inch = 0.25 miles)

Figure 3.1 Milford water distribution system.

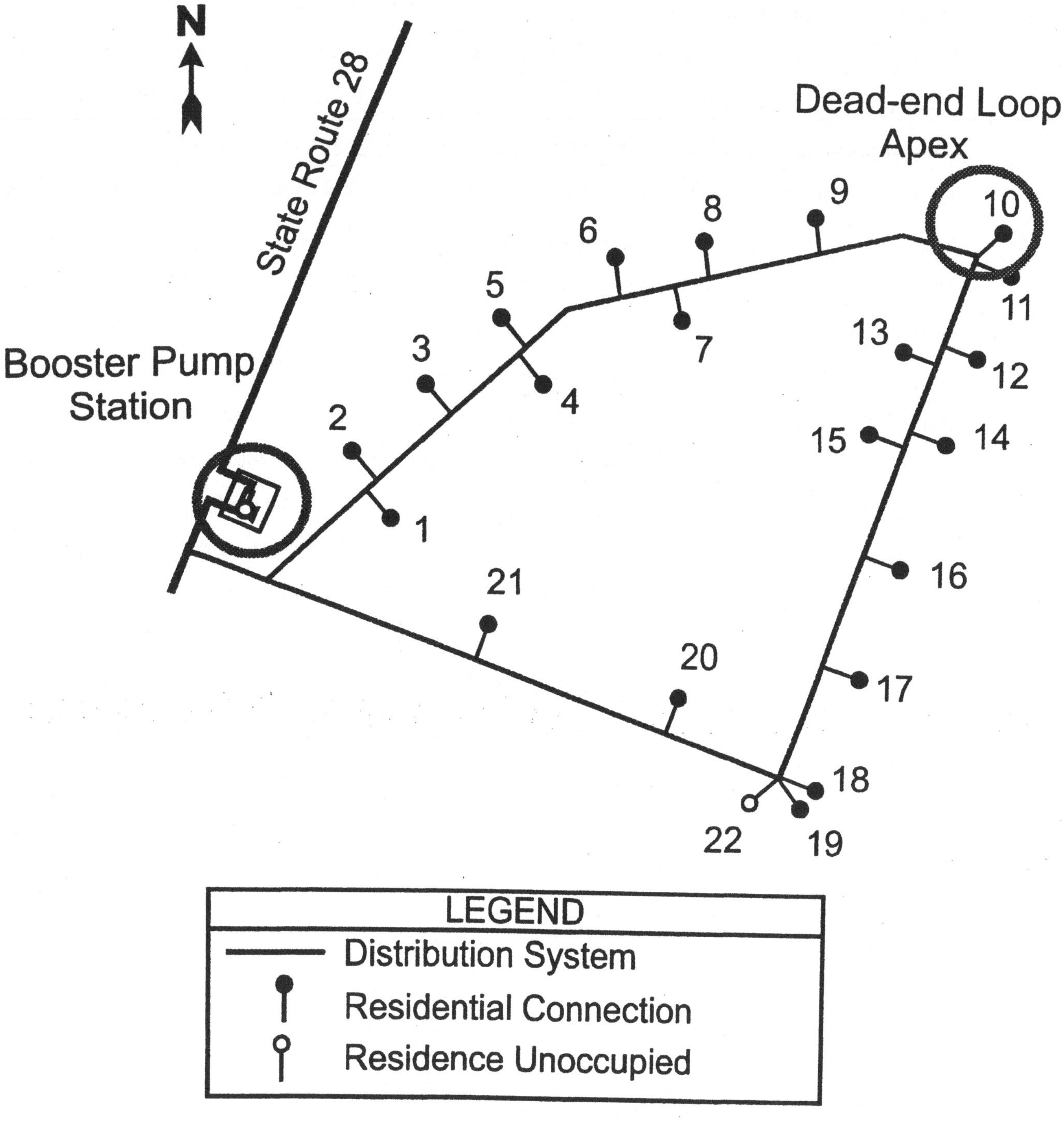

Figure 3.2 Dead-end loop study site.

FIELD METHODS AND MATERIALS

Water Demand Monitoring

Instrument packs to monitor residential water use were installed on the service line at secure indoor locations in each of the 21 homes on the loop[(1)]. Residential water consumption including indoor (hot and cold) and outdoor water use as well as leaks were recorded around-the-clock. The instrument package included a Rustrak Ranger II data logger, 4-29 mA impulse sensor, count and dump input pod, COM-504 modem connection, and 2,400 baud Hayes modem (see Figure 3.3). Data loggers were coupled with an input sensor inserted between the base and register head of a 5/8 inch T-10 Neptune water meter.

The water meters had a nominal accuracy of 100% (±1.5%) over a range of 0.5 to 20 gpm (Schlumberger, 1992). The input sensor monitored revolutions of a magnet fixed to a positive displacement nutating disc in the measuring chamber of the meter. At 1-second intervals, the sensor transmitted a signal to the data logger where the readings were converted to a flow rate and stored. Most recordings were zero. The nutating disc generated 231 signals per gallon (1 signal per cubic inch). Water demands over 0.26 gpm generated a continuous signal; demands under 0.26 gpm were recorded as an interrupted signal.

Water demands were downloaded nightly via dedicated modems linked to a computer at the University of Cincinnati. Pilot testing of the water demand monitoring units began at two residences in Autumn 1995. By February 1996, all 21 homes were on-line. Full-scale water demand monitoring continued through November 1997 [(2)]. During this 22-month period (669 days), over 1.2 billion water demand signals were collected from the residences on the dead-end loop [(3)].

---

(1) Participating homes are listed in Table 3.1. Six homes originally with outdoor pit meters were fitted with indoor meters. Home 19 was built and occupied during the study.

(2) The water quality monitoring units were deployed in March 1997. In order to provide simultaneous measurements of water demands and water quality at the study site, the water demand monitoring units were operated for 10 months beyond the original 12-month period.

(3)

$$\left(\frac{1\ reading}{sec \cdot home}\right) \times \left(\frac{86{,}400\ sec}{day}\right) \times (669\ days) \times (21\ homes) = 1.2\ billion\ readings$$

Table 3.1. Residential Participants in Dead-end Loop Study

Home No	Home Owner's Name	Service Address	1995 Avg Use (gallon/month)	Number of Occupants	Download Time	Miscellaneous Notes
1	Family #01	12 Valley View Circle	2,400	2	01:30	
2	Family #02	11 Valley View Circle	4,200	2	02:10	leaks(?)
3	Family #03	13 Valley View Circle	4,900	3	01:40	
4	Family #04	16 Valley View Circle	5,600	2	03:25	leaks(?)
5	Family #05	15 Valley View Circle	5,600	3	01:50	pit meter; corrosion
6	Family #06	21 Valley View Circle	4,000	2	02:45	pit meter, swimming pool
7	Family #07	20 Valley View Circle	6,800	4	03:55	
8	Family #08	25 Valley View Circle	4,900	3	02:40	swimming pool, line cut w/ rake
9(*)	Family #09	29 Valley View Circle	2,900	2	06:15 (*)	swimming pool, moved in 3/97
10	Family #10	46 Oakcrest Drive	5,800	3	03:45	pit meter, CMU at DLA
11	Family #11	40 Oakcrest Drive	1,900	2	02:20	leaks(?)
12	Family #12	34 Oakcrest Drive	6,200	2	03:15	pit meter, cut w/ rake
13(*)	Family #13	37 Oakcrest Drive	3,500	3	06:10 (*)	pit meter
14	Family #14	32 Oakcrest Drive	2,500	1	03:10	swimming pool
15	Family #15	35 Oakcrest Drive	4,700	2	02:30	pit meter
16	Family #16	26 Oakcrest Drive	3,200	2	03:35	
17(*)	Family #17	22 Oakcrest Drive	5,700	2	06:05 (*)	
18	Family #18	1 Promont Court	na	4	02:50	new home
19	Family #19	3 Promont Court	na	2	04:05	home built during study
20	Family #20	19 Oakcrest Drive	7,000	4	02:00	
21(*)	Family #21	17 Oakcrest Drive	7,900	4	06:00 (*)	

Notes: (1) Modem was accessed for a 10-minute download (* = called 3 times/day).
(2) Homes 18 and 19 were built after 1995.
(3) Home 10 held the continuous monitoring unit for water quality at the dead-end loop apex (CMU at DLA).

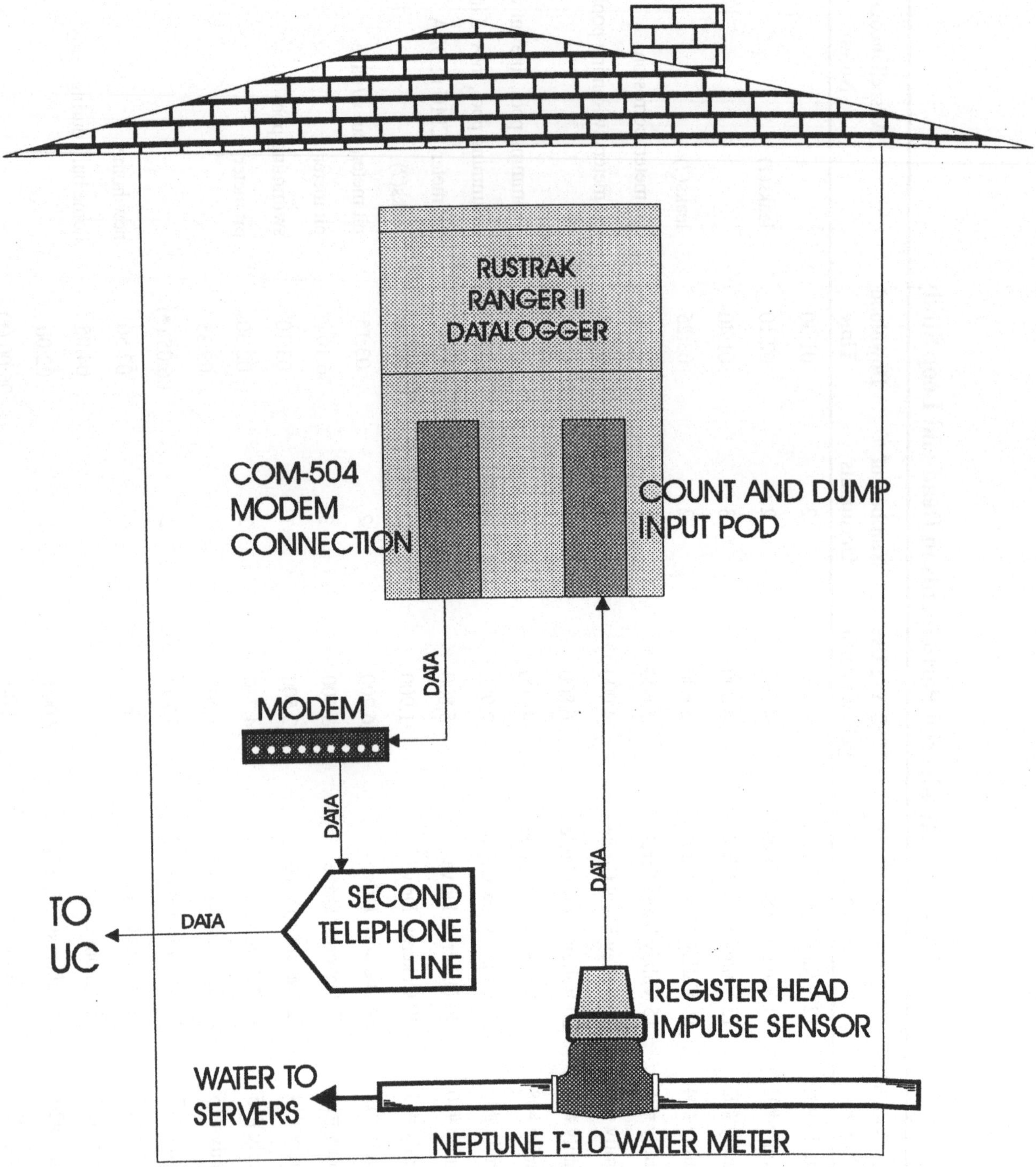

Figure 3.3 Instrument plan to monitor residential water demands in each home.

Water Quality Monitoring

Water quality parameters (described in Chapter 2) were monitored at four points in the municipal distribution system: the water treatment plant (WTP), the Wallace Street tank (WST), the booster pump station (BPS) and in a home at the dead-end loop apex (DLA). These sampling locations provided convenient and strategic points of access to the municipal distribution system. In addition, the City of Milford provided daily records of water levels in the Wallace Street storage tank, production flow rates and quality measurements of the finished water at the treatment plant.

Continuous Monitoring Units

Continuous monitoring units (CMU, see Figure 3.4) were placed at two locations (BPS and DLA) to track fluctuations in physical and chemical water quality parameters through the dead-end loop. The CMUs measured temperature, turbidity, conductivity, ORP, pH, DO, NO_3 and free residual chlorine. Readings at the BPS provided a baseline against which to compare water quality measurements taken at the DLA. The DLA was selected to maximize travel time from the loop entrance. Based on hydraulic modeling, travel time between the BPS and DLA averaged about 76 hours during the Winter (Schade, 1996).

The first CMU was installed on the 8-inch mainline at the BPS approximately 120 feet downstream of the entrance to the dead-end loop. The mainline tap included a pressure reducer with gage to maintain pressures under 10 psi and minimize stresses on the fittings and probes in the CMU. The second CMU was installed on the service line upstream of the water heater in the basement of the home at the DLA. A ½ inch copper bypass line, fitted with a solenoid valve and backflow preventer, supplied water to the CMU.

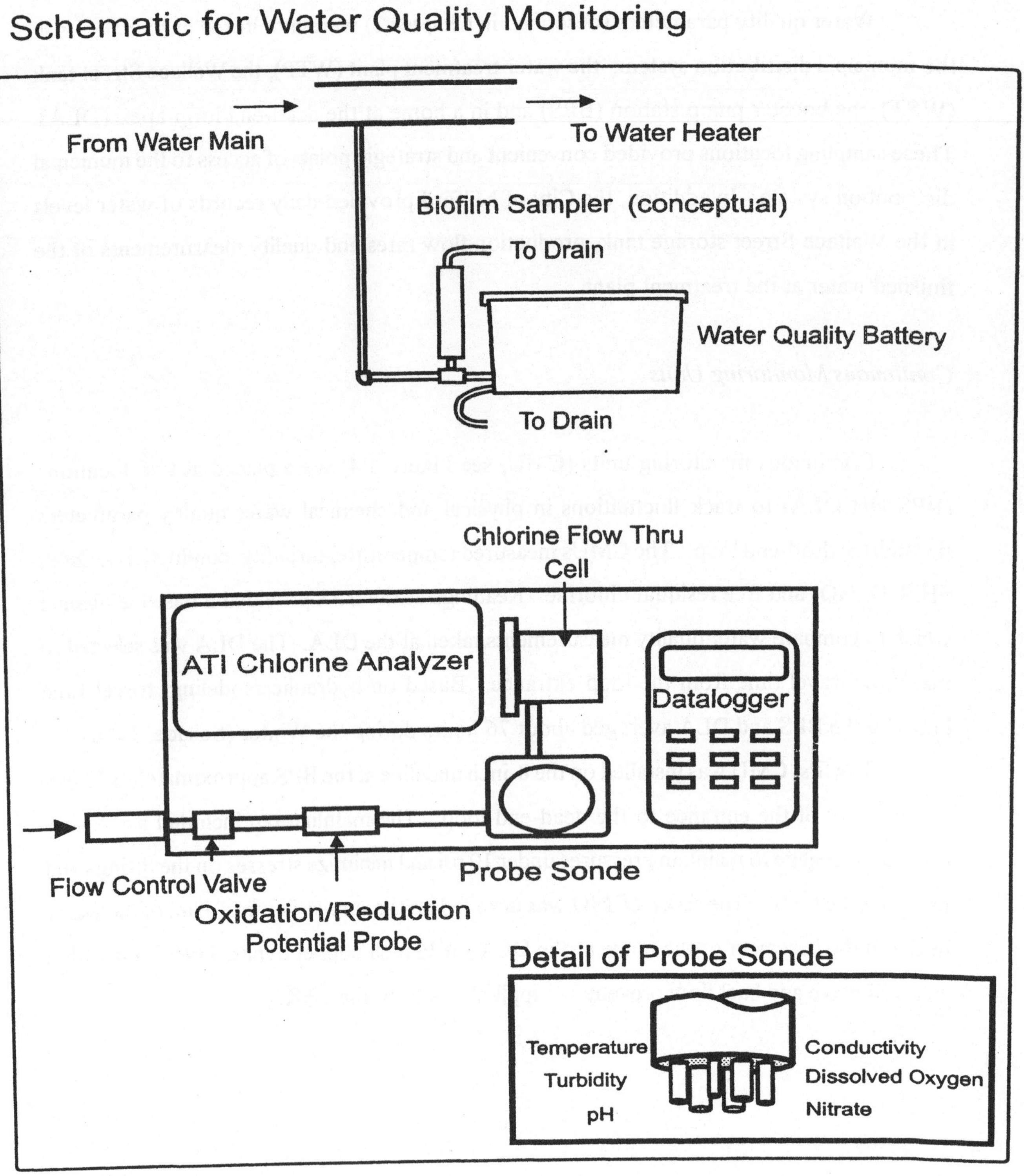

Figure 3.4 Instrument plan (CMUs) to monitor water quality at BPS and DLA.

At both sampling locations, a continuous bleed of approximately 0.26 gpm (1 L/min) was drawn from a tap set in the water main. From this bleed, a flow of 0.16 gpm (0.6 L/min) was piped to the CMU to satisfy minimum flow requirements of the chlorine analyzer. The balance of the flow was diverted to an on-line bio-reactor (i.e., modified Robbins Device, described below).

Both CMUs were constructed and operated in an identical manner. Drinking water entered the CMUs at a rate of 0.16 gpm and flowed through a ½ inch Teflon tube connected to a polyvinyl chloride tee housing a submerged Sensorex ORP probe. Water then passed through a retention cell which housed a Solomat Neotronics, Inc., 803PS water quality probe sonde. The sonde measured temperature, turbidity, conductivity, pH, DO, and NO_3 with a multi-channel WP4007 Data Logger that took recordings at 10-minute intervals. Water then passed through an acrylic constant head flow-through cell where free residual chlorine was measured using an Analytical Technology Inc., polarographic A15/62 membrane electrode. Following the chlorine measurement, flow from the CMU was discharged to an open drain (Carter 1998).

Grab Samples

Weekly grab samples were collected at the WTP, WST, BPS and DLA locations and analyzed by the US Environmental Protection Agency (USEPA) Laboratory in Cincinnati for heterotrophic plate count (HPC) bacteria, total coliform (TC), assimilable organic carbon (AOC), and total organic carbon (TOC). All samples were taken after thorough flushing of the connection to the service line to capture a representative parcel from the bulk fluid. HPC samples were collected in sterile 250 mL polypropylene bottles containing sodium thiosulphate [1 mL 10% (w/v) solution] to neutralize oxidation reactions that might change the nature of the sample during transportation. AOC samples were collected in a 1000 mL glass bottle containing 5 mL of sodium thiosulphate. TOC samples were placed in a 50 mL glass vial and treated with 1 mL of phosphoric acid.

Modified Robbins Device

The experimental apparatus at the BPS and DLA also included on-line bio-reactors (called Modified Robbins Devices or MRDs, see Figure 3.5) designed to monitor biofilm growth. Similar to reactors used by Bezanson *et al.* (1992) and Manz *et al.* (1993), the MRDs consisted of a 4-inch diameter clear schedule 40 PVC pipe, 52 inches in length. Both MRDs contained 60 sampling ports, in 20 groups of three, staggered lengthwise at a spacing of 1.5 inches and offset by an angle of 60°. The three sampling ports in each group were spaced equally around the circumference of the MRD. Each port held an acrylic plug with a removable CPVC tip, which provided 0.196 in^2 of surface area for biofilm growth. Both MRDs were mounted in a horizontal position and sealed in a wooden box to prevent exposure to light. The flow rate through the MRDs was 0.1 gpm (0.4 L/m), yielding a nominal residence time of approximately 28 minutes. The MRDs discharged to an open drain in the basement floor.

During installation at the BPS and DLA, the MRD pipe section was disinfected with a 10 percent chlorine solution for a period of 24 hours. The CPVC tips were autoclaved and packaged in foil until aseptic placement in the field. On a weekly basis, one group of three plugs was extracted from the MRDs at both locations (the BPS and the DLA). The tips of the acrylic plugs were removed with sterile steel tongs and immediately placed into separate 50 mL plastic vials containing 10 mL of sterile phosphate buffer solution and 0.5 mL of sodium thiosulfate to quench the residual free chlorine. The plug tips were transported immediately to the USEPA Laboratory in Cincinnati where they were sonicated for 30 seconds to dislodge attached microorganisms into a phosphate buffered solution. The cell suspensions were then cultured on R2A, PCA, and TSA-SB media for biofilm plate counts.

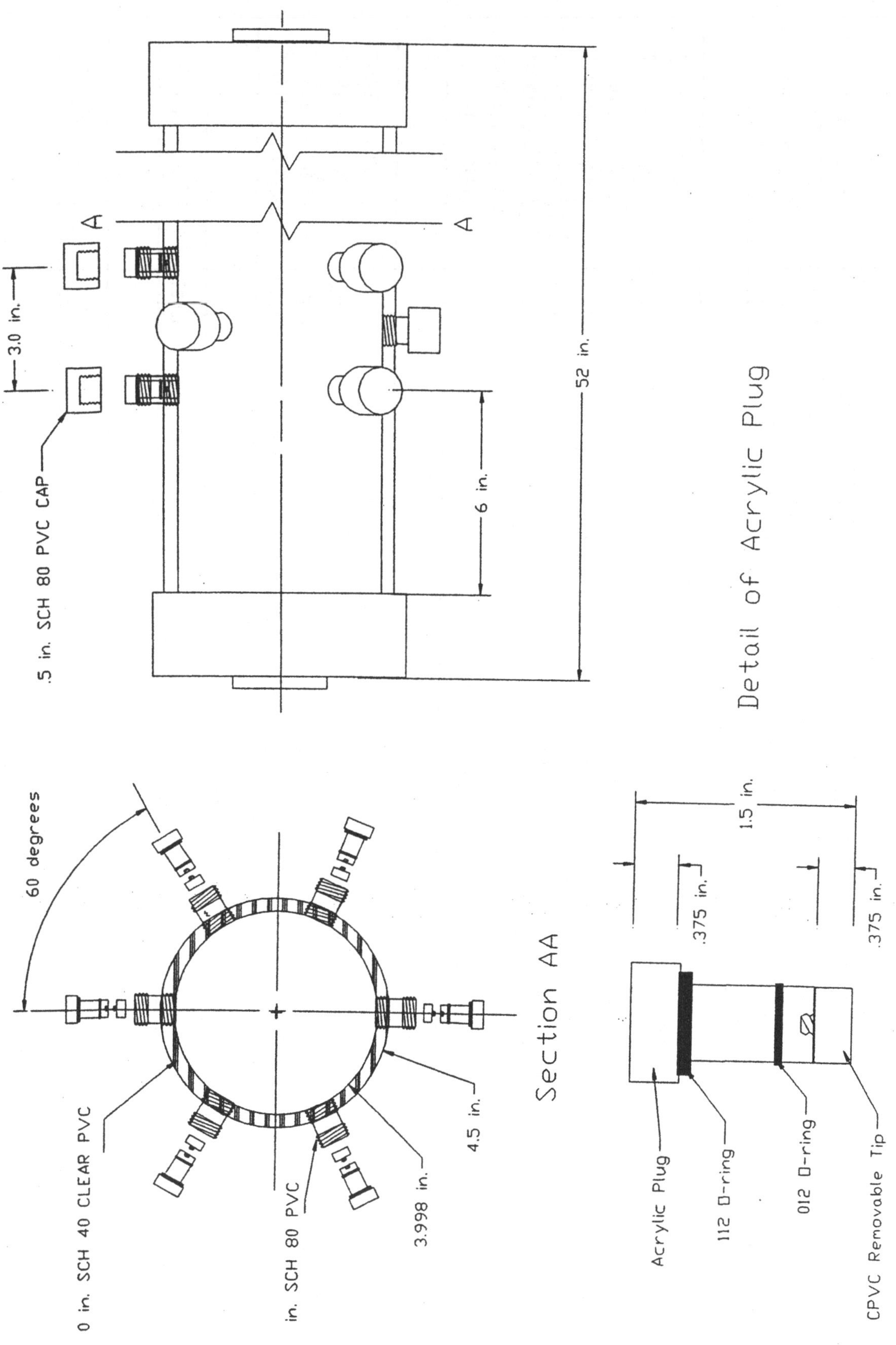

Figure 3.5 Modified Robbins Device and sampling plug.

Table 3.2 Experimental Matrix for Field Sampling and Lab Analysis

Parameter	Reporting Units	Analytical Method (APHA)	Detection Limit	Location and Frequency		Instrument (*)	Period of Observation
				WTP, WST	BPS, DLA		
❖ Water Quality: Continuous Monitoring at 0.16 gpm							
temperature	°C	platinum electrode	–5	–	6 / hr	1	3/8/97 – 11/18/97
turbidity	NTU	2130 B	0.10	–	6 / hr	1	3/8/97 – 11/18/97
conductivity	µS/cm	2510 B	0.50	–	6 / hr	1	3/8/97 – 11/18/97
ORP	mV	Ag/AgCl	0.00	–	6 / hr	2	3/8/97 – 11/18/97
pH	SU	4500–H^+ B	0.01	–	6 / hr	1	3/8/97 – 11/18/97
DO	mg/L	4500–O G	0.00	–	6 / hr	1	3/8/97 – 11/18/97
NO_3	mg/L	4500–NO_3 D	0.10	–	6 / hr	1	3/8/97 – 11/18/97
Free Cl_2	mg/L	polarograph membrane	<0.05	–	6 / hr	3	3/8/97 – 11/18/97
❖ Water Quality: Grab Samples							
TOC	mg/L	5310 D	0.10	2 / mo	2 / mo	4	1/8/97 – 11/18/97
AOC	mg-C/L	9217 B	0.005	2 / mo	2 / mo	visual	1/8/97 – 11/18/97
T coliform	+ or –	9222 B	0.0	1 / wk	1 / wk	visual	1/8/97 – 11/18/97
HPC	CFU/mL	9215 C	0.0	1 / wk	1 / wk	visual	1/8/97 – 11/18/97
❖ Water Quality: Modified Robbins Device at 0.10 gpm							
biofilm	CFU/cm^2	see Chapter 3	0.0	–	1 / wk	visual	7/7/97 – 11/18/97
❖ Water Demands							
water use	gpm	see Chapter 4	0.10	1 per sec at 21 homes		5	2/1/96 – 12/1/97

(*)Instruments: (1) Solomat Neotronics 803PS and WP4007 Data Logger (Norwalk, CT 06850).
(2) Sensorex ORP Probe
(3) Analytical Technology, Inc., Model A15/62 Free Chlorine Monitor (Oaks, PA 19456).
(4) 700 O.I. TOC Analyzer
(5) Neptune T-10 5/8" water meter and Rustrac Ranger II data logger linked to PC modem.

LABORATORY METHODS AND MATERIALS

The experimental matrix for field sampling and lab analysis is summarized in Table 3.2. Additional details on parameter measurement and probe calibration are provided below.

Physical Water Quality Parameters

Temperature – Water temperature was measured in °C using a platinum electrode.

Turbidity – Continuous turbidity measurements were taken by a probe immersed in the water stream and reported as Nephlometric Turbidity Units (NTU) per Standard Methods (APHA 2130 B). The probe was calibrated once per week using a zero and 100 NTU resin standard and cleaned with Kimwipes and Type II deionized water.

Conductivity – Conductivity was measured according to Standard Methods (APHA 2510 B). The probe was calibrated once a week using a one point method. The calibration fluid held a magnitude of 1500 μS and was supplied by Solomat, Inc.

Oxidation Reduction Potential – ORP was measured using a Sensorex electrode immersed in a steady flow of 0.16 gpm from the service line. No calibration was necessary for this probe; it was cleaned once a month using non-abrasive wipes and distilled water.

Chemical Water Quality Parameters

pH – The pH of the sample bleed was measured per Standard Methods (APHA 4500-H+ B). The probe was calibrated weekly using low and high points of 4.00 and 7.00, respectively.

Dissolved Oxygen – DO was measured according to the Standard Methods for galvanic membranes (APHA 4500-O G). The probe was calibrated once per week using the two point linear method, an automatic feature of the WP4007 Solomat Data Logger. The calibration points were 0 and 9.06 ppm, representing anoxic and saturation conditions at 20 °C.

Nitrate–Nitrate was measured using an electrode with a semi-porous inert membrane according to Standard Methods (APHA 4500-NO3- D). The electrode was calibrated once a week using low and high calibration points of 1 and 10 ppm, respectively.

Free Residual Chlorine–Free chlorine was measured using the Analytical Technology, Inc., A15/62 free chlorine monitor. This unit provided continuous on-line monitoring of the residual in the range of 0 to 2 ppm. The sensing element of the analyzer used a polarographic membrane housed in a constant head flow through cell. The ATI analyzer was calibrated weekly to a standard sample measured independently by a Hach Pocket Colorimeter. The membrane was cleaned with type II deionized water weekly and replaced along with the electrolyte reference solution monthly or as the sensitivity fell below 40 percent.

Total Organic Carbon – Total organic carbon was measured with a 700 O.I. Analytical Total Organic Carbon Analyzer based on Standard Methods for Wet Oxidation (APHA 5310 D).

Biological Water Quality Parameters

Assimilable Organic Carbon – AOC detection used two organisms (van der Kooij et al., 1982): *Pseudomonas fluorescens* (P-17 strain) and *Spirillum* (NOX strain). The P-17 strain is often isolated from drinking water and grows on small amounts of a diverse substrates. The NOX strain detects degradable organics not utilized by the P-17 strain. Water samples first were pasteurized to inactivate indigenous microbes, then inoculated with the P-17 and NOX strains. Samples were then incubated at 15 °C for up to 20 days without shaking, followed by plating (0, 3, 5, 7 days) to track the peak density of the organisms. Results were compared against standard growth curves (P-17 strain on acetate and NOX strain on oxalate) and reported as micrograms of acetate and oxalate equivalences per liter according to Standard Methods (APHA 9217 B). Note that there is some risk in overestimating AOC using this method. The two indicator organisms may consume a common nutrient with equal vigor thereby resulting in a higher total cell density and AOC concentration (Geldreich, 1996).

Total Coliforms – Coliform bacteria are indicators of recent fecal contamination or other undesirable water quality changes in water distribution and storage systems. The presence of coliforms was tested using the Total Coliform Membrane Filter Procedure per Standard Methods (APHA 9222 B). The test required a sample of bulk fluid from the distribution network which was then sent through a filter. This filter was then placed on a M-Endo LES medium and the bacteria were allowed to grow. The number of bacterial colonies were counted and used to determine the number of coliforms per volume of bulk fluid.

Heterotrophic Plate Count – The heterotrophic plate count analysis, formerly known as the standard plate count, was carried out per Standard Methods for the spread plate method (APHA 9215 C). Beginning in July 1997, sheep's blood agar and plate count agar were used to increase the substrate to which the samples were exposed.

Biofilm Growth – The MRD samples were sonicated for 30 seconds at setting 10 to transfer the attached organisms from the plug tip to a phosphate buffered solution. The cell suspension was then applied to plates containing R2A nutrient agar for enumeration and identification. The plating technique used here was the same as that used for the HPC analysis, including the switch to sheep's blood agar and plate count agar. These substrates increased the number of microorganisms which could be subcultured for identification.

Quality Assurance/Quality Control

Weekly calibration of the Solomat probes and the ATI unit was performed in accordance with applicable standard methods, U.S. EPA approved methods, and/or manufacturer's recommended procedures. A volumetric flow calibration was done periodically to detect any drift in the flow diverted through the monitoring devices.

ARCHIVING RESIDENTIAL WATER DEMANDS

Water use was recorded every second at each residence. Over 1.8 million readings were taken *each day* from the neighborhood of 21 homes on the dead-end loop. This level of detail afforded unprecedented resolution of residential water use patterns, but it also presented enormous data management challenges. Fortunately, most of the water use recordings were redundant (usually zero) and, consequently, could be eliminated without loss of information. A seven step process was used to transform the raw input signals into archived residential water pulses (see Figure 3.6). In the process, three basic sets of water demand data were generated, as summarized in Table 3.3.

Steps 1 and 2, involving data retrieval and format conversion, were largely automated. These data represent water demands at the "Point-of-Entry"to the residence (i.e., measured on the service line, see Figure 2.3). In step 3, readings showing zero water demand were deleted and flow measurements at home 10 were reduced. This flow adjustment was necessary to account for the steady bleed (Q = 0.26 gpm) diverted through the water quality monitoring unit in the basement of home 10 at the DLA (details are presented in Chapter 4). Steps 4, 5 and 6, comprise the heart of the data reduction process and are based on algorithms developed and demonstrated by Buchberger and Wells (1996). Some of the key points will be mentioned here briefly, but the interested reader should refer to the original paper for complete details.

In step 4, the non-zero flow data at each residence were filtered using a 5-point moving average scheme to convert fluctuating water use measurements into "smooth signals". For example, Figure 3.7 illustrates the smoothing process applied to three raw data signals. In most cases, the smoothed signal reduced to a single rectangular pulse. In some cases, however, the smoothing process generated complex patterns resembling Strings of Random Blocks (SORBs). The presence of SORBs might indicate multiple busy servers in the home (see Figure 2.3). Hence, all SORBs were analyzed in a subsequent step for coincident water use.

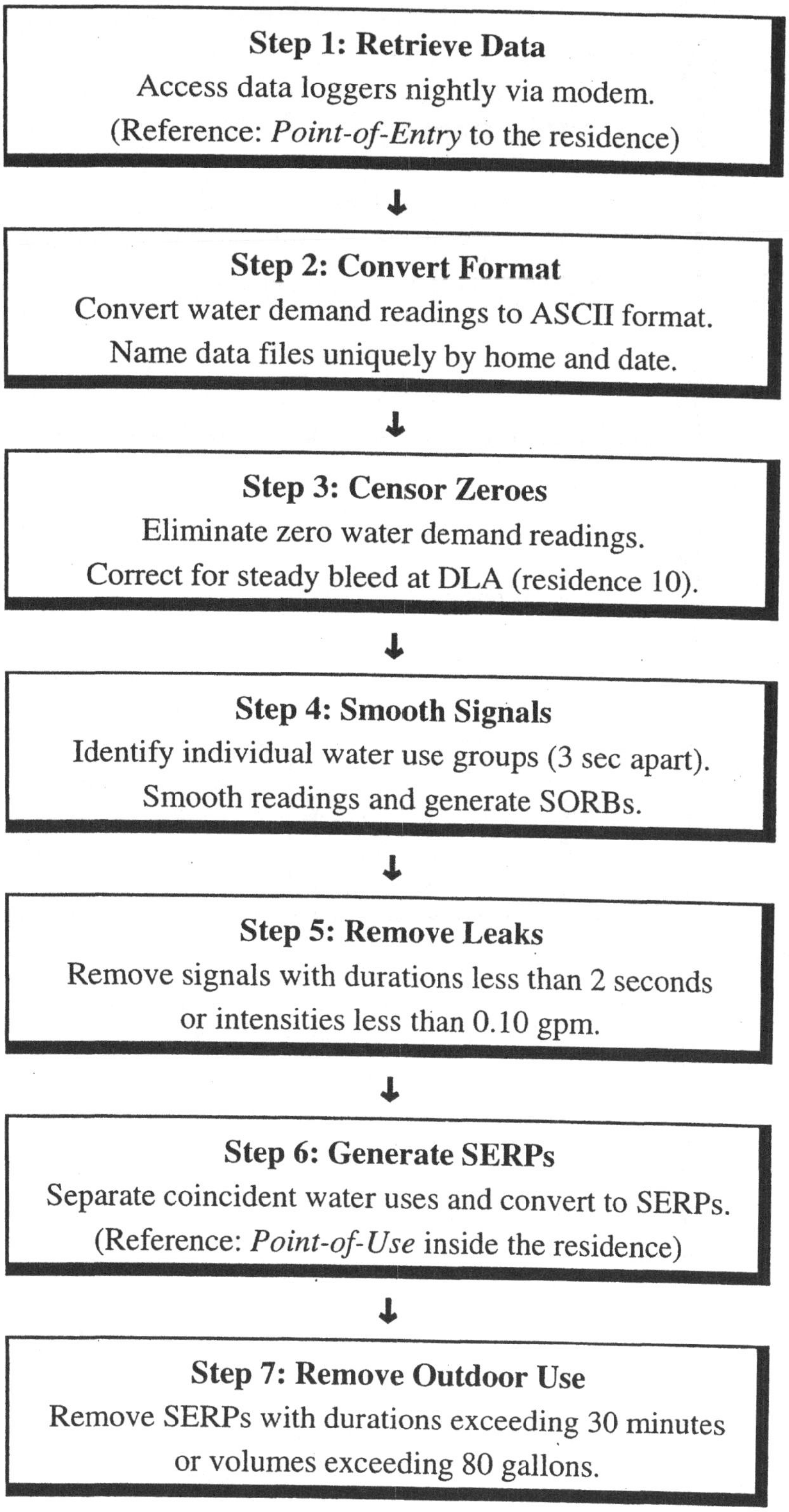

Figure 3.6 Data Reduction Flowchart for Residential Water Use.

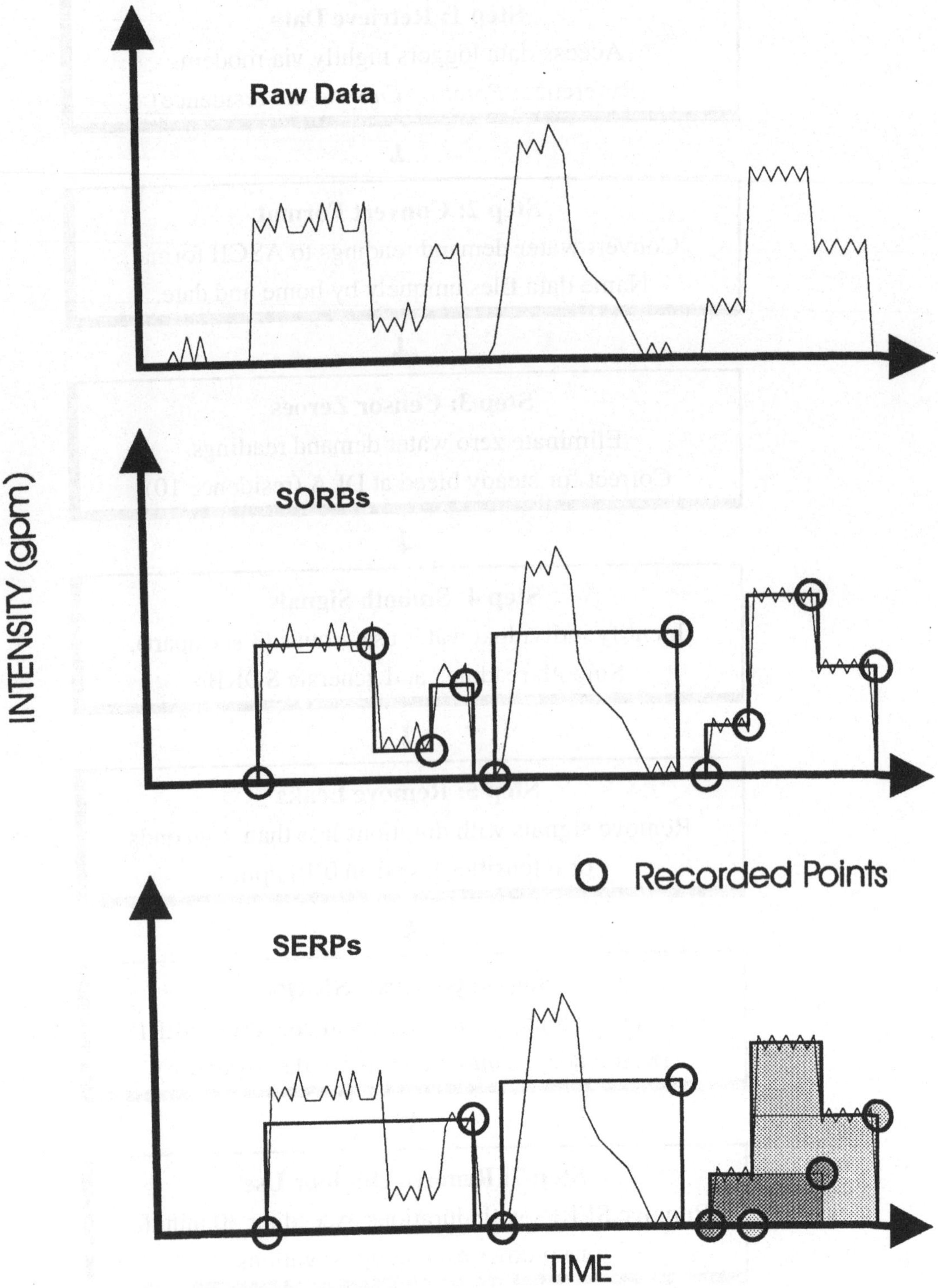

Figure 3.7 Signal smoothing process.

In step 5, leaks and ultra-low demands were censored from the smoothed signals(4). Differentiating between leaks and legitimate low flows was problematic because residential demands under 0.26 gpm registered as an intermittent signal on the Rustrac Ranger data logger(5). Hence, it was not always obvious when a genuine low flow demand began or ended. In other cases, however, clear patterns emerged. For instance, a few homes had many water pulses with durations of exactly one second (i.e., "blips"), spaced at regular intervals during long periods of otherwise zero flow. In some cases, these blips comprised over 20 percent of the total pulse count for the residence, but usually less than 0.2 percent of the total water use(6). These 1-second blips were interpreted as slow leaks(7). Ultimately, a smoothed water demand signal was classified as a leak if it met either of the following two guidelines:

- ❖ very short duration (under 2 seconds) or
- ❖ very low intensity (under 0.10 gpm).

As a general outcome, when applied to the data set, these criteria effectively deleted any signal with a total volume less than about 0.01 gallons (see Figure 3.8). Some legitimate water demands were probably censored, but their contribution to the total amount of water used at any residence was negligible.

---

(4) This was necessary because the PRP hypothesis for residential water use does not include background leaks; if leaks are present, their effect must be incorporated separately.

(5) For example, a steady demand of 0.0625 gpm (1 cup per minute) would generate one non-zero signal every four seconds.

(6) In general, 1-second blips accounted for less than 5 percent of the recorded pulse count at all homes on the dead-end study site. Some blips may have been caused by ice makers.

(7) A barely detectable leakage rate of 1 gal/day would generate over 6,000 (see below) spurious 1-second signals during a 30 day reporting period, assuming that residential servers were idle 90 percent of the time. The remaining 10 percent of the time, water intentionally used during busy periods would mask the leak signal.

$$\left(\frac{231\ blips}{gallon}\right)\times\left(\frac{1\ gallon}{day}\right)\times\left(\frac{30\ days}{month}\right)\times(0.90\ idle) = 6{,}237\ \frac{blips}{month}$$

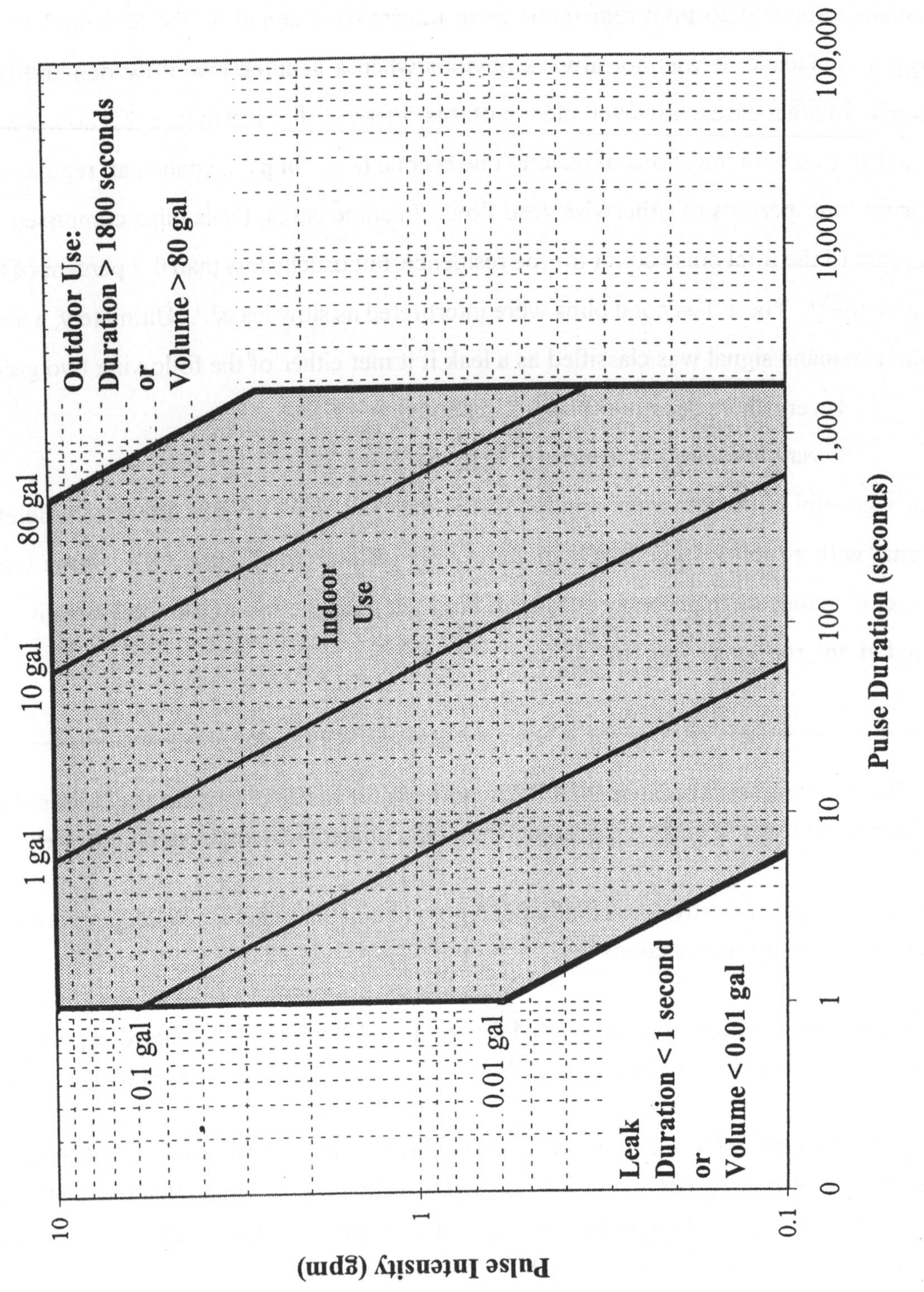

Figure 3.8 Domains for leaks, indoor water use and outdoor water use.

In step 6, SORBs from the smoothing operation (step 4) were scanned for coincident water use with an algorithm by Wells (1994). If coincident use were not detected, the SORB was converted directly to a Single Equivalent Rectangular Pulse or "SERP". However, if coincident water uses were confirmed, they were separated and then expressed as SERPs.

The SERPs preserved the average intensity, duration, frequency, volume, and timing of the parent signal. Using only three parameters (starting time, average intensity and pulse duration), SERPs provide a compact scheme to accurately catalog complex water demands. Figure 3.9 illustrates two common SORBs found in the water demand data base and shows how SERPs were derived from them. Converting a SORB into a SERP effectively transferred water demand information from the "Point-of-Entry" to the "Point-of-Use"(see Fig 2.3). SERPs that extended from one day to the next were split into two pulses at midnight.

In step 7 very large pulses were screened[(8)]. A water use was classified as "outdoor" if its duration exceeded 30 minutes or if its volume exceeded 80 gallons (see Figure 3.8). These thresholds were based on analyses of water demands during the winter months, when virtually all residential water use is indoor.

Steps 6 and 7 sometimes had an undesirable side effect. The algorithm to detect coincident water use was developed using data from the winter season when water use is indoor and brief. In the presence of long-duration outdoor water demands, the search algorithm often performed poorly. Very few (short-duration) indoor pulses were separated from (long-duration) outdoor pulses. Hence, when outdoor demands were converted to SERPs in step 6, coincident indoor demands were subsumed in the conversion. When outdoor demands subsequently were screened in step 7, some legitimate indoor demands were unintentionally censored, leaving gaps in the indoor data at some homes. This loss of information slightly reduced estimates of total indoor water consumption for the summer season. Fortunately, these missing data points had very little (if any) effect on testing the validity of the PRP hypothesis for residential water use.

[(8)] The PRP model for residential water use treats joint indoor and outdoor demands as the superposition of two individual PRP processes. Therefore, to test the PRP hypothesis, it is necessary to separate indoor and outdoor demands. Chapter 6 has more details.

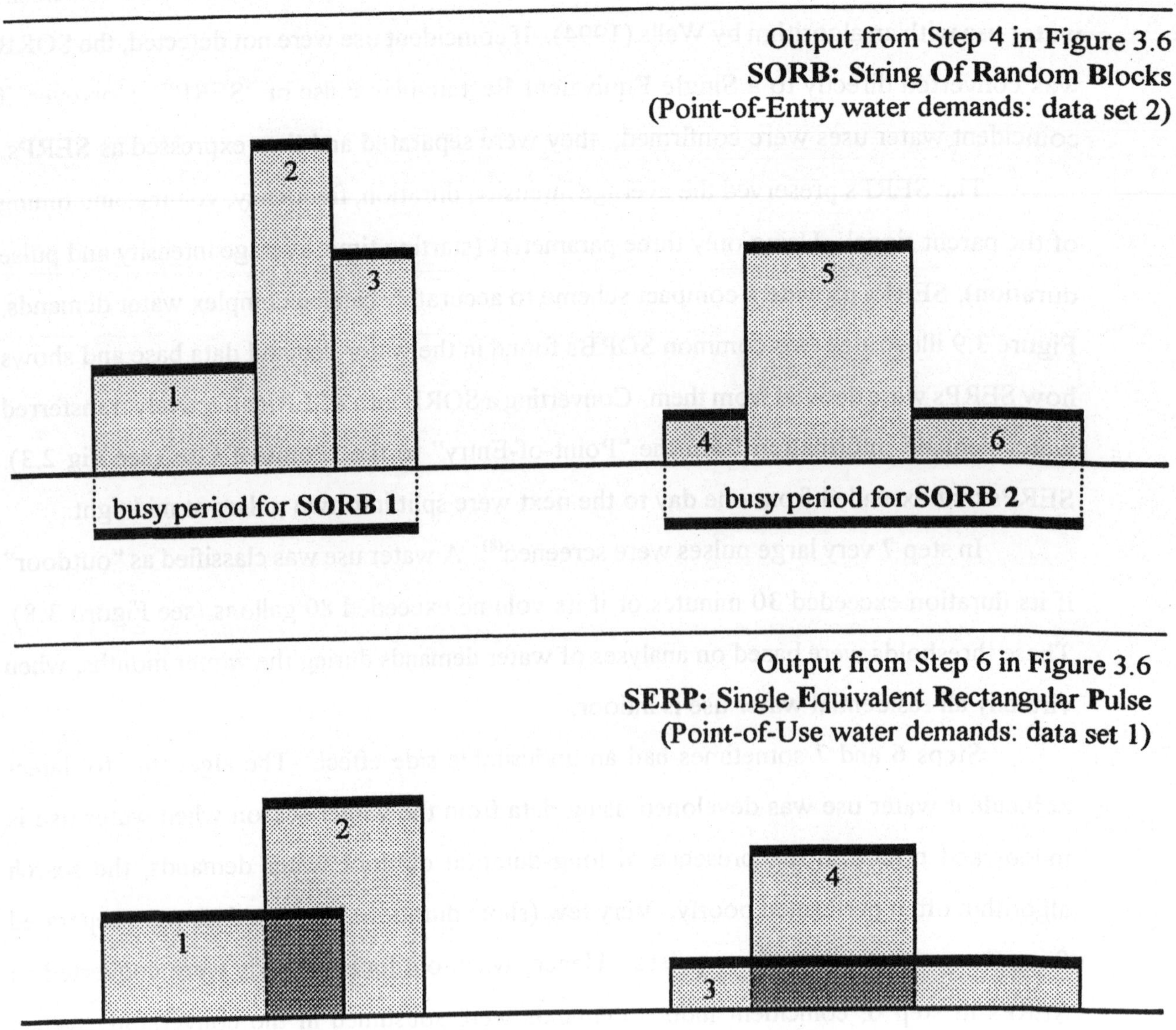

Figure 3.9 Conversion of SORBs to SERPs.

These plots show two busy periods with water demand (y-axis) versus time (x-axis). The upper plot shows two common examples of "Strings Of Random Blocks" (SORBs), both with 3 distinct water demand intensity levels. In the lower plot, each **SORB** is decomposed into 2 "Single Equivalent Rectangular Pulses" (SERPs). This step transforms information measured at the Point-of-Entry into plausible water demands at the Point-of-Use. The volume of water contained in the 2 **SORBs** is identical to the volume of water represented by the 4 **SERPs**. The dark areas in the lower plot correspond to the volume of coincident water use (joint demands at 2 servers) in the residence. Each **SERP** was archived with the following information: home, date, pulse number, start time, duration, intensity and volume (see Table 3.4).

Eliminating zero readings, censoring residential leaks and converting demands to SERPs effectively reduced the total number of recorded data points by nearly 99.9 percent. Nonetheless each day of monitoring at the dead-end loop added over 1,700 demands to the water use files. Following the flowchart outlined in Figure 3.6, these residential water demands were archived as three types of data sets (listed in Table 3.3).

The archiving process was a slow painstaking exercise, requiring nearly 3 years [9]. The final product contains 365,000 residential water demands spanning 214 consecutive days of use at 21 homes. This collection represents one of the most detailed and accurate continuous high resolution records (1 second interval) of residential water use assembled. It is available for researchers, practitioners, consultants, regulators and others who may be investigating demands on water distribution systems.

Table 3.3
Archived Residential Water Demand Files
April 1, 1997 through October 31, 1997

Data Set No	Point of Reference and Hierarchy Level	Water Demand Pattern	Coincident Pulses Separated?	Indoor or Outdoor Demands
1	Point-of-Use, H-1	SERP	yes	indoor only
2	Point-of-Entry, H-2	SORB	no	indoor only
3	Point-of-Use, H-1	SERP	yes	indoor & outdoor

(a) Point-of-Use values were inferred from measurements on service line into home.
(b) Point-of-Entry values were obtained from measurements made on service line to home.
(c) SORB: Strings Of Random Blocks and SERP: Single Equivalent Rectangular Pulse.

---

[9] The effort required to groom and archive water demand data that were free of anomalies, gaps, and inconsistencies was considerably more challenging than anticipated. Difficulties were exacerbated by the sheer size of the data base and the protracted field monitoring program (see footnote 2 in Chapter 2). On average, about 100 hours *per home* were required to archive the water demands for the neighborhood.

The residential water demand data sets were archived as text files, each uniquely identified by residence and by month. The entire collection of water demand data for the seven month period April 1, 1997 to October 31, 1997 was saved on a compact disc (57 megabytes of storage). Appendix A contains information on how to order a copy of this CD. An example of the information contained in each of the data sets appears in Tables 3.4, 3.5 and 3.6 which show water demands measured at home 12 during the second week of July 1997 [(10)].

These residential water demand data were used to carry out three types of analysis: (i) generate descriptive statistics of indoor and outdoor residential water use (using data set 1 and data set 3); (ii) compare PRP theoretical predictions of busy home properties (e.g., probability density function, autocorrelation function, state transitions) and flow regime characteristics (e.g., probability density function and moments) against sample statistics obtained from field observations (using data sets 1 and 2); and (iii) drive an event-based network hydraulic model to track movement of discrete volume elements in order to estimate travel time in the dead-end loop (using data set 3).

[(10)] The pulse counts do not coincide across the three Tables, because each data set has slightly different elements and sample size:

- ❖ Table 3.4 has indoor SERPs (Data Set 1, Point of Use).
- ❖ Table 3.5 has indoor SORBs (Data Set 2, Point of Entry).
- ❖ Table 3.6 has indoor and outdoor SERPS (Data Set 3, Point of Use).

The date and time of each pulse are unique and can be used to identify common water demands among the three Tables.

Table 3.4
Archived Residential Water Demands
Example from Data Set 1 for Home 12
(Point-of-Use: Indoor SERPs only)

(1) month	(2) date	(3) pulse	(4) time	(5) duration	(6) intensity	(7) volume
..	..	..	..	..	..	..
7	8	1142	26861	7	0.36	0.04
7	8	1143	33124	4	0.54	0.04
7	8	1144	35683	6	0.42	0.04
7	8	1145	39017	24	1.43	0.57
7	8	1146	39046	68	2.80	3.17
7	8	1147	39357	9	0.42	0.06
7	8	1148	40170	9	0.42	0.06
7	8	1149	42586	54	3.41	3.07
7	8	1150	42601	15	1.20	0.30
7	8	1151	43683	6	0.42	0.04
..	..	..	..	..	..	..

Legend:

column (1): month	=	Month of year (1=January, 2=February,....)
column (2): date	=	Day of month (1, 2, 3,...., 31)
column (3): pulse	=	Pulse number since start of month (1, 2, 3,...)
column (4): time	=	Seconds since midnight (1, 2, 3,...., 86400)
column (5): duration	=	Pulse duration (seconds)
column (6): intensity	=	Pulse intensity (gpm)
column (7): volume	=	Pulse volume (gal) [col(5)×col(6)/60]

Shaded area highlights two **SERPs** extracted from the **SORB** identified in Table 3.5.

Table 3.5
Archived Residential Water Demands
Example from Data Set 2 for Home 12
(Point-of-Entry: Indoor SORBs only)

(1) month	(2) date	(3) pulse	(4) time	(5) duration	(6) intensity	(7) volume
..	..	..	..	..	..	..
7	8	1174	26861	7	0.36	0.04
7	8	1175	33124	4	0.54	0.04
7	8	1176	35683	6	0.42	0.04
7	8	1177	39017	24	1.43	0.57
7	8	1178	39046	68	2.80	3.17
7	8	1179	39357	9	0.42	0.06
7	8	1180	40170	9	0.42	0.06
7	8	1181	42586	15	3.41	0.85
7	8	1182	42601	15	4.61	1.15
7	8	1183	42616	24	3.41	1.36
7	8	1184	43683	6	0.42	0.04
..	..	..	..	..	..	..

Legend:

column (1): month	=	Month of year (1=January, 2=February,....)
column (2): date	=	Day of month (1, 2, 3,...., 31)
column (3): pulse	=	Pulse number since start of month (1, 2, 3,...)
column (4): time	=	Seconds since midnight (1, 2, 3,...., 86400)
column (5): duration	=	Pulse duration (seconds)
column (6): intensity	=	Pulse intensity (gpm)
column (7): volume	=	Pulse volume (gal) [col(5)×col(6)/60]

Shaded area highlights a **SORB** that yields two **SERPs** identified in Table 3.4.

Table 3.6
Archived Residential Water Demands
Example from Data Set 3 for Home 12
(Point-of-Use: Indoor and Outdoor SERPs)

(1) month	(2) date	(3) day	(4) season	(5) pulse	(6) total	(7) time	(8) dur	(9) int	(10) volume	(11) loc	(12) status
..	..	..	..	..	..	..	..	..	..	..	..
7	8	2	2	1146	28	26861	7	0.36	0.04	1	1
7	8	2	2	1147	29	28292	1070	4.51	80.41	2	1
7	8	2	2	1148	30	33124	4	0.54	0.04	1	1
7	8	2	2	1149	31	35683	6	0.42	0.04	1	1
7	8	2	2	1150	32	39017	24	1.43	0.57	1	1
7	8	2	2	1151	33	39046	68	2.80	3.17	1	1
7	8	2	2	1152	34	39357	9	0.42	0.06	1	1
7	8	2	2	1153	35	40170	9	0.42	0.06	1	1
7	8	2	2	1154	36	42586	54	3.41	3.07	1	2
7	8	2	2	1155	37	42601	15	1.20	0.30	1	2
7	8	2	2	1156	38	43683	6	0.42	0.04	1	1
..	..	..	..	..	..	..	..	..	..	..	..

Legend:

column (1): month	=	Month of year (1=January, 2=February,....)
column (2): date	=	Day of month (1, 2, 3,...., 31)
column (3): day	=	Day of week (1=Monday, 2=Tuesday,....)
column (4): season	=	Season (1=winter, 2=summer)
column (5): pulse	=	Pulse number since start of month (1, 2, 3,...)
column (6): total	=	Total number of pulses for the day (1, 2, 3,...)
column (7): time	=	Seconds since midnight (1, 2, 3,...., 86400)
column (8): duration	=	Pulse duration (seconds)
column (9): intensity	=	Pulse intensity (gpm)
column (10): volume	=	Pulse volume (gal) [col(8)×col(9)/60]
column (11): location	=	Pulse location (1=indoor, 2=outdoor)
column (12): status	=	Pulse status (1=single pulse, 2=coincident pulse)

Shaded area highlights an outdoor water use.

CHAPTER 4
RESIDENTIAL WATER DEMAND ANALYSIS

OVERVIEW

This chapter covers data recovery and water use statistics, including a comparison against findings from the recent AwwaRF report on residential end uses of water (Mayer *et al*, 1999).

DATA RECOVERY

During the water demand monitoring phase, there were three possible outcomes at a home on any given day, namely:

- Full recovery–all daily water demands were successfully recorded.
- Partial recovery–some daily water demands were successfully recorded.
- No recovery–all daily water demands were lost (i.e., none recorded).

Table 4.1 summarizes these three outcomes at the 21 homes along the dead-end loop for the 36-week period from March 12, 1997 to November 18, 1997. This corresponds to the time during which calibrated water quality CMUs were in place. Table 4.2 shows the break down of data recovery at all 21 homes on a weekly and monthly basis.

The overall success rate for full-day complete data recovery across the entire dead-end loop is 91.5%. This compares favorably to the 91.8% data logger success rate reported in the REUW study. On a monthly basis, overall data recovery ranged from just over 88% in June to nearly 95% in May. At 15 of the 21 homes on the dead-end loop, complete daily data recovery was very high, averaging 96.2% (column (7) in Table 4.1).

Most of the days with "partial recovery" (178 out of 266 days or 67%) in Table 4.1 occurred at the four homes (9, 13, 17, 21) instrumented with older data loggers. Owing to their limited storage capacity (64K with CPU version 4.31 versus 256K with CPU version 4.56, see Figure 3.3), the older units required three data downloads per day. These units malfunctioned about once per week and could not be reset over the phone; hence, whenever communication problems arose, it was necessary to visit the residence to get the unit back on-line. As a consequence, the average full-day success rate of the four homes with the older data loggers was about 80% (column (7) in Table 4.1).

Table 4.1 Residential Water Demand Recovery
March 12, 1997 through November 18, 1997.

(1) Home on Dead-end Loop	(2) Days Home Was Occupied	(3) Days with Full Data Recovery	(4) Days with Partial Recovery	Days with No Recovery		(7) Full Data Recovery (%)
				(5) Replicated	(6) Lost	
1	252	243	4	5	0	96.4
2	252	242	5	5	0	96.0
3	252	241	4	7	0	95.6
4	252	242	4	6	0	96.0
5	252	221	4	27	0	87.7
6	203	125	9	6	63	61.6
7	252	245	2	5	0	97.2
8	252	236	10	6	0	93.7
9(*)	196	150	45	1	0	76.5
10	252	249	3	0	0	98.8
11	252	244	3	5	0	96.8
12	252	237	12	3	0	94.0
13(*)	252	204	47	1	0	81.0
14	252	242	6	4	0	96.0
15	252	243	3	6	0	96.4
16	252	244	4	4	0	96.8
17(*)	252	203	43	6	0	80.6
18	252	244	5	3	0	96.8
19	252	245	5	2	0	97.2
20	252	239	5	8	0	94.8
21(*)	252	208	43	1	0	82.5
Total	5,187	4,747	266	111	63	Average 91.1 %
% Total	100	91.52	5.13	2.14	1.21	

Notes:
Col (2) : There were 252 days during the period March 12 through November 18, 1997.
Col (6) : Nine weeks in June and July at Home 6 have no data.
Col (7) = Col (3) / Col (2)

(*) Home with "older" data logger (limited storage and 3 downloads per day).

Table 4.2 Number of Complete Days of Residential Water Demand Data Recovery During 36 Week Period from Wednesday March 12, 1997 through Tuesday November 18, 1997

Home No	March			April				May					June				July					August				September				October					Nov		①	②	③	④
	1	2	3	4	5	6	7	8	9	10	11	12	13	14	15	16	17	18	19	20	21	22	23	24	25	26	27	28	29	30	31	32	33	34	35	36				
1	7	4	7	6	6	7	7	7	5	7	7	7	7	7	7	7	7	6	7	7	7	7	7	7	6	7	7	7	7	7	7	7	7	7	7	7	243	4	5	96.4
2	7	4	7	6	6	7	7	7	7	7	7	7	7	6	7	7	7	7	6	6	7	7	7	7	6	7	7	7	7	7	7	7	7	7	6	7	242	5	5	96.0
3	7	4	7	6	7	6	7	6	7	7	7	6	5	7	7	7	7	7	7	7	7	7	7	7	6	7	7	7	7	6	7	7	7	7	7	7	241	4	7	95.6
4	7	4	7	6	7	7	7	6	7	7	7	7	7	7	6	7	7	6	7	7	7	7	7	7	6	7	7	7	7	6	7	7	6	7	7	7	242	4	6	96.0
5	7	4	7	6	5	6	7	6	7	7	7	7	7	7	7	1	0	0	6	7	7	7	7	7	6	7	7	7	7	6	7	7	7	7	7	7	221	4	27	87.7
6	⊠	⊠	⊠	⊠	⊠	⊠	⊠	4	7	7	7	3	0	0	0	0	0	0	0	0	0	7	7	7	6	6	4	6	7	7	7	7	7	7	5	7	125	9	6	61.6
7	7	5	7	7	6	7	6	7	7	7	7	7	7	7	7	7	6	7	6	7	7	7	7	7	6	7	7	7	7	7	7	7	7	7	7	7	245	2	5	97.2
8	4	4	7	7	7	7	7	7	7	7	7	7	7	7	6	7	7	7	6	7	7	7	7	7	6	6	4	6	7	7	7	7	7	7	5	7	236	10	6	93.7
9(*)	⊠	⊠	⊠	⊠	⊠	⊠	⊠	⊠	2	5	6	7	6	7	6	5	7	6	5	6	7	6	6	7	4	6	3	2	4	2	3	7	6	6	6	7	150	45	1	76.5
10	7	5	7	7	7	7	7	7	7	7	7	7	7	7	7	7	7	7	7	7	7	7	7	7	6	7	7	7	7	7	7	7	7	7	7	7	249	3	0	98.8
11	7	4	7	7	7	7	7	7	7	7	7	7	7	5	5	7	7	7	7	7	7	7	7	7	6	7	7	7	7	7	7	7	7	7	7	7	244	3	5	96.8
12	1	3	7	6	6	7	7	7	7	7	7	7	7	7	7	7	7	7	7	7	7	7	7	7	6	7	7	7	7	7	7	6	6	7	7	7	237	12	3	94.0
13(*)	6	4	6	7	7	5	5	7	4	6	6	7	6	6	5	5	5	6	6	6	7	6	6	7	4	5	4	3	6	4	4	7	6	6	7	7	204	47	1	81.0
14	7	4	7	6	7	7	7	7	7	7	7	7	7	3	7	7	7	7	7	7	7	7	7	7	7	7	7	7	7	7	7	7	6	7	6	7	242	6	4	96.0
15	7	5	7	7	6	7	7	6	7	7	7	7	7	7	7	7	7	7	6	7	7	7	7	7	6	7	7	7	7	5	7	7	6	7	7	7	243	3	6	96.4
16	7	4	7	6	7	7	7	6	7	7	7	7	7	7	7	7	7	7	7	7	7	7	7	7	6	7	7	7	7	5	7	7	7	7	7	7	244	4	4	96.8
17(*)	6	4	7	6	6	6	6	6	5	6	6	7	6	7	4	4	6	5	4	4	6	7	6	7	4	6	4	2	7	5	5	7	6	6	7	7	203	43	6	80.6
18	7	4	7	6	6	7	7	7	7	7	7	7	7	7	7	7	7	7	7	7	7	7	7	7	6	7	7	7	7	6	7	7	7	7	6	7	244	5	3	96.8
19	7	4	7	6	7	7	7	7	7	7	7	7	7	7	7	7	7	7	7	7	7	7	7	7	6	7	7	7	7	6	7	7	7	7	6	7	245	5	2	97.2
20	7	6	7	7	6	7	7	6	7	7	7	7	7	7	7	7	7	6	7	7	7	6	7	7	4	6	7	7	7	7	7	7	7	6	7	4	239	5	8	94.8
21(*)	6	3	7	6	7	6	7	7	6	6	6	7	6	7	5	5	6	6	6	7	7	5	6	6	4	6	4	3	5	4	5	6	6	6	6	7	208	43	1	82.5
Sum	332 / 399			500 / 532				691 / 728					519 / 588				654 / 735					548 / 588				535 / 588				687 / 735					281/294		4747	266	111	91.1
%	83.2			94.0				94.9					88.3				89.0					93.2				91.0				93.5					95.6					

Notes: Col ① is number of **complete days** of data recovery at a home [sum of weeks 1 through 36 at each home].
Col ② is number of **partial days** of water data recovery at a home: [Col ② = 252 – Col ① – Col ②, except at homes 6 and 9].
Col ③ is number of **replicated days** of water demand data at a home [data copied from same home, same day, but different week].
Col ④ is percent of complete days recovered at a home. [Col ④ = Col ① / 252, except at homes 6 and 9]
Cells with shading are excluded from computations of residential water use statistics.
Cells with ⊠ signify times that home was either not occupied or not participating.
(*) signify homes with older data loggers having limited storage capacity (downloaded 3 times per day).

Occasionally no water demand data were recovered from an occupied home during a 24-hour period. This happened 174 out of 5,187 home-days or 3.4% of the total. The primary reasons for days with no data recovery were:

- ❖ modem connection severed (line cut by landscapers, corroded by moisture, etc)
- ❖ campus PC crashed (occurred August 30, 1997; affected all homes)
- ❖ home not accessible (homes 5 and 6 during June and July).

On most days with no data recovery, water demands from the same home and same day (but different week) were replicated and substituted for the missing data [(1)]. For example, water demands measured at Home 15 on Sunday July 20 were replicated and used to fill the missing record at Home 15 on Sunday July 27. As shown in column (5) of Table 4.1, data replication occurred on 111 out of 5,187 home-days (2.14% of the total). This step preserved a continuous, consistent data record for homes on the dead-end loop. When days with replicated data are included, the overall recovery rate approaches 94%. If the on-line portions of partial days are counted, then the overall data recovery rate exceeds 97% for the entire dead-end loop during the 36 week period from mid-March to mid November 1997.

Some unique difficulties arose at homes 5, 6 and 10. Communication with home 5 was lost for 21 days in late June and early July 1997 (see Table 4.2). Three weeks passed before access could be coordinated to correct the problem (the copper modem connections had been corroded by moisture from a nearby dryer vent). Aside from this incident, home 5 had a nearly complete record of water use. Therefore water demand data at home 5 from mid-June and late-July were replicated and used to "patch" this 21 day period of missing data.

At home 6, seven weeks of water demand data were completely lost in June and July 1997. In this case, property access needed to restore the modem connection was denied. As a compromise, a new wire was run from the data logger at home 8 to the outdoor pit meter at home 6. Home 6 was back on line in August. However, considering the extended down time at home 6, it was decided not to patch the missing 63 days with replicated data. This period is a gap in the data base at home 6 and labeled "lost data" in column (6) of Table 4.1.

[(1)] In general, the effect of data replication on the computed mean monthly water use statistics is quite small (less than ±2 percent in this study). See Appendix 4.1 for details.

At home 10 (apex of dead-end loop), a bleed was established to maintain a continuous flow of about 0.26 gpm past the water quality probes in the CMU (see Figure 3.4). In principle, it is straightforward to estimate the actual demand at home 10 as the difference between the total flow and the background bleed. In practice, however, this was not a simple exercise because the bleed rate fluctuated slightly due to pressure variations, temperature changes and other perturbations in the supply line.

When the bleed rate at home 10 was *under-estimated*, the computed water demands contained very long sequences of low flows. In subsequent analysis, these low-end flows were often mistaken for either leaks or outdoor use. Conversely, when the bleed rate was *over-estimated*, some of the computed water demands took negative values. While negative flows were not permissible from a physical point of view, this latter case proved easier to remedy and ultimately provided the basis for estimating water demands at home 10.

In the end, the best strategy for estimating water demands at home 10 was to over-correct the continuous readings by assuming an artificially high bleed rate (~ 0.33 gpm). All resulting negative water demands were then censored and the remaining demands were adjusted by adding 0.07 gpm (0.33 – 0.26) to the pulse intensities. Admittedly, some water pulses with low intensities were lost in this step; however, because low intensity pulses constitute a very minor portion of residential water consumption, these lost data are not considered to be serious. Statistics of the resulting water demands indicate that the derived data set provides a reasonable picture of water use at home 10.

RESIDENTIAL WATER USE STATISTICS

Point-of-Use water demands from data sets 1 and 3 (see Table 3.2) were used to compute sample statistics for indoor and outdoor residential water use at each of the 21 homes along the dead-end loop for the 31 week period spanning April 1 to October 31, 1997. The shaded portions in Table 4.2 identify the weeks which were not included in the statistical computations. The beginning period (three weeks in March) and ending period (two weeks in November) were omitted from the calculations for all 21 homes because they are partial months. Home 6 had no occupants in April and subsequent access problems (described above) in June and July. Home 9 was off-line until early May due to a change in ownership.

Water demand statistics for each home along the dead-end loop are presented in three ways, each corresponding to one of the following subsections:

- **Full-Period Summary**: overview of indoor and outdoor use at each home.
- **Monthly Summaries**: statistics for indoor and outdoor use at each home.
- **Previous Studies**: comparison against other water demand studies.

Full-Period Summary

Tables 4.3 and 4.4 provide summaries of indoor and outdoor[(2)] water demands, respectively, at each of the 21 homes along the dead-end loop for the 31 week period spanning April 1 to October 31, 1997. Included for each home are total and average values for pulse count, pulse volume, pulse duration, pulse intensity, and the utilization factor (key parameter in PRP model, see Chapter 2). Although outdoor demands represented only about 0.12% of the total pulse count, outdoor use comprised over 17% of the water consumption at the study site (Figure 4.1). The cumulative distributions of residence average values for indoor and outdoor use are plotted on normal probability paper in Figures 4.2 and 4.3.

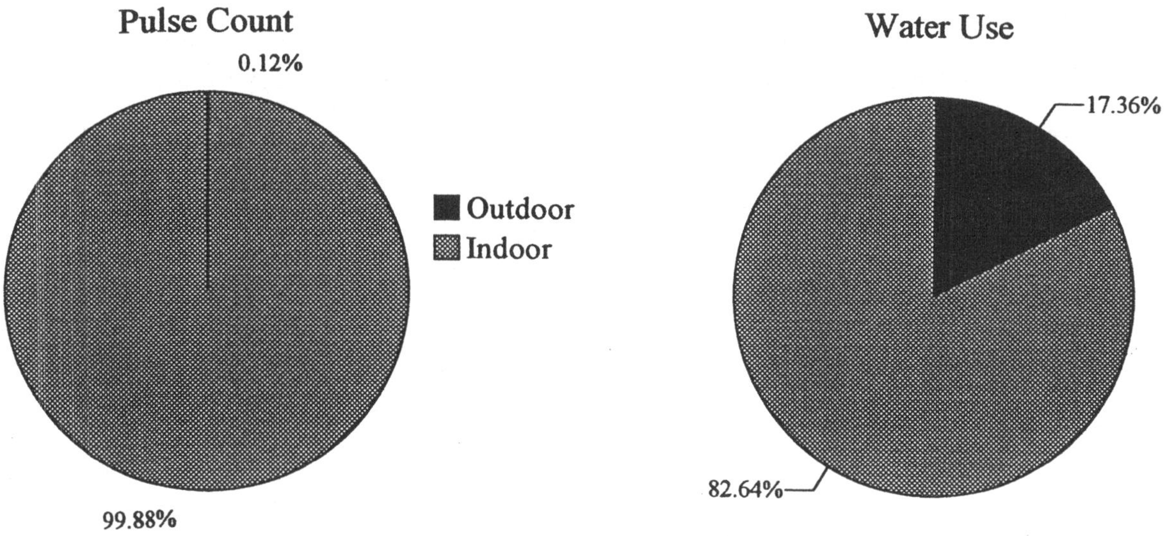

Figure 4.1 Indoor and outdoor use as measured by pulse count and water consumption.

---

(2) Water demands with a volume exceeding 80 gallons or a duration exceeding 30 minutes were classified as outdoor use (step 7 in Fig 3.6 and Fig 3.8). This cut-off between indoor and outdoor water use, while somewhat arbitrary, agrees well with fixture summaries given in the REUW report for showers (Figs 5.11 and 5.12) and washing machines (Fig 5.14).

Table 4.3 Indoor Water Use (SERPs) at Dead-End Loop for Period April 1, 1997 through October 31, 1997

Home No	Pulse Count		Pulse Volume (gal)		Pulse Duration (min)		Pulse Intensity (gpm)	Utilization Factor
	Total	Avg (min^{-1})	Total	Avg	Total	Avg		
1	13,897	0.0451	17,283	1.24	8,378	0.60	2.06	0.0272
2	12,500	0.0406	25,731	2.06	9,540	0.76	2.70	0.0310
3	14,951	0.0485	27,856	1.86	13,263	0.89	2.10	0.0430
4	60,845	0.1974	37,941	0.62	18,672	0.31	2.03	0.0606
5	13,176	0.0428	35,923	2.73	20,113	1.53	1.79	0.0653
6	12,819	0.0724	12,951	1.01	5,850	0.46	2.21	0.0330
7	20,913	0.0679	33,966	1.62	13,018	0.62	2.61	0.0422
8	17,304	0.0562	40,045	2.31	13,495	0.78	2.97	0.0438
9	9,924	0.0375	27,742	2.80	10,430	1.05	2.66	0.0394
10	18,430	0.0598	35,125	1.91	17,179	0.93	2.04	0.0557
11	10,419	0.0338	14,667	1.41	6,706	0.64	2.19	0.0218
12	24,209	0.0786	39,209	1.62	17,179	0.71	2.28	0.0557
13	11,768	0.0382	32,580	2.77	14,360	1.22	2.27	0.0466
14	16,561	0.0537	15,265	0.92	7,972	0.48	1.91	0.0259
15	13,645	0.0443	31,492	2.31	15,334	1.12	2.05	0.0498
16	15,760	0.0511	18,075	1.15	7,803	0.50	2.32	0.0253
17	9,301	0.0302	20,981	2.26	9,150	0.98	2.29	0.0297
18	24,616	0.0799	41,520	1.69	26,909	1.09	1.54	0.0873
19	5,326	0.0173	20,906	3.93	11,859	2.23	1.76	0.0385
20	17,198	0.0558	49,487	2.88	15,577	0.91	3.18	0.0505
21	20,985	0.0681	43,349	2.07	17,487	0.83	2.48	0.0567
Sum	364,547	1.2192	622,094	41.17	280,274	18.64	47.44	0.9290
GA-1	17,359	0.0581	29,624	1.961	13,346	0.888	2.26	0.0442
GA-2				1.706		0.769	2.22	

Notes: (1) All homes based on 214 days (308,160 min) except home 6 (123 days) and home 9 (184 days).
(2) GA is "global average"; see Table 4.5 for equations.

Table 4.4 Outdoor Water Use (SERPs) at Dead-End Loop for Period April 1, 1997 through October 31, 1997

Home No	Pulse Count		Pulse Volume (gal)		Pulse Duration (min)		Pulse Intensity (gpm)	Utilization Factor
	Total	Avg (day^{-1})	Total	Avg	Total	Avg		
1	17	0.0794	3,165	186	1,361	80	2.33	0.00442
2	16	0.0748	5,665	354	1,155	72	4.90	0.00375
3	4	0.0187	286	72	619	155	0.46	0.00201
4	1	0.0047	118	118	20	20	5.90	0.00006
5	42	0.1963	5,665	135	1,782	42	3.18	0.00578
6	5	0.0406	924	185	547	109	1.69	0.00309
7	1	0.0047	335	335	39	39	8.59	0.00013
8	16	0.0748	12,676	792	2,002	125	6.33	0.00650
9	90	0.4891	51,968	577	7,088	79	7.33	0.02675
10	9	0.0421	1,645	183	1,005	112	1.64	0.00326
11	7	0.0327	4,147	592	787	112	5.27	0.00255
12	71	0.3318	16,960	239	5,400	76	3.14	0.01752
13	7	0.0327	297	42	341	49	0.87	0.00111
14	29	0.1355	5,908	204	3,233	111	1.83	0.01049
15	9	0.0421	536	60	362	40	1.48	0.00117
16	26	0.1215	7,210	277	2,012	77	3.58	0.00653
17	1	0.0047	323	323	124	124	2.60	0.00040
18	33	0.1542	8,006	243	2,905	88	2.76	0.00943
19	6	0.0280	1,792	299	646	108	2.77	0.00210
20	14	0.0654	1,767	126	435	31	4.06	0.00141
21	21	0.0981	1,257	60	727	35	1.73	0.00236
Sum	425	2.0719	130,650	5,402	32,590	1684	72.44	0.11082
GA-1	20.24	0.0987	6,221	257.2	1,552	80.2	3.45	0.00528
GA-2				307.4		76.7	4.01	

Notes: (1) All homes based on 214 days (308,160 min) except home 6 (123 days) and home 9 (184 days).
(2) GA is "global average"; see Table 4.5 for equations.

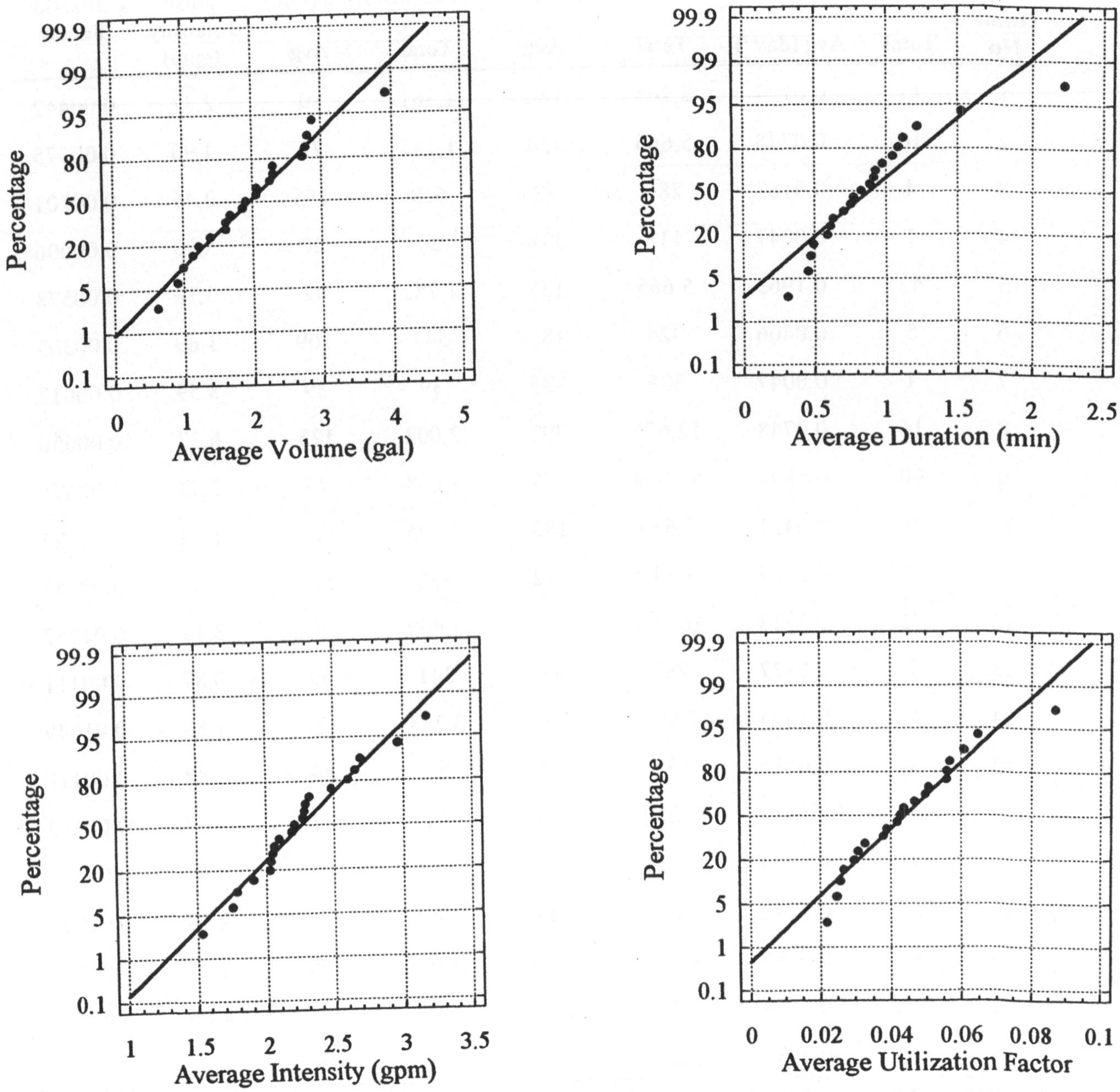

Figure 4.2 Probability distribution of residence averaged pulse volumes, pulse durations, pulse intensities and utilization factors for indoor water use.

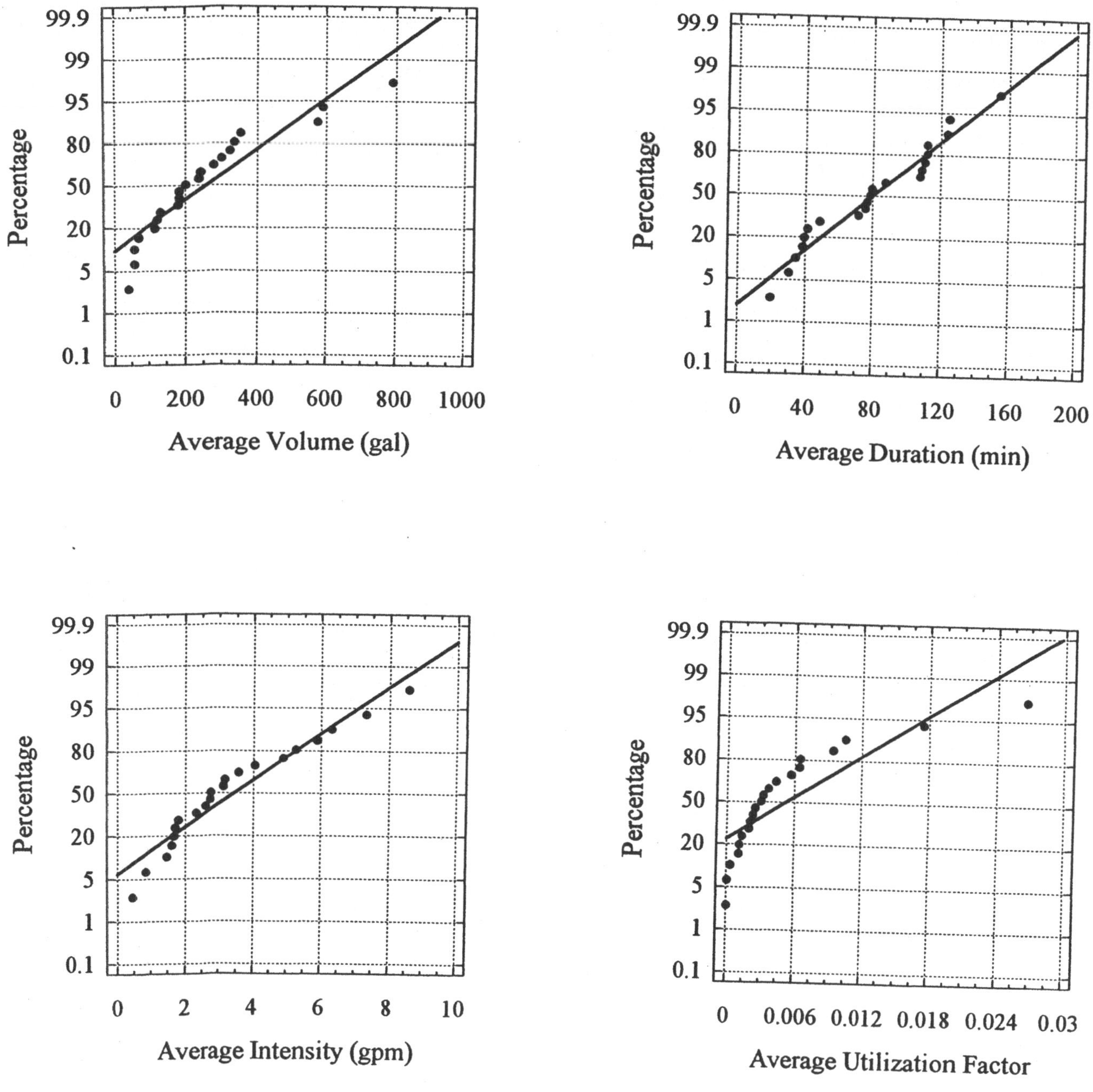

Figure 4.3 Probability distribution of residence averaged pulse volumes, pulse durations, pulse intensities and utilization factors for outdoor water use.

The sum near the bottom of Tables 4.3 and 4.4 is the total of each column. These values represent the collective behavior of all 21 homes in the neighborhood (hierarchy level H-3, Table 2.3). During the period April 1 to October 31, a total of 364,973 individual uses (indoor and outdoor combined) drawing 752,744 gallons of water were recorded. The vast majority of the demands (364,548 or 99.88%) were deemed indoor use.

Average Demand

To characterize water use habits for the entire neighborhood, two types of global average values are given at the bottom of Tables 4.3 and 4.4 for pulse volume, duration and intensity. The first global average ("GA-1") is the *arithmetic mean* of the 21 home averages. In Table 4.3, GA-1 of the indoor pulse volumes is 41.17 gallons per pulse/21 = 1.961 gallons per pulse. This estimate gives equal weight to each of the 21 homes on the loop.

In reality, however, each home does not contribute equally to water use in the neighborhood. As a consequence, the arithmetic mean (given by GA-1) does not conserve mass. To illustrate, when GA-1 is multiplied by the total number of indoor pulses, the result gives 714,879 gallons (1.961 gallons per pulse × 364,548 pulses) which is 15% greater than the total indoor demand of 622,094 gallons. The reason for this "discrepancy" is that there is a mild but significant negative correlation (R^2=0.28) between total pulse count and average pulse volume at the study site (for example, compare homes 4 and 19). Appendix 4.2 presents further details about estimating a neighborhood sample average.

The second global average value ("GA-2") provides a more representative picture of neighborhood water use and also preserves mass. The GA-2 statistic is a *weighted mean* of the 21 home averages. Total pulse count (i.e., arrival rate) serves as the weighting factor for volumes and durations; total pulse duration is the weighting factor for intensity. In Table 4.3, for example, GA-2 of the indoor pulse volumes is given by (622,094 gallons) / (364,547 pulses) = 1.706 gallons per pulse. This estimate is 13% less than the simple arithmetic average given by GA-1. Clearly, the choice of averaging schemes can have a measurable effect on the outcome. The equations and weighting factors used for both averaging schemes (GA-1 and GA-2) are summarized in Table 4.5.

Table 4.5
Summary of Equations Used to Compute Global Average Values Listed in Tables 4.3 and 4.4 for Pulse Volume, Duration and Intensity

Feature	Average Pulse Volume	Average Pulse Duration	Average Pulse Intensity
Entries in Rows 1-21	$\frac{v_i}{n_i}$	$\frac{t_i}{n_i}$	$\frac{v_i}{t_i}$
Column Sum	$\sum_{i=1}^{21}\frac{v_i}{n_i}$	$\sum_{i=1}^{21}\frac{t_i}{n_i}$	$\sum_{i=1}^{21}\frac{v_i}{t_i}$
Global Average-1	$\frac{1}{21}\sum_{i=1}^{21}\frac{v_i}{n_i}$	$\frac{1}{21}\sum_{i=1}^{21}\frac{t_i}{n_i}$	$\frac{1}{21}\sum_{i=1}^{21}\frac{v_i}{t_i}$
Wtg Factor for GA-1	1/21	1/21	1/21
Comment for GA-1	simple arithmetic mean	simple arithmetic mean	simple arithmetic mean
Global Average-2	$\frac{\sum_{i=1}^{21} v_i}{\sum_{i=1}^{21} n_i}$	$\frac{\sum_{i=1}^{21} t_i}{\sum_{i=1}^{21} n_i}$	$\frac{\sum_{i=1}^{21} v_i}{\sum_{i=1}^{21} t_i}$
Wtg Factor for GA-2	$\frac{n_i}{\sum_{i=1}^{21} n_i}$	$\frac{n_i}{\sum_{i=1}^{21} n_i}$	$\frac{t_i}{\sum_{i=1}^{21} t_i}$
Comment for GA-2	arrival weighted mean	arrival weighted mean	duration weighted mean

Notation:
n_i = total number of individual pulses at home i
v_i = total volume of water used at home i
t_i = total duration of individual pulses at home i

Appendix 4.2 provides additional insight on the differences between GA-1 and GA-2.

Utilization Factor

Values of the non-dimensional utilization factor (ρ) given in Table 4.3 and 4.4 are computed as the ratio of total pulse duration to total observation time. For instance, at home 10, the utilization factor for indoor use is ρ = (17,179 minutes)/(308,160 minutes) =0.0557. In this sense, the utilization factor can be viewed as the percentage of time or probability that the home is busy (i.e., using water). This interpretation is valid provided the utilization factor is small (i.e., $\rho<0.10$). All the household averaged values listed in Tables 4.3 and 4.4 are within this approximate threshold. Further discussion on the significance of ρ is presented in Chapter 6 with confirmation of the PRP hypothesis for residential water use.

Per Capita Demand

Expressing residential water use on the basis of per capita-days allows meaningful comparisons among water consumption rates at different locations. Table 4.6 summarizes daily per capita water use at the study site. The average indoor water use (without leaks) is 55 gallons per capita per day (gpcd). This usage is consistent with values in the 1999 REUW study which reported average indoor water use (without leaks) ranging from 51 to 70 gpcd (see REUW Table 5.8). In addition, at the dead-end loop study site, there was an average of 2.57 occupants per household, close to the value of 2.71 residents per household at 14 cities cited in the 1999 REUW study (Table 4.10, pg 76).

With per capita-day water use, there again is an issue of how best to estimate a representative global average value. As before, two averaging approaches are used in Table 4.6. The simple arithmetic average of 56.8 gpcd, given by GA-1, does not preserve mass(3). The weighted average of 55.0 gpcd, given by GA-2 (with weighting factor based on "capita-days" and listed in Table 4.6), does properly conserve mass.

Table 4.7 compares some the key results for household averaged indoor and outdoor water use at the study site. Indoor water use is a high frequency, short duration event with moderate intensities and small volumes. By contrast, outdoor water use tends to be a low frequency, long duration signal with relatively high intensities and large volumes.

---

(3) The 1999 REUW report uses the GA-1 approach to estimate daily per capita water use.

Table 4.6 Per Capita Water Use at Dead-End Loop for Period April 1, 1997 through October 31, 1997

Home No	Persons	Capita Days		Pulse Volume (gal)			Daily Per Capita Use (gpcd)		
		Cap-Day	Weight	Indoor	Outdoor	Combined	Indoor	Outdoor	Combined
1	2	428	0.0378	17,283	3,165	20,448	40.4	7.4	47.8
2	2	428	0.0378	25,731	5,665	31,396	60.1	13.2	73.4
3	3	642	0.0567	27,856	286	28,142	43.4	0.4	43.8
4	2	428	0.0378	37,941	118	38,059	88.6	0.3	88.9
5	3	642	0.0567	35,923	5,665	41,588	55.9	8.8	64.8
6	2	246	0.0217	12,951	924	13,875	52.6	3.8	56.4
7	4	856	0.0757	33,966	335	34,301	39.7	0.4	40.1
8	3	642	0.0567	40,045	12,676	52,721	62.4	19.7	82.1
9	2	368	0.0325	27,742	51,968	79,710	75.4	141.2	216.6
10	3	642	0.0567	35,125	1,645	36,770	54.7	2.6	57.3
11	2	428	0.0378	14,667	4,147	18,814	34.3	9.7	44.0
12	2	428	0.0378	39,209	16,960	56,169	91.6	39.6	131.2
13	3	642	0.0567	32,580	297	32,877	50.7	0.5	51.2
14	1	214	0.0189	15,265	5,908	21,173	71.3	27.6	98.9
15	2	428	0.0378	31,492	536	32,028	73.6	1.3	74.8
16	2	428	0.0378	18,075	7,210	25,285	42.2	16.8	59.1
17	2	428	0.0378	20,981	323	21,304	49.0	0.8	49.8
18	4	856	0.0757	41,520	8,006	49,526	48.5	9.4	57.9
19	2	428	0.0378	20,906	1,792	22,698	48.8	4.2	53.0
20	4	856	0.0757	49,487	1,767	51,254	57.8	2.1	59.9
21	4	856	0.0757	43,349	1,257	44,606	50.6	1.5	52.1
Sum	54	11,314	1.000	622,094	130,650	752,744	1,192	311.3	1,503
GA-1	2.57	538.8	0.0476	29,624	6,221	35,845	56.8	14.8	71.6
GA-2							55.0	11.5	66.5

Notes: (1) All homes based on 214 days except home 6 (123 days) and home 9 (184 days).
(2) GA is "global average"

Table 4.7 Summary of Average Residential Water Use at 21 Homes Along Dead-End Loop for Period April 1, 1997 through October 31, 1997

Indoor Water Use
(N=364,547 pulses for all 21 homes)

Pulse Feature (Units)	Minimum of Means	Maximum of Means	Arithmetic Mean (1)	Weighted Mean (2)	Median of Means
Volume (gal)	0.62	3.93	1.96	1.71	1.91
Duration (min)	0.31	2.23	0.89	0.77	0.83
Intensity (gpm)	1.54	3.18	2.26	2.22	2.21
Utilization factor	0.022	0.087	0.044	0.044	0.043
Use (gal per cap-day)	34.3	91.6	56.8	55.0	52.6
Arrivals (pulse per cap-day)	12.4	142.2	36.2	32.2	27.0

Outdoor Water Use
(N=425 pulses for all 21 homes)

Pulse Feature (Units)	Minimum of Means	Maximum of Means	Arithmetic Mean (1)	Weighted Mean (2)	Median of Means
Volume (gal)	42	792	257	307	204
Duration (min)	20	155	80	77	79
Intensity (gpm)	0.46	8.59	3.45	4.01	2.77
Utilization factor	0.00006	0.02675	0.00528	0.00528	0.00309
Use (gal per cap-day)	0.30	141.2	14.8	11.5	4.2
Arrivals (pulse per cap-day)	0.0012	0.2446	0.0458	0.0376	0.0210

Notes: (1) Arithmetic mean is GA-1 from Tables 4.3, 4.4 and 4.6
(2) Weighted mean is GA-2 from Tables 4.3, 4.4 and 4.6

Monthly Summary

Monthly Pulse Totals

Appendix B summarizes monthly totals for pulse counts, pulse volumes, and pulse duration for indoor, outdoor and combined water use at each of the 21 homes in the study site. This information shows the monthly distribution of indoor and outdoor water use data presented in Tables 4.3 and 4.4. As expected, residential indoor water use is quite uniform from month to month across the study site. For instance, over the 7 month period, the arrival rate for indoor water use averaged 32.2 pulses per capita-day (ppcd), ranging from a low of 30.4 ppcd in May to a high of 33.4 ppcd in September. In contrast, outdoor water use exhibits a strong seasonal pattern with most of the outdoor activity occurring during the hot dry summer months of July and August.

Monthly Pulse Statistics

Appendix C contains detailed monthly statistical summaries for the duration, intensity and volume of indoor single equivalent rectangular pulses (**SERPs**) for each of the 21 homes at the study site. A typical entry for home 2 from June 1997 appears in Table 4.8. These statistics were computed from Point-of-Use data set 1 (see Table 3.3) and are needed to estimate parameters in the Poisson Rectangular Pulse model for residential water use.

Table 4.8 Excerpt from Appendix C:
Summary Statistics for Indoor Water Demands, June 1-30, 1997

Home 2, Sample Size N=1688	Duration (sec)	Intensity (gpm)	Volume (gal)
Mean	48.0	2.78	2.22
Median	9.0	1.87	0.16
Mode	7.0	1.35	0.05
Standard Deviation	101.8	1.81	5.59
Minimum	2.0	0.11	0.01
Maximum	816.0	8.05	47.91
Skewness	3.9	0.64	4.17
Coefficient of Variation	2.1	0.65	2.51
Sum	81041.0	-.-	3753.59

Figures 4.4, 4.5 and 4.6 show the sample average, standard deviation, and coefficient of variation (standard deviation/average) for the duration, intensity and volume of the indoor SERPs for each of the 21 homes on a month by month basis (from statistics in Appendix C). In some cases, there is substantial variation in the pulse statistics among the homes. For example, consider the differences in mean volume or mean duration between homes 4 and 19. Interestingly, however, the coefficient of variation for pulse duration, intensity and volume is relatively stable month after month across the study site.

At any given home, the sample statistics are remarkably consistent from month to month, especially those for SERP intensity. This indicates that future efforts to continuously monitor residential water demands probably can obtain representative estimates of indoor water use by sampling periods of one month or less. This also suggests that computer simulation of indoor (and outdoor) water demands at individual homes may be achieved with relatively parsimonious probability models. In view of these findings, a simple method for simulating residential water demands based on the PRP model is outlined in Chapter 6.

Figure 4.7 shows the household utilization factor (ρ) and household amplification factor ($\eta=\rho/[1-\exp(-\rho)]$) for each of the 21 homes on a month by month basis. Both of these dimensionless parameters, instrumental in the PRP model, are discussed in Chapter 6. As with the SERP statistics, values of ρ and η are very consistent from month to month at a given residence. The utilization factor does not exceed 10% at any home. Across the entire study site (with N=21 homes) the average value for a single household is $\overline{\rho} \approx 4.5\%$.

The 31-day period from May 11 to June 10, 1997 is used extensively in Chapter 6 to test the PRP hypothesis for residential water use. Appendices D, E, and F, provide a close look at the SERPs from this 31-day test period. Appendix D shows the hourly distribution of SERP intensity, duration and volume. Clusters of data accentuate the diurnal nature of domestic water demands and signal the use of certain water fixtures in the home. As noted in a previous study (Buchberger and Wells 1996), the log-normal distribution provides a reasonable description of SERP intensity, duration and volume at most of the residences. Up to 35% or more of indoor demands have durations under 10 seconds. This, underscores the high-frequency, short-duration nature of indoor residential water use.

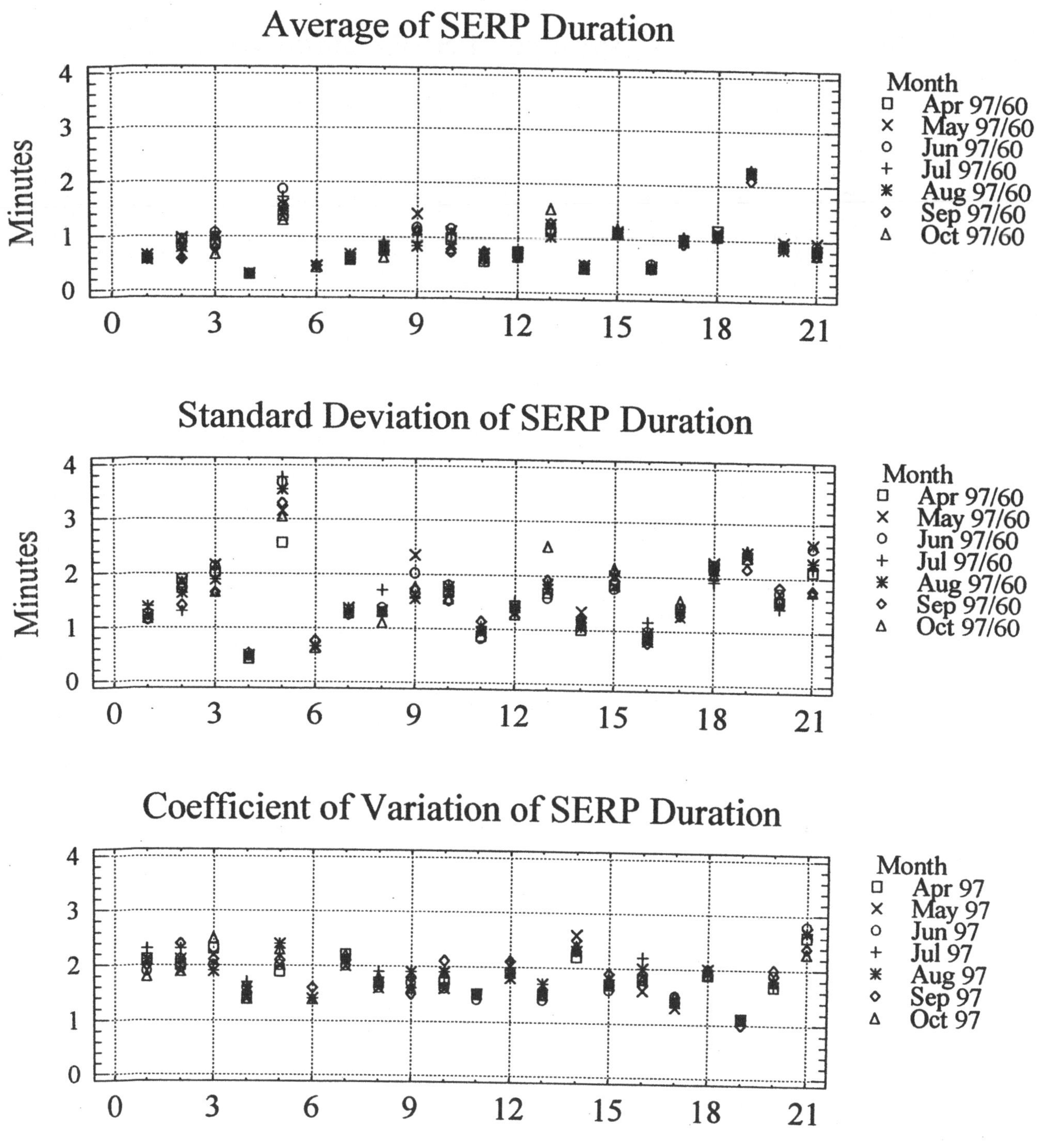

Figure 4.4 Monthly statistics for SERP duration at 21 homes.

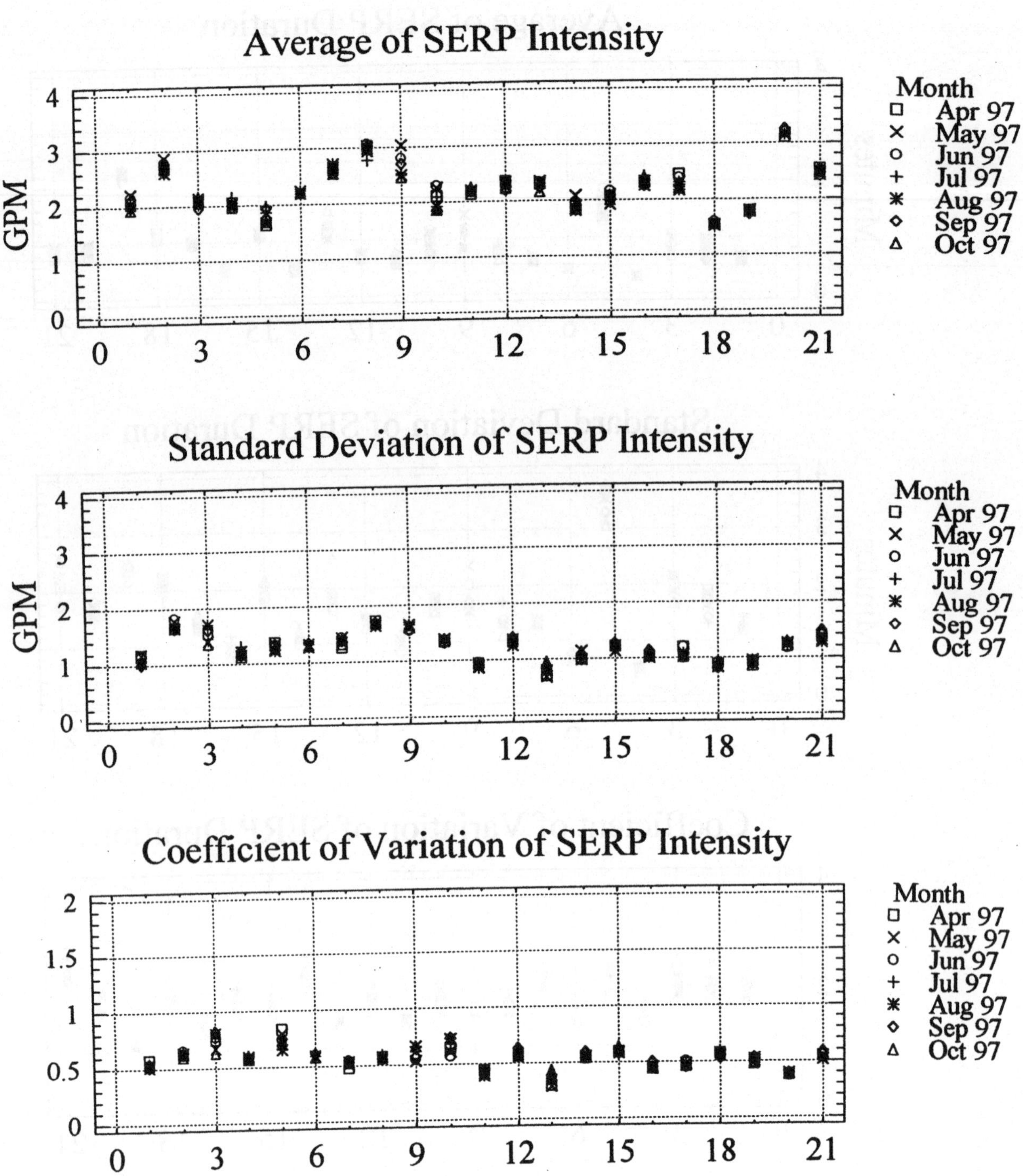

Figure 4.5 Monthly statistics for SERP intensity at 21 homes.

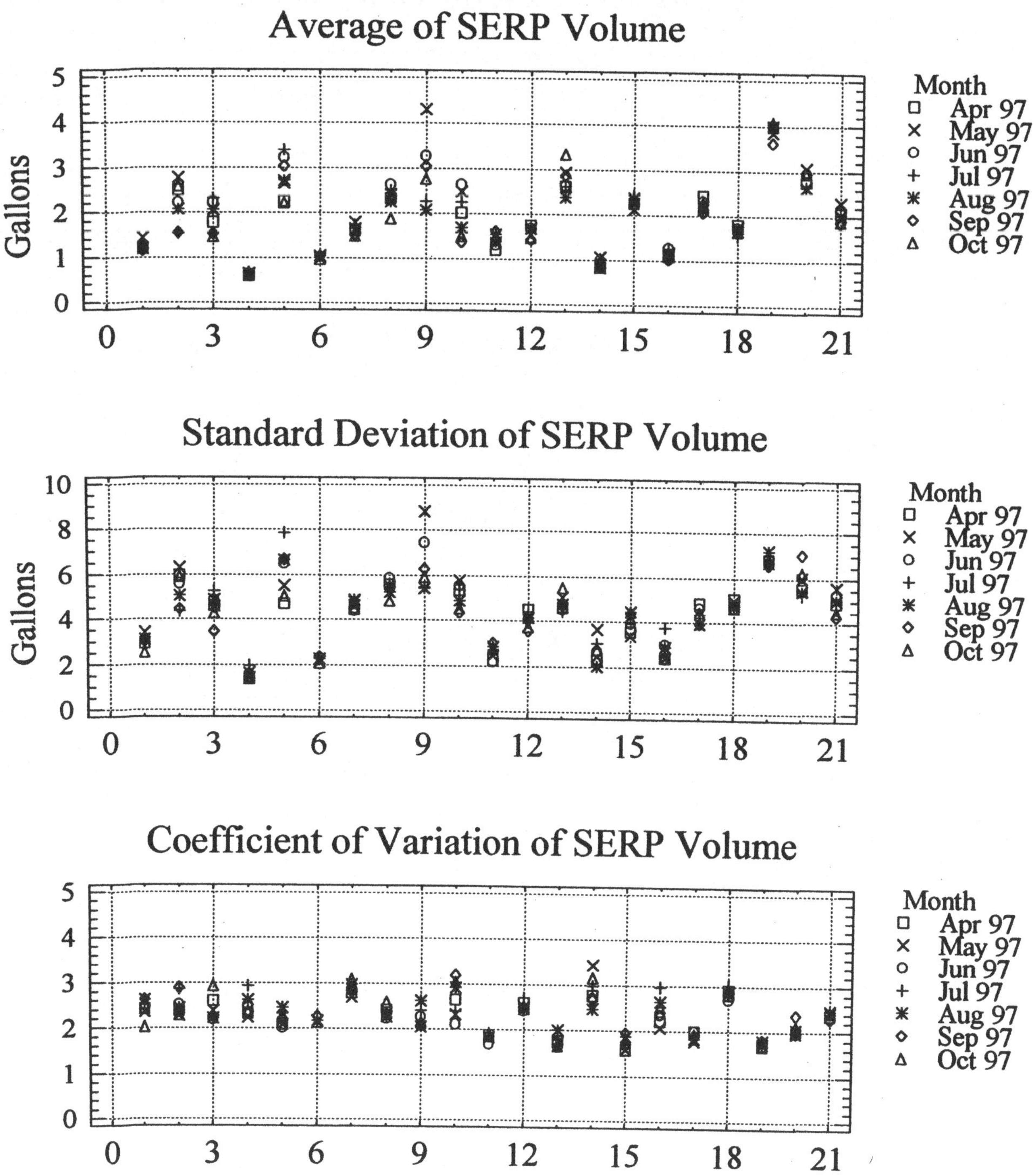

Figure 4.6 Monthly statistics for SERP volume at 21 homes.

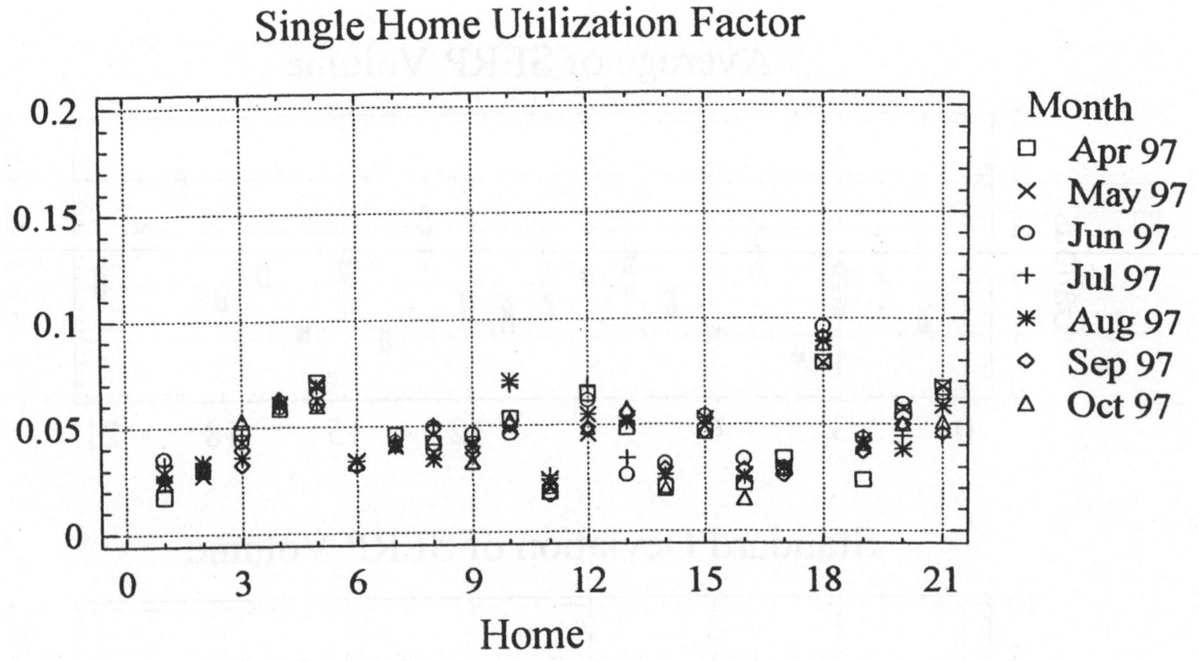

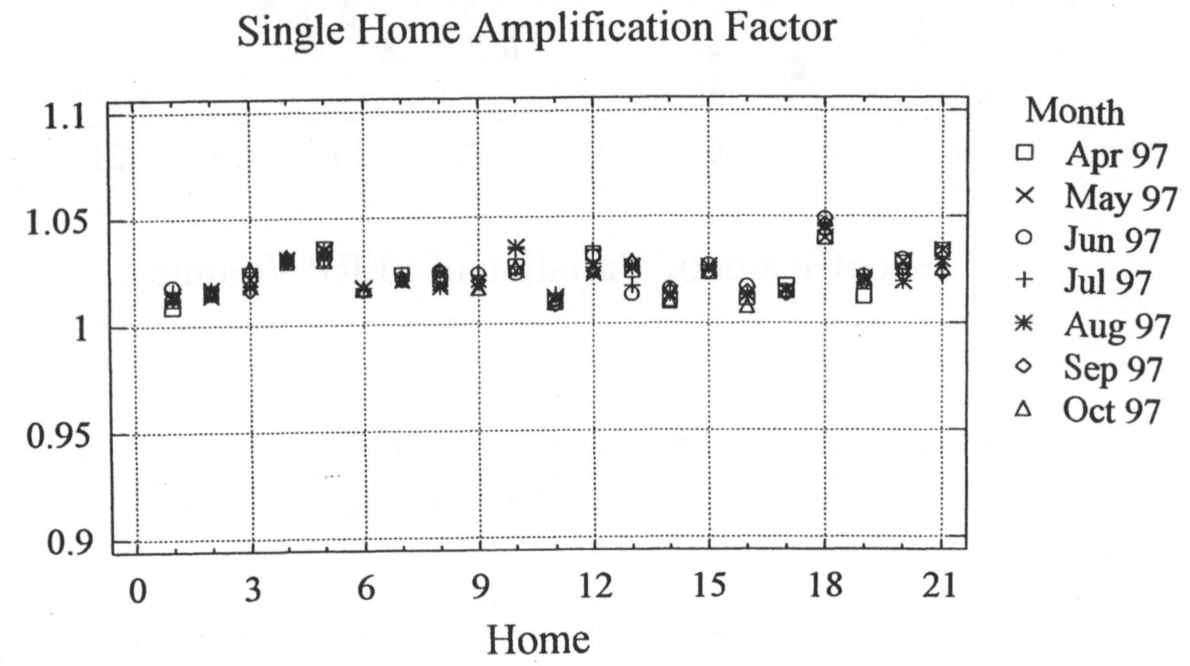

Figure 4.7 Monthly utilization factor and amplification factor at 21 homes.

Appendix E shows 3-D box plots for SERP intensity, duration and volume. The strong correlation between pulse volume and pulse duration is clear (volume is the product of intensity and duration). The ACFs in Appendix F indicate that total daily pulse counts and total daily water volumes are essentially independent from day to day. The normal distribution provides a good fit to the *daily* data, especially the daily pulse count (i.e., arrivals).

Previous Studies

Several benchmarks are mentioned here to further corroborate the reasonableness of residential water demands recorded at the dead-end loop from April 1 to October 31, 1997.

Point No 1: Assuming that no outdoor water use occurs at the study site from November through March, then the annual average water consumption during 1997 at the loop is approximately 61.7 gpcd [(4)] for indoor and outdoor use combined. This rate agrees with the 1995 water demand of 60.9 gpcd estimated from City of Milford billing records [(5)].

Point No 2: Ten studies on residential water use in the United States and Canada, dating back to the mid-1960s, are listed Table 4.9. Although sample size, study duration, and sampling frequency vary greatly among these studies, computed rates of indoor water use (excluding leaks) generally fall in the range of 50 to 65 gpcd. The estimate of indoor water use from the current study (55 gpcd) is comfortably between these limits. Additionally, the average number of occupants per household at the dead-end loop study site is 2.57. This is very similar to the value of 2.71 residents per household averaged across 14 cities in the 1999 REUW study (Table 4.10, page 76).

[(4)] From Table 4.6, the relevant rates are 66.5 gpcd (indoor and outdoor use) for 7 months and 55.0 gpcd (indoor only) for 5 months. The annual average is given by

$$[(66.5 \text{ gpcd} \times 7 \text{ mo}) + (55.0 \text{ gpcd} \times 5 \text{ mo})] / 12 \text{ mo} = 61.7 \text{ gpcd}$$

[(5)] In 1995, there were 19 homes on the dead-end loop and the total annual consumption was 1,076,400 gallons. Assuming 2.55 persons per household, this gives

$$\frac{1{,}076{,}400 \text{ gallons per yr}}{365 \text{ days per yr} \times 19 \text{ homes} \times 2.55 \text{ persons per home}} = 60.9 \text{ gpcd}$$

Table 4.9 Ten Studies on Residential Water Use

Investigator and Date	Study Location	Interval Between Readings	Sample Size		Indoor Water Use (gpcd)	
			no of homes	days per home	including leaks	excluding leaks
Linaweaver et al, 1966	41 residential areas in US	1 min	~ 10,950	(1–3 yrs)	~ 59	~ 53
Brown and Caldwell, 1984	6 Cities across US	recorded by user	210	14	na	66.2
Aher et al, 1991	Oakland, California	2 sec	25	60	na	40.0
Bowen et al, 1993	5 Cities across US	1 hour	242(*)	~ 42	72.4(*)	na
Anderson et al, 1993	Tampa Bay, Florida	??	25	??	na	50.7
Buchberger and Wells, 1996	Cincinnati, Ohio	1 sec	4	273 – 365	na	58.5
DeOreo et al, 1996	Boulder, Colorado	10 sec	16	21	58.8	52.7
Aquacraft 1998	Westminster, Colorado	10 sec 10 sec	20 20	21 21	63.5 50.6	62.0 47.3
Mayer et al, 1999 (REUW)	12 Cities in US/Canada	10 sec	1188	28	69.3	59.8
This Study 2002	**Cincinnati, Ohio**	**1 sec**	**21**	**252**	**na**	**55.0**

(*) excludes Altamonte Springs, Fla.
(**) Before and after water conserving devices were installed

WATER DEMANDS FROM WINTER 1996

Pilot testing of the water demand monitoring units began in Autumn 1995 (see Chapter 3, page 20). All homes were on-line by the end of January 1996. Hence, February and March 1996 represent the first two months with complete coverage of residential water demands at the dead-end loop. The 2-month 1996 winter period has two advantages over the 7-month 1997 period: (i) all water demands occurred indoors and (ii) the continuous bleed at the loop apex was absent[(6)]. Hence, it was not necessary to separate indoor and outdoor demands nor to extract the bleed rate from home 10 at the apex.

Water demands from February and March 1996 were processed to generate SERPs using an algorithm very similar to the flowchart in Figure 3.6. However, at the time, the criterion used to censor low flow events was not stringent. As a consequence, the SERP data for winter 1996 included thousands of tiny pulses which, according to the criteria in Figure 3.8, would be classified and then deleted as leaks. To illustrate, the hourly distributions of pulse intensities, pulse durations and pulse volumes from February and March 1996 are shown in Figures 4.8, 4.9 and 4.10, respectively. These plots show many SERPs with durations under 2 seconds or volumes under 0.02 gallons.

The net effect is that the pulse count (arrival rate) for the Winter 1996 data is elevated and, hence, the average pulse duration is biased low. As a consequence, the SERPs from February and March 1996 were not archived nor used to characterize statistical properties of residential indoor water use. Fortunately, total water consumption is not appreciably affected by the low flow censor threshold. Since travel time essentially depends on total consumption (or more precisely on the time required for total water consumption to displace a given pipe volume), the 1996 SERPs were used in a hydraulic model to estimate the winter season travel time distribution to all homes on the loop. Details on the travel time analysis are presented in Chapter 5 of this report.

---

(6) The water quality CMU at the DLA was not deployed until January 1997.

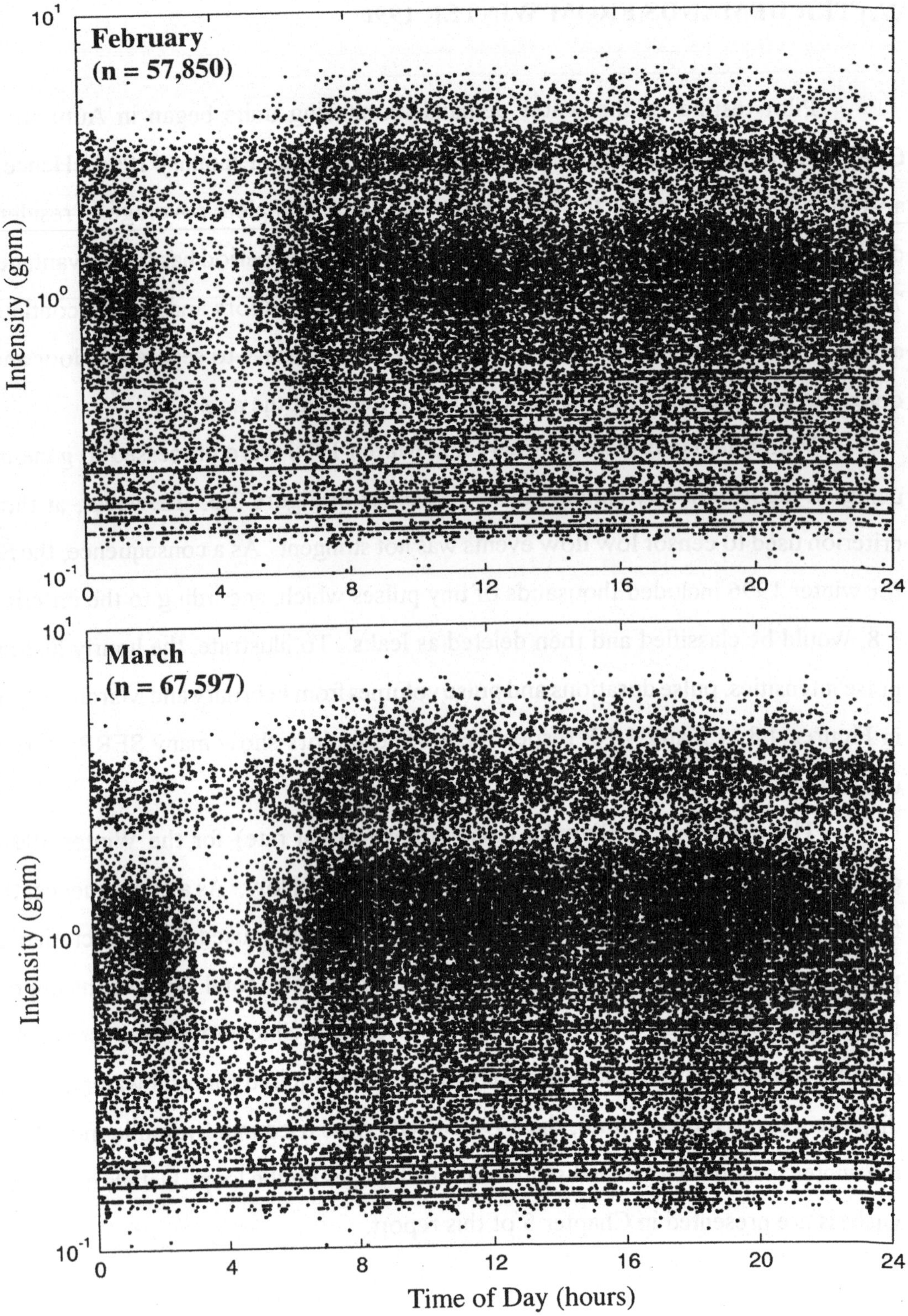

Figure 4.8 Daily distribution of pulse intensities in February 1996 (top) and March 1996 (bottom) at the dead-end loop.

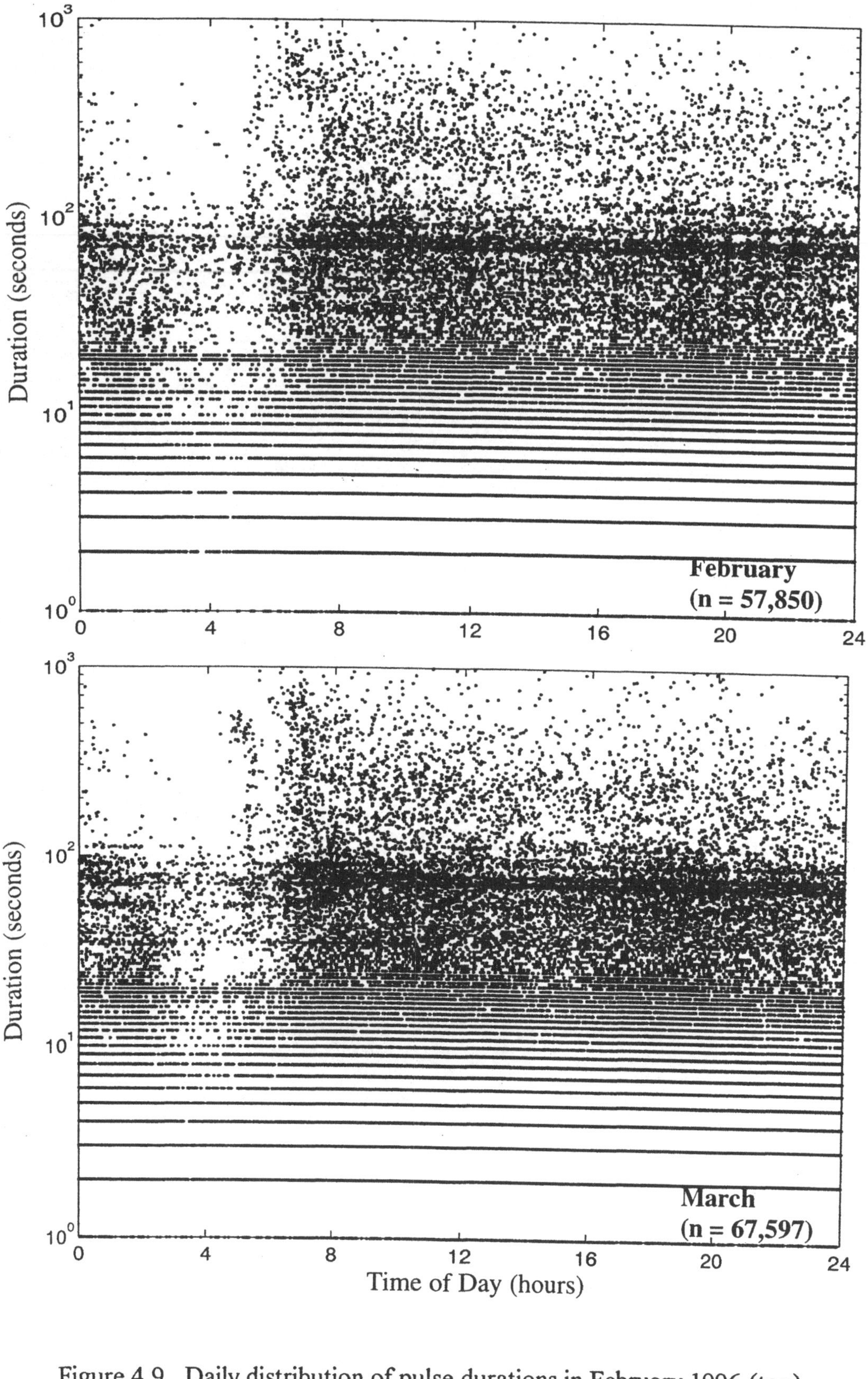

Figure 4.9 Daily distribution of pulse durations in February 1996 (top) and March 1996 (bottom) at the dead-end loop.

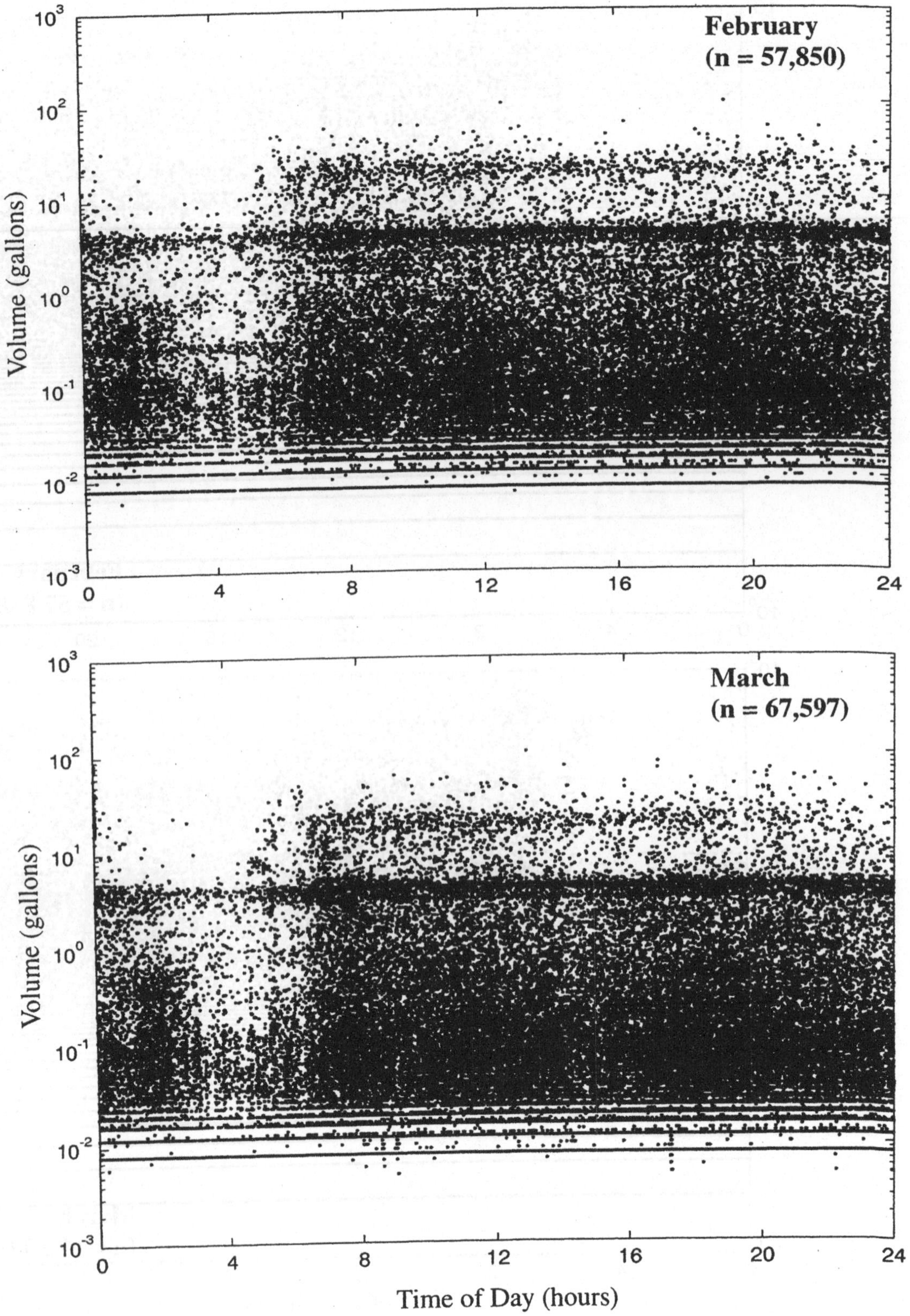

Figure 4.10 Daily distribution of pulse volumes in February 1996 (top) and March 1996 (bottom) at the dead-end loop.

SUMMARY (1997 Demand Data)

Nearly 365,000 residential water demands were recorded in a neighborhood of 21 homes along a dead-end loop during a seven month period from April to November 1997. The estimated data recovery rate approached 92%. All water demands were converted to single equivalent rectangular pulses (SERPs) and archived as digital text files on a compact disc (see Appendix A). With a time resolution of one second, these water demand data represent an unprecedented level of detail and constitute one of the most extensive records of residential water use ever compiled.

While outdoor use comprised only 0.12% of the total pulse count, it accounted for nearly 18% of the total water volume. The intensity, duration and volume of the indoor pulses is well described by a log-normal probability distribution. About 20% of all indoor pulses have durations under five seconds; about 35% have durations under ten seconds. This underscores the observation that indoor water use is a high frequency, short duration event with moderate intensity and small volumes. By contrast, outdoor water use tends to be a low frequency, long duration signal with relatively high intensity and large volumes.

There is remarkable consistency in the statistics of residential indoor water use from month to month at a given home and across the dead-end loop. The average indoor water use at the study site is 55 gallons per capita day (without leaks). This value is consistent with other studies of residential water use in North America and it agrees well with historical water use at the study site.

Considering indoor and outdoor uses together, there are about 1.2 demands per minute at the loop. During 1997, the mean daily demand (both indoor and outdoor uses) from all 21 homes on the loop is roughly 752,744 gal / 214 days = 3,517 gallons per day. Noting that the feeder stub and dead-end loop have a combined storage volume of 2,700 gallons, the flow-weighted average residence time in the pipe network at the study site is about 18 hours[(7)].

[(7)] The stub and loop traverse 1840 feet with a 6 inch cast iron pipe. This provides a storage capacity of 2,700 gal. The mean residence time is $T=V/Q = (2700 \text{ gal})/(3517 \text{ gal per day}) = 0.768 \text{ days} = 18.4 \text{ hours}$.

Appendix 4.1: Effect of Data Replication on Demand Statistics

Over a period of N days, suppose data are obtained on K days but are missing for M=N-K days. Observed values are replicated to fill in missing data. How does the inclusion of replicated data affect the sample mean and variance reported for the full period of N days? (In this study, typically, $M \approx 1$ day/month).

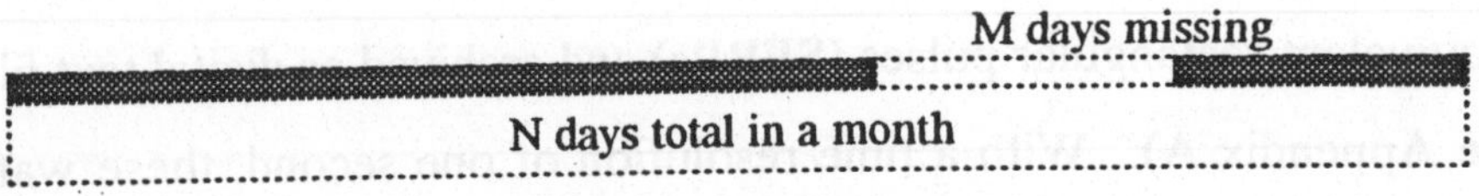

Let X_K denote observed data available for K=N-M days.
Let X_M denote replicated data for M=N-K days to fill in the missing values.
Let X_N denote the full period of data for N=K+M days, including M days of replicated data.

The nth moment of X_N is given by:

$$E[X_N^n] = \frac{(N-M)E[X_K^n] + (M)E[X_M^n]}{N} \tag{4.1.1}$$

Dividing by $E[X_K^n]$ gives a ratio of moments for the adjusted full period to the observed period,

$$\frac{E[X_N^n]}{E[X_K^n]} = 1 - \frac{M}{N}\left(1 - \frac{E[X_M^n]}{E[X_K^n]}\right) \tag{4.1.2}$$

The case n=1, gives the ratio of means,

$$\frac{E[X_N]}{E[X_K]} = 1 - \frac{M}{N}\left(1 - \frac{\mu_M}{\mu_K}\right) \tag{4.1.3}$$

Hence, sample means for K days and N days will be equal if there are no missing days (M=0) or if the mean of replicated data matches the mean of observed data ($\mu_M = \mu_K$). Noting Var[X] = $E[X^2] - E[X]^2$, equations (4.1.2) and (4.1.3) lead to

$$\frac{Var[X_N]}{Var[X_K]} = 1 - \frac{M}{N}\left[\left(1 - \frac{\sigma_M^2}{\sigma_K^2}\right) - \left(1 - \frac{M}{N}\right)\left(\frac{1}{\Theta_K^2}\right)\left(1 - \frac{\mu_M}{\mu_K}\right)^2\right] \tag{4.1.4}$$

where $\Theta = \sigma/\mu$. Variances will be equal if there are no missing days (M=0) or if the mean and variance of replicated data match the mean *and* the variance of observed data ($\mu_M = \mu_K$ and $\sigma_M^2 = \sigma_K^2$). Except for home 5, on average there was M=0.5 replicated days per home per month (see Table 4.1). As illustrated in Figure 4.11, with less than one (M<1) day of replication per month, it is unlikely that the estimate of the sample mean and variance will shift by more than 2% and 4%, respectively.

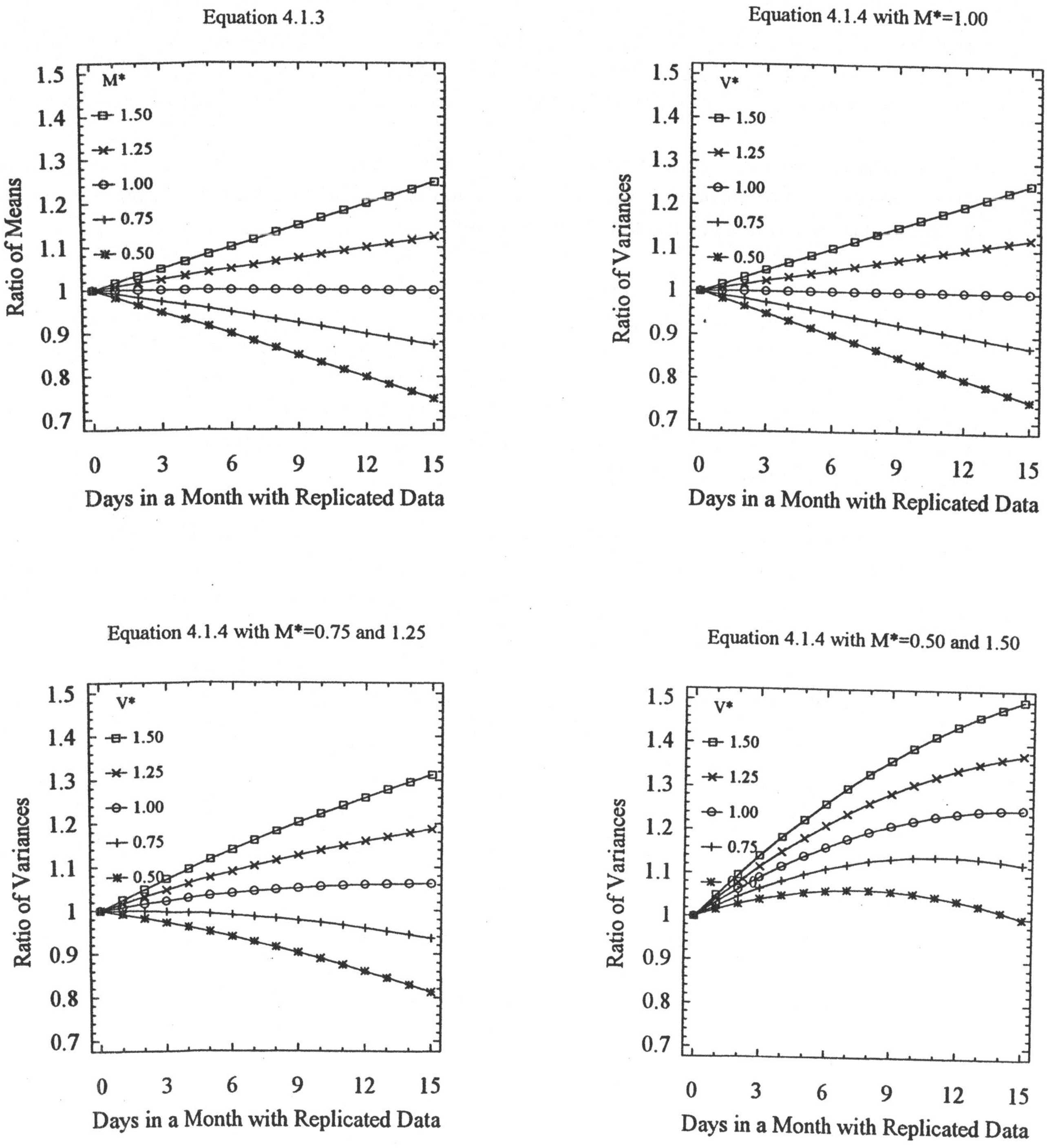

Figure 4.11 Effect of data replication on estimate of sample mean and sample variance for one month (N=30 days) of residential water demands. Plots are based on a coefficient of variation, Θ_K=0.50. $M^* = \mu_M / \mu_K$ and $V^* = \sigma_M^2 / \sigma_K^2$.

Appendix 4.2: Computing a Neighborhood Average

In Tables 4.3 and 4.4 two values (called GA-1 and GA-2) are given for the neighborhood average intensity, duration and volume of a SERP. Similarly in Table 4.6, two values are given for the neighborhood average daily per capita water use. Which value is appropriate?

A simple comparison in Table 4.10 highlights the difference between the two averages. The entries in columns (2) through (4) are measured quantities. The entries in columns (5) through (7) are computed values, each expressed as a ratio.

GA-1 is the simple arithmetic average of entries in each column[8]. It gives equal weight to each of the three homes, although each home does not contribute equally to water use. As a consequence, the computed average values do not conserve mass. For example, note that 12 occupants at 3 homes used a total of 750 gallons. However, (70 gal/cap)×(12 people)=840 gallons>750 gallons. In this example, the GA-1 approach overestimates the neighborhood consumption by 12 percent. The simple arithmetic average is *not appropriate* for estimating the average of a "busy-time" quantity.

GA-2 is a weighted average where weights are based on the relative contributions of each home to the total neighborhood water use (see weights in Table 4.5). The GA-2 approach preserves mass, (62.5 gal/cap)×(12 people)=750 gallons, and therefore, it is the appropriate method for estimating the average of a "busy-time" quantity (see Chapter 6).

Table 4.10 Comparison of two methods for computing neighborhood average.

(1) Home	(2) Number of Residents	(3) Number of Pulses	(4) Total Volume (gallons)	(5) Pulse per Capita (pulse/cap)	(6) Gallons per Capita (gal/cap)	(7) Volume per Pulse (gal/pulse)
1	2	90	180	45	90	2.0
2	4	300	300	75	75	1.0
3	6	180	270	30	45	1.5
Sum	12	570	750	150	210	4.5
GA-1	4	190	250	50	70	1.5
GA-2	4	190	250	47.5	62.5	1.31
GA-2/GA-1	1	1	1	0.950	0.893	0.873

Note: Col (5) = Col (3) / Col (2)   Col (6) = Col (4) / Col (2)   Col (7) = Col (4) / Col (3)

[8] GA-1 is the method used to compute average values for 12 study sites in the 1999 REUW report.

The see more clearly the distinction between both approaches, note that the entries in columns 5,6 and 7 of Table 4.10 can be expressed as a ratio Z=X/Y.

Let X_i represent the quantity in the numerator of the ratio for home i (for instance, X is pulses, gallons and gallons in columns 5, 6 and 7, respectively), and

Y_i represent the quantity in the denominator of the ratio for home i (for instance, Y is resident, resident and pulses in columns 5, 6 and 7, respectively)

The general expression for computing the average of N items is,

$$\overline{Z} = \sum_{i=1}^{N} w_i Z_i = \sum_{i=1}^{N} w_i \left(\frac{X_i}{Y_i} \right) \tag{4.2.1}$$

where w_i is a weighting factor with $\Sigma w_i = 1$. In the case of GA-1, the weighting factor is $w_i = N^{-1}$ so Eq 4.2.1 becomes

$$GA \cdot 1 = \frac{1}{N} \sum_{i=1}^{N} \left(\frac{X_i}{Y_i} \right) \tag{4.2.2}$$

which shows GA-1 is the simple arithmetic *average of ratios*. For GA-2, the weighting factor is

$$w_i = \frac{Y_i}{\sum_{i=1}^{N} Y_i}$$

When combined with Eq 4.2.1 this leads to

$$GA \cdot 2 = \frac{\sum_{i=1}^{N} X_i}{\sum_{i=1}^{N} Y_i} \tag{4.2.3}$$

which indicates that GA-2 is a *ratio of averages*. It can be shown

$$\frac{GA \cdot 2}{GA \cdot 1} = 1 + \rho_{Z,Y} \Theta_Z \Theta_Y \tag{4.2.4}$$

where $\Theta = \mu/\sigma$ is the coefficient of variation. Hence, the two estimates for the sample mean are equal only if the correlation between Z and Y is zero, $\rho_{Z,Y} = 0$. The simplest way to achieve this is for $Y_i = Y_j$. In this example, each home would need the same number of residents or the same number pulses. Since the ratio in Eq 4.2.4 < 1 (as appears in Table 4.10), it implies $\rho_{Z,Y} < 0$.

CHAPTER 5
FLOW REGIMES AND TRAVEL TIMES

OVERVIEW

Residential water demands from Winter 1996 are used in two analyses. First, the demands are used as input to a network model to simulate pipe flow through the dead-end loop. Second, results from the hydraulic simulation are analyzed to quantify flow behavior (regimes, directions and reversals) and to estimate bi-directional travel times along the pipe loop. Portions of this material are drawn from the MS thesis by Trent Schade (1996).

NETWORK MODEL

A network hydraulic model was developed to simulate the movement of water parcels through the dead-end loop. The closed loop was represented by 22 links and 22 nodes (see Figure 3.2). The node at the entrance to the loop served as the inflow supply point. The remaining 21 junctions were demand nodes, each corresponding to a home on the dead-end loop.

Using measured nodal demands, the hydraulic model calculated the rate and direction of flow through each pipe link along the dead-end loop. Because residential water demands could change at any second, the calculations were carried out with a 1-second time step. Despite the unsteady flow conditions, inertial effects were not considered. This is not a particularly serious oversight because the actual flow rates through the dead-end loop were quite low [(1)].

The continuity condition for the network model requires that the total inflow (Q_{in}) and the total outflow (Q_{out}) are equal at each node and across the entire loop or

$$\sum_{node}(Q_{in}) = \sum_{node}(Q_{out}) \tag{5.1}$$

---

(1) Flow through the dead-end loop was either stagnant or laminar over 97% of the time, especially during the winter months (see Tables 5.2 – 5.3 and Figures 5.1 – 5.2).

Similarly, energy conservation requires that the net headloss around the pipe loop be zero. For a single loop having a fixed pipe diameter, the energy condition can be expressed,

$$\sum_{i=1}^{N} L_i \left[a\, Q_i^{\,b} \right] = 0 \qquad (5.2)$$

where N is the total number of pipe links around the loop, L_i is the length of link i, Q_i is the flow through link i, a is a direction indicator ($a = +1$ if flow is clockwise, $a = -1$ if flow is counter-clockwise) and b is a flow exponent [(2)].

Initial flows in each pipe link were estimated from the known residential water demands (most of which were zero) at each node, taking care to satisfy Eq 5.1. Headlosses along each link were computed from the assumed initial flow rates; minor losses were neglected. If the computed head losses did not satisfy Eq 5.2, a correction, ΔQ, was added to each of the assumed flow rates and the head losses were re-estimated (see Schade, 1996 for details). In this manner, the computed flows quickly converged to a solution for the current iteration. The demand data for the next time interval were then read, initial flows were estimated again, and the process was repeated until all demands at all nodes were considered. Output from the network hydraulic model provided the rate and the direction of flow in each link on a second by second basis for the full simulation period (typically one month). These results were then used to calculate statistics of the flow through each pipe link and to estimate water travel times to each home.

HYDRAULIC CONDITIONS

Flow Regimes

Flow through any pipe link in the dead-end loop can be classified into one of four regimes (stagnant, laminar, transitional or turbulent) according to the Reynolds Number,

$$\mathbf{Re} = \frac{4Q}{\pi D \nu} \qquad (5.3)$$

---

(2) For laminar flow, b=1. For turbulent flow, b=1.85 with the Hazen-Williams formula or b=2 with the Darcy-Weisbach formula. Friction factor is assumed constant.

In Eq 5.3, D is the pipe diameter and ν is the kinematic viscosity of water. Table 5.1 summarizes the range of discharge rates for each of the four flow regimes at several different water temperatures in a six inch pipe. The range for laminar (and transition) flow is largest in the cool winter months when the kinematic viscosity of water is highest [(3)].

Using flow rates from the network hydraulic model, the relative frequency of the four flow regimes was computed for each pipe link. Results summarized in Tables 5.2 and 5.3 (plotted in Figs 5.1 and 5.2) indicate that during winter 1996, the most prevalent regime was stagnation (no flow, ~53% of the time) followed by laminar flow (~46% of the time). There were occasional instances of transitional flow, especially in the links near the loop entrance. Practically speaking, however, turbulent flow did not occur in the loop. This is noteworthy, because existing water quality models for distribution systems assume that *quasi-steady continuous turbulent flow* exists throughout the network. At this site (and most other dead-ends zones), the actual flow condition is most likely to be *random intermittent laminar flow*.

The relative frequency of stagnation is the same for all links around the dead-end loop. By continuity, if one link is stagnant, then all must be [(4)]. The incidence of laminar, transition, and turbulent flow, however, varies with location around the loop. Homes closest to the loop entrance can experience slightly higher Reynolds numbers than homes near the apex. To see why this is so, consider concurrent and nearly equal demands occurring only at two homes on opposite branches of the loop (e.g., homes 3 and 18). The demand for home 3 will be delivered mainly via the left branch while that for home 18 will arrive mainly through the right branch. The section of the loop between homes 3 and 18 will carry a residual flow to satisfy continuity and energy conservation, but the rate will be less than the flow through the left and right branches going directly to homes 3 and 18, respectively.

---

(3) Owing to the random intermittent nature of the residential water use, it is unlikely that any sequence of demands will be sustained for a period sufficient to establish a fully developed laminar flow (with parabolic velocity profile) in the pipe loop. Nonetheless, an operational definition is adopted here: whenever $0<\mathbf{Re}<2000$, the flow in the link is called "laminar".

(4) This contrasts sharply to flow behavior in a dead-end stem (which are much more common than dead-end loops in water distribution systems). In a dead-end stem, only the pipe links upstream of a demand experience flow. Hence, the relative frequency of stagnation varies from link to link along a dead-end stem, increasing in the downstream direction.

Table 5.1 Rate of Water Discharge Through a Six Inch Pipe. (Q is gpm)

Kinematic Viscosity of Water (ft²/sec)	Regime (1): Stagnant Conditions Re = 0	Regime (2): Laminar Flow 0<Re<2,000	Regime (3): Transition Flow 2,000<Re<4,000	Regime (4): Turbulent Flow Re>4,000
		Winter Season (Water Temperature = 50°F)		
1.41×10^{-5}	$Q = 0$	$0 < Q < 5.0$	$5.0 < Q < 10.0$	$Q > 10.0$
		Summer Season (Water Temperature = 70°F)		
1.06×10^{-5}	$Q = 0$	$0 < Q < 3.7$	$3.7 < Q < 7.5$	$Q > 7.5$
		Annual Average (Water Temperature = 60°F)		
1.22×10^{-5}	$Q = 0$	$0 < Q < 4.3$	$4.3 < Q < 8.6$	$Q > 8.6$

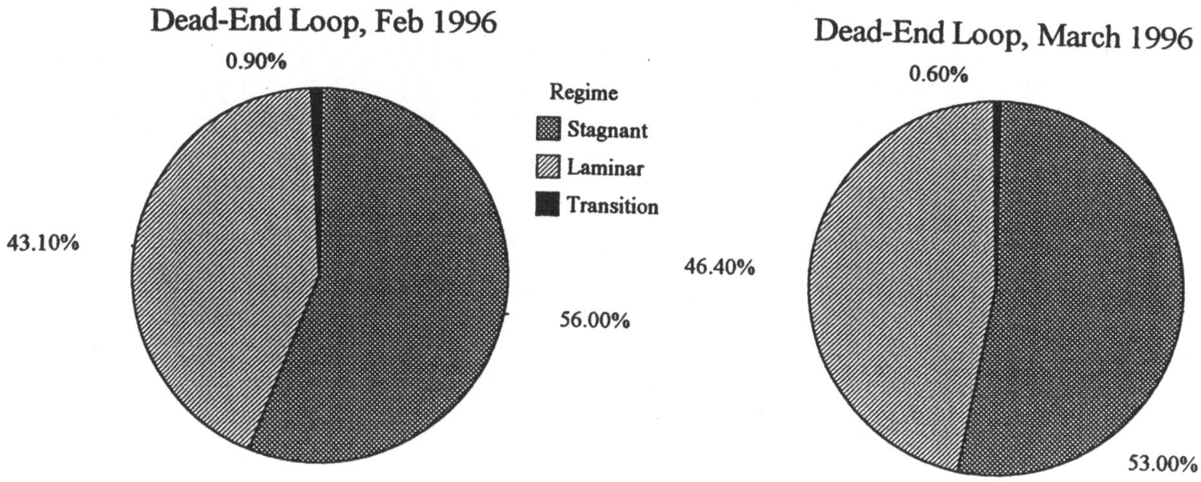

Figure 5.1 Proportion of stagnant, laminar and transition flow observed in February 1996 (left) and March 1996 (right) at the dead-end loop.

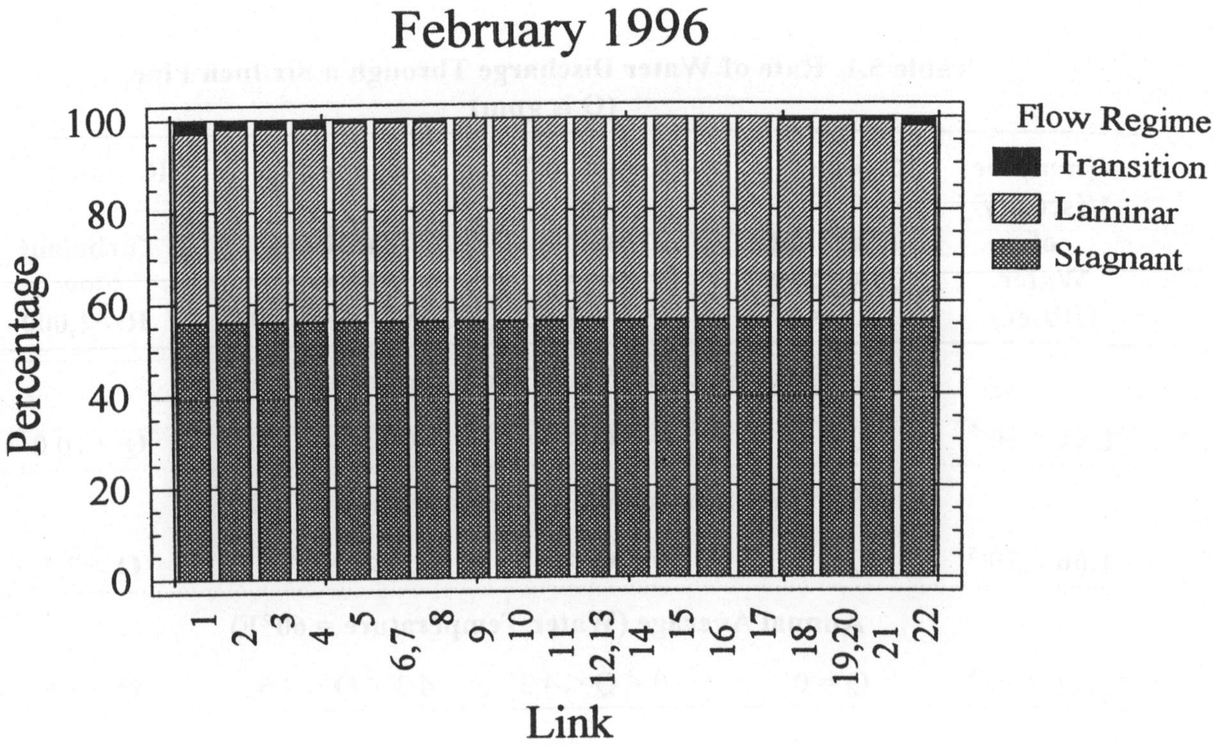

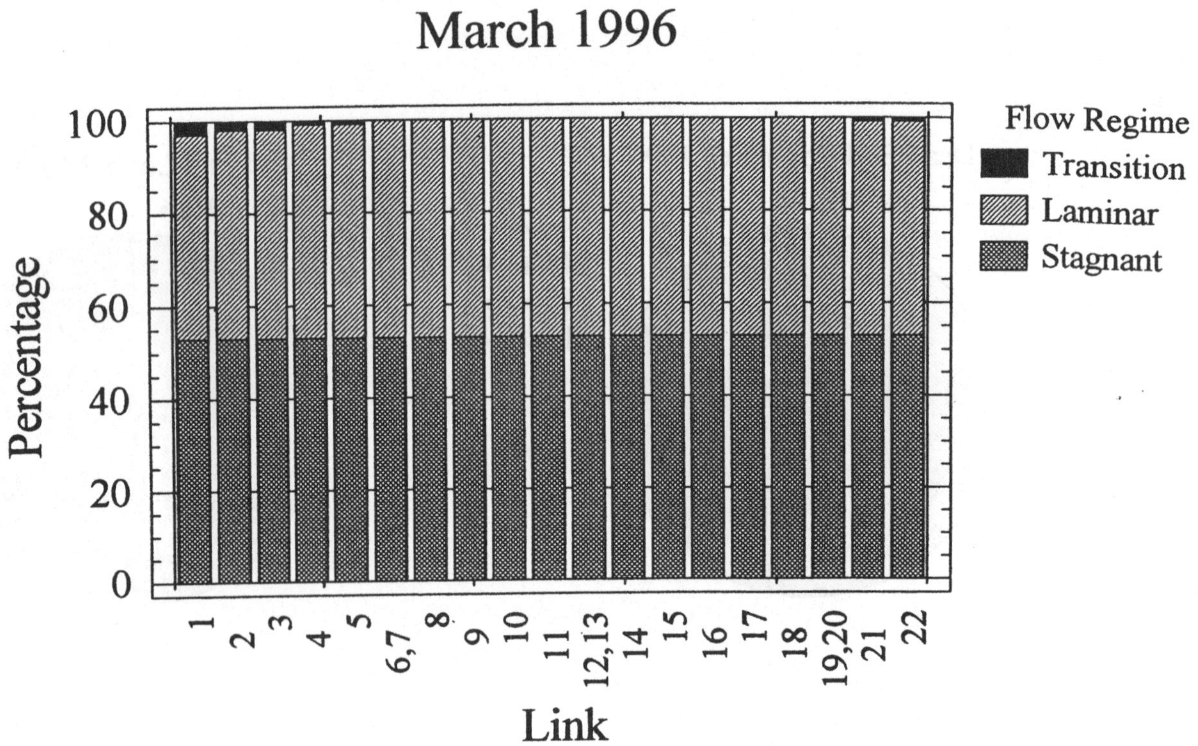

Figure 5.2 Variation of flow regime with pipe link in February 1996 (top) and March 1996 (bottom) at the dead-end loop.

Table 5.2 Flow Regimes in Dead-end Loop during February 1996 (Entries are percent of time)

Pipe Link	Stagnant Condition	Laminar Flow	Transition Flow	Turbulent Flow
1	55.52	41.48	2.98	0.02
2	55.52	41.51	2.95	0.02
3	55.52	42.08	2.39	0.01
4	55.52	42.43	2.04	0.01
5	55.52	42.81	1.66	0.01
6	55.52	43.62	0.86	0
7	55.52	43.62	0.86	0
8	55.52	43.97	0.51	0
9	55.52	44.18	0.30	0
10	55.52	44.23	0.25	0
11	55.52	44.25	0.23	0
12	55.52	44.33	0.15	0
13	55.52	44.33	0.15	0
14	55.52	44.30	0.18	0
15	55.52	44.28	0.20	0
16	55.52	44.17	0.31	0
17	55.52	44.04	0.44	0
18	55.52	43.88	0.60	0
19	55.52	43.75	0.73	0
20	55.52	43.75	0.73	0
21	55.52	43.21	1.26	0.01
22	55.52	42.45	2.01	0.02

Notes: (1) Adapted from Schade (1996) assuming T=50 °F in February 1996.
(2) See Table 5.1 for discharge rates corresponding to each flow regime.

Table 5.3 Flow Regimes in Dead-end Loop during March 1996 (Entries are percent of time)

Pipe Link	Stagnant Condition	Laminar Flow	Transition Flow	Turbulent Flow
1	52.92	44.58	2.49	0.01
2	52.92	44.70	2.37	0.01
3	52.92	45.51	1.57	0
4	52.92	45.74	1.34	0
5	52.92	46.08	1.00	0
6	52.92	46.68	0.40	0
7	52.92	46.68	0.40	0
8	52.92	46.79	0.29	0
9	52.92	46.99	0.09	0
10	52.92	47.04	0.04	0
11	52.92	47.06	0.02	0
12	52.92	47.07	0.01	0
13	52.92	47.07	0.01	0
14	52.92	47.03	0.05	0
15	52.92	47.02	0.06	0
16	52.92	46.99	0.09	0
17	52.92	46.90	0.18	0
18	52.92	46.75	0.33	0
19	52.92	46.69	0.39	0
20	52.92	46.69	0.39	0
21	52.92	46.45	0.63	0
22	52.92	45.60	1.47	0.01

Notes: (1) Adapted from Schade (1996) assuming T=50 °F in February 1996.
(2) See Table 5.1 for discharge rates corresponding to each flow regime.

Flow Direction

Water moves through a dead-end stem in one direction only [6], along a path from the entrance to the point of consumption, in response to downstream demands. By contrast, flow in the dead-end loop always moves in two directions, clockwise and counter-clockwise. Both conditions occur simultaneously in the dead-end loop because any single point of demand can be supplied from two directions.

Suppose a dead-end loop, with a fixed diameter and a total length L, has one active node with a demand Q. Let Q_{CW} be the rate of flow in the clockwise direction (see Figure 5.3). From conservation of energy and mass, the fraction of the total demand traveling in clockwise direction is given by

$$\frac{Q_{CW}}{Q} = \frac{1}{1 + \left(\frac{L_{CW}}{L_{CCW}}\right)^b} \tag{5.4}$$

Here b is a flow regime exponent and L_{CW} and L_{CCW} are the lengths of the flow paths along the loop in the clockwise and counter-clockwise directions, respectively. For laminar flow, b=1 and Eq 5.4 becomes,

$$\frac{Q_{CW}}{Q} = \frac{L_{CCW}}{L_{CW} + L_{CCW}} = \frac{L_{CCW}}{L} = L* \tag{5.5}$$

while for turbulent flow assuming a constant friction factor with b=0.5, Eq 5.4 is

$$\frac{Q_{CW}}{Q} = \frac{\sqrt{L_{CCW}}}{\sqrt{L_{CW}} + \sqrt{L_{CCW}}} = \frac{\sqrt{L*}}{\sqrt{L*} + \sqrt{1 - L*}} \tag{5.6}$$

---

[6] Exceptions can occur but they are very rare. An extraordinary event (e.g., fire demand, pipe burst, etc) near the entrance of a dead-end stem could cause water to flow "backwards" out of the pipe stub.

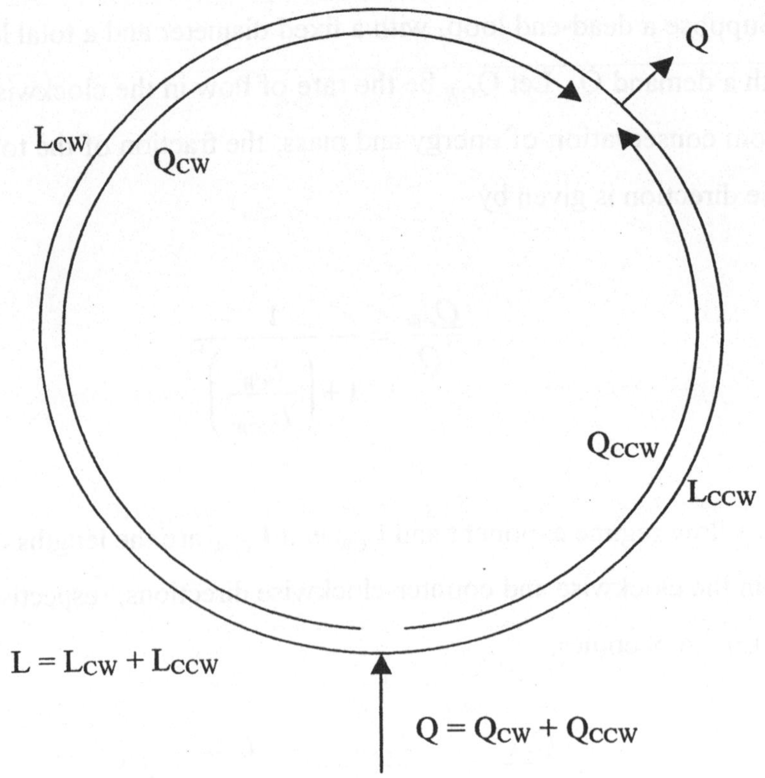

Figure 5.3 Definition sketch for flow pattern resulting from one demand on the dead-end loop. (CW = clockwise; CCW = counter-clockwise).

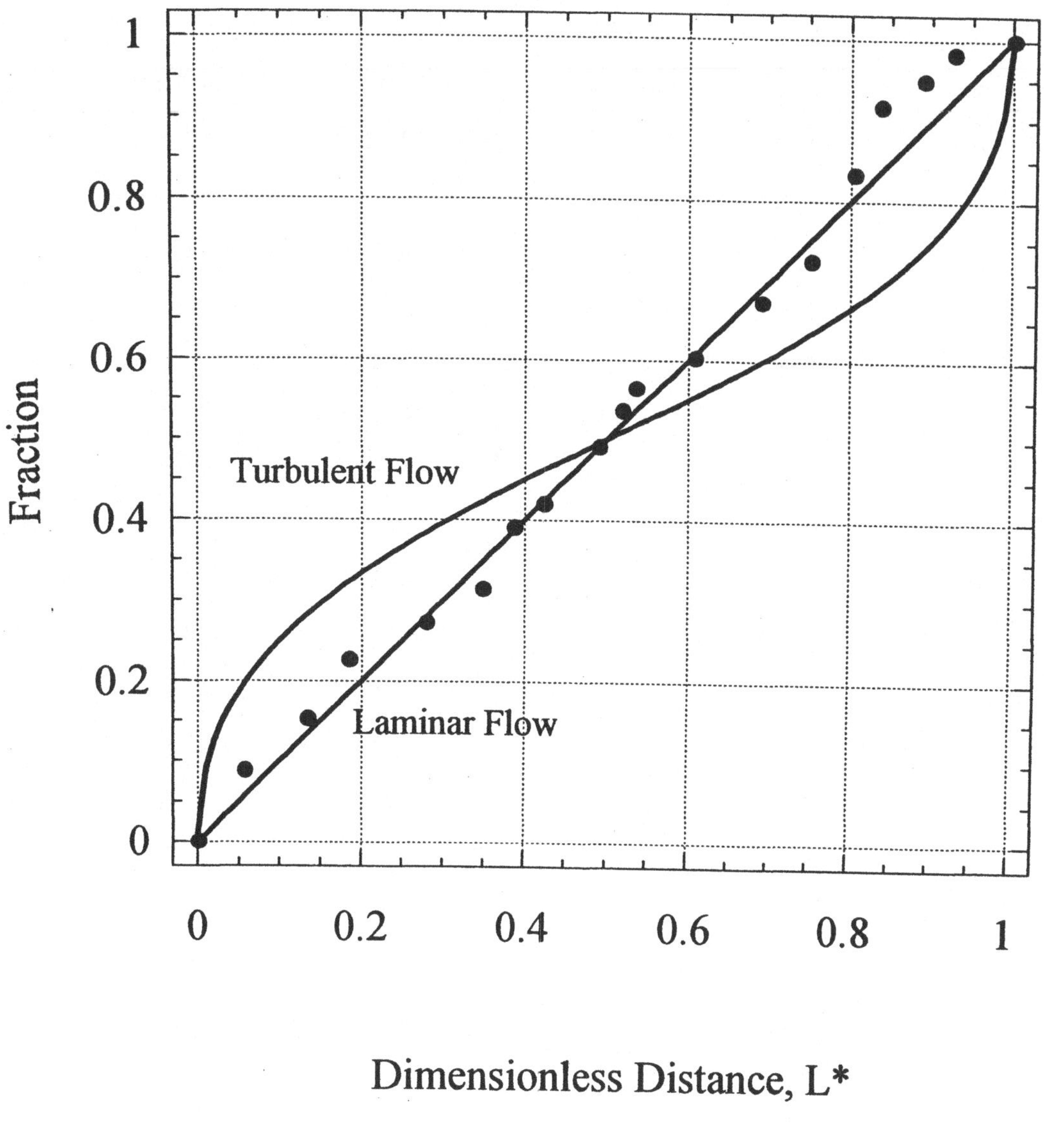

Figure 5.4 Fraction of clockwise flow in dead-end loop.

Table 5.4 Percent of Time that Flow is Clockwise at Dead-End Loop During February and March 1996.

Link	Nodes	Length (ft)	L_{CCW} (ft)	$L^*=L_{CCW}/L$	Q_{CW}/Q
1	0, 1	208	--	100.0	100.0
2	1, 2	17	1257	92.7	98.2
3	2, 3	81	1207	89.0	94.9
4	3, 4	68	1133	83.6	91.6
5	4, 5	17	1090	80.4	83.2
6	5, 6	73	1045	75.1	72.5
7	6, 7	40	989	75.1	72.5
8	7, 8	69	935	69.0	67.3
9	8, 9	155	823	60.7	60.3
10	9, 10	40	725	53.5	56.4
11	10, 11	1	704	51.9	53.6
12	11, 12	27	690	49.0	49.0
13	12, 13	79	637	49.0	49.0
14	13, 14	48	574	42.3	41.9
15	14, 15	54	523	38.6	39.0
16	15, 16	48	472	34.8	31.4
17	16, 17	139	379	27.9	27.2
18	17, 18	117	250	18.4	22.6
19	18, 19	1	191	13.3	15.2
20	19, 20	40	171	13.3	15.2
21	20, 21	151	76	5.6	8.8
22	21, 0	299	--	0.0	0.0
		Σ 1772			

Notes: (1) L_{CCW} is based on the midpoint of the pipe link. The total apparent loop length, L = 1265 + 91 = 1356 feet; 91 ft is the difference (299–208) in the lengths of the left and right feeder links to the first nodes on the left and right branches of the loop.
(2) Q_{CW}/Q is weighted average of results from hydraulic model for February and March 1996.

In Eqs 5.5 and 5.6, L* is a dimensionless distance representing the percentage of the total loop where the flow is counter-clockwise. Equations 5.5 and 5.6 are plotted in Figure 5.4 for values of L* ranging from 0 to 100 percent. For both laminar and turbulent flow cases, the majority of the flow takes the shortest route from the entrance to the point of demand. When the demand occurs at L*=0.50, the supply is split evenly between the left and right branches for both cases. Under laminar conditions (where energy loss varies with average velocity), the fraction of clockwise flow is directly proportional (equal) to L*. Under turbulent conditions (where energy loss depends on the average velocity squared), the fraction of clockwise flow varies in a nonlinear manner with L*. The turbulent fraction of clockwise flow exceeds the laminar percentages when 0<L*<0.50 (left branch of the loop) but falls below the laminar percentages when 0.50<L*<1.00 (right branch of the loop).

Also included in Figure 5.4 are results from the hydraulic model on the frequency of clockwise flow through all 22 pipe links at the dead-end study site during February and March 1996 (three links with unoccupied homes were combined with adjacent links). The data points plotted in Figure 5.4 (also summarized in Table 5.4) agree very closely with the line for the laminar regime. This further corroborates the observation that flows with low Reynold's numbers prevailed at the study site. Deviations between the model data points and the theoretical laminar line can be attributed primarily to random variations in the magnitude and the timing of residential water demands at the dead-end loop.

Flow Reversals

A flow reversal occurs whenever the direction of water movement through the loop changes. Reversals are rare in links close to the loop entrance where either clockwise flow (left branch) or counter-clockwise flow (right branch) predominate (see Table 5.4). Near the apex, however, flow in either direction is equally likely and, consequently, flow reversals are common.

Consider an idealized dead-end loop having nodes with equal demands uniformly distributed around the perimeter. It can be shown (see Appendix 5.1 for details) that the number of laminar flow reversals (N_{FR}) during time interval T at dimensionless location L* (and, by symmetry, 1–L*) along the loop has a mean value given by

$$E[N_{FR}] = 4\lambda T(L^*)(1-L^*)(1-P_0)^2 \tag{5.7}$$

where λ is the average customer arrival rate for the entire loop and P_0 is the probability of stagnation.

Equation 5.7 is plotted in Figure 5.5 for typical arrival rates (λ=1,700 demands per day) and stagnation conditions (P_0 = 0.50) observed at the dead-end loop study site. Even though the study site does not exactly match the ideal conditions of the hypothetical loop, the overall trends in Figure 5.5 can provide some insight into the general behavior of flow at the study site. The links near the loop entrance have very few flow reversals, but the number increases rapidly away from the entrance. The links near the loop apex ($0.4<L^*<0.6$) can expect to experience between 360 and 480 flow reversals per day (mean ± 3 standard deviations). The high incidence of flow reversals has implications for travel times in the dead-end loop as seen in the next section.

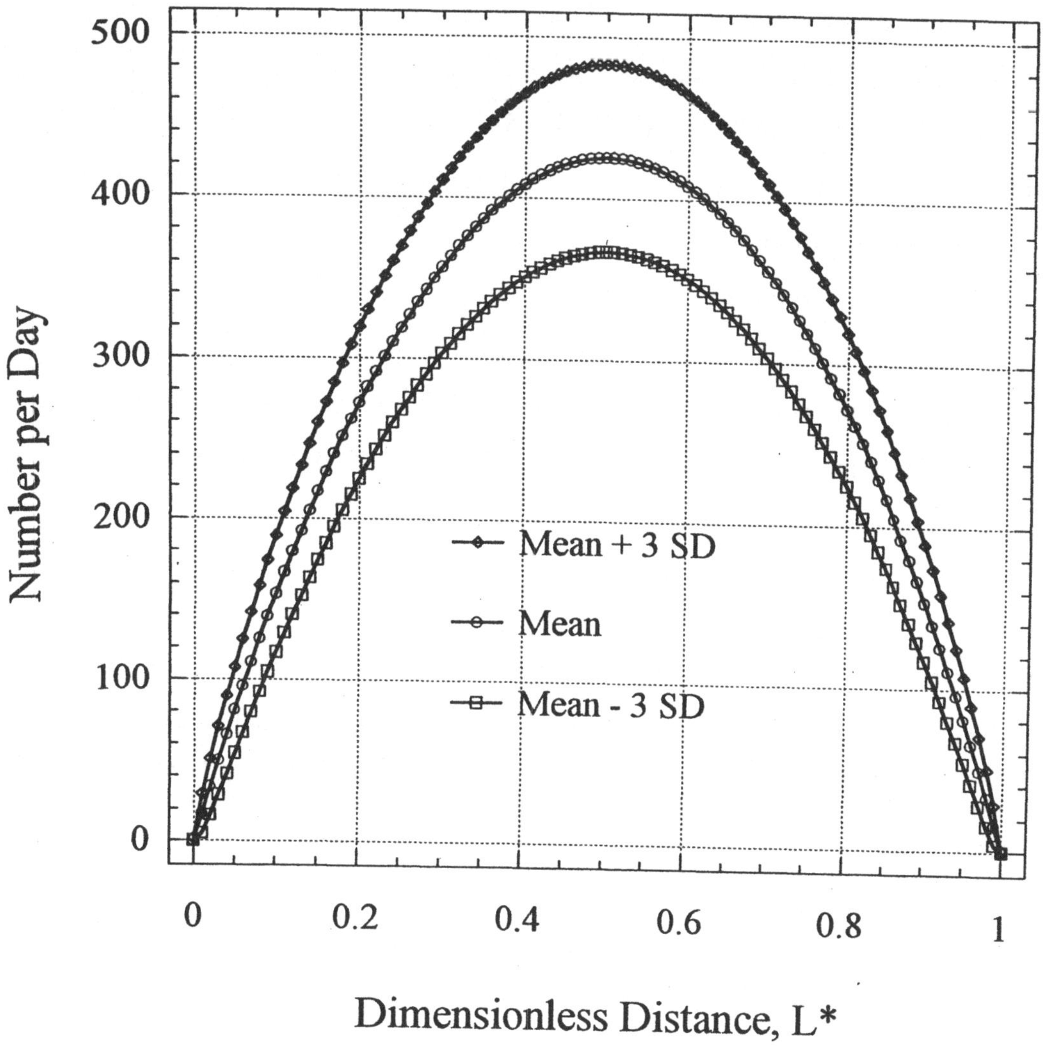

Figure 5.5 Laminar flow reversals in dead-end loop. (1700 daily demands and 50% stagnation).

TRAVEL TIME

Tracking Water Parcels

Travel times from the loop entrance to all homes along the dead-end main were estimated by tracking the age of small parcels of water moving through the supply line to the demand nodes. Each of the 22 pipe links was segmented into discrete cells having a constant volume, $V_{CELL} = 0.10$ gallons. The entire dead-end loop required over 27,000 cells. Water within each cell was assumed to be completely mixed and mass exchange between adjacent cells was not permitted [7].

The cells were translated through the pipe links in response to residential demands imposed at the nodes. The rate, direction and duration of flow in each pipe link was obtained from the hydraulic network model. Water aged from the time the cell entered the dead-end loop until the time it arrived at a busy home. The age of the water in the cell upon arrival at a busy home is the travel time.

---

[7] The discrete cell method is a convenient and expedient approach normally reserved for one-dimensional turbulent flow where the velocity distribution is nearly uniform over the cross section of the pipe. It might also offer a reasonable approximation for unsteady laminar flow, provided the developing velocity distribution is relatively flat (i.e., does not achieve its steady-state parabolic distribution). Based on a stagnation probability of P_0=54% and a loop-wide arrival rate of λ=1,800 demands per day, the average duration of a typical busy period (i.e., interval with Q>0) in the loop is about 41 seconds (see Appendix 6.3 for details). In a 6 inch pipe at 50 °F, the dimensionless Szymanski time for a developing laminar velocity profile is

$$\Delta T_B = \frac{\nu t}{r^2} = \frac{(1.41 \times 10^{-5}\ \mathrm{ft/sec})(41\ \mathrm{sec})}{(0.25\ \mathrm{ft})^2} = 0.009$$

With ΔT_B=0.009, the velocity at the pipe centerline attains less than 4% of its steady-state maximum value (see Figure 4.1-3 in Bird, Steward and Lightfoot, 1960). For the conditions that prevail at the study site, it appears that intermittent laminar flow moves through the dead-end loop in sporadic bursts with an undeveloped velocity profile that more closely resembles a plug flow uniform distribution rather than the steady-state parabolic distribution. Hence, the discrete cell approach might be suitable for flow conditions that exist in the dead-end loop.

The rate of flow in the dead-end loop changed on average about every 25 seconds [(8)]. Hence, tracking and updating water age was done on an event basis. In this context, an event is defined as a period of steady flow (including zero flow) in the loop; a new event corresponds to any change in the flow (caused by the start or finish of a residential water demand). Whenever the flow rate changed, the duration of the new discharge was noted and this amount of time was added to the age of each cell in the loop.

If the loop were idle during an event (Q=0, no home is using water), then all cells were aged but not moved. If the loop were busy during an event (Q>0, at least one home is using water), then the cells were aged and transported to active nodes. The number of cells advanced along a pipe link during an event is given by

$$\eta_{LINK} = \frac{Q_{LINK}\ T}{V_{CELL}} \tag{5.8}$$

where Q_{LINK} is the flow rate (gpm) through the link and T is the duration (min) of the event and V_{CELL} is fixed at 0.10 gallons. When $Q_{LINK} = 0$, no cells advance.

All cells between the same two active nodes move the same distance (because the event flow rate is constant along these links). The number of discrete cells withdrawn at an active node on the loop during an event of duration T is given by

$$\eta_{NODE} = \frac{Q_{NODE}\ T}{V_{CELL}} \tag{5.9}$$

[(8)] Each residential water demand causes the flow rate to change twice (once when the pulse begins and again when the pulse ends, assuming concurrent pulses neither start nor stop at the same instant). Hence 365,000 water demands occurring at the loop over a period of 7 months (214 days = 308,160 min) gives

$$\frac{(365{,}000 \text{ pulses})(2 \text{ flow changes per pulse})}{308{,}160 \text{ minutes}} = 2.37 \text{ flow changes per minute}$$

or

1 flow change every 25 seconds

where Q_{NODE} is the intensity of the residential water demand. Fractional portions of cells are held at the node until enough flow accumulates for a complete cell to be withdrawn.

By continuity, the total inflow at the loop entrance must equal the total outflow from the residential nodes for each busy event, or

$$\eta_{LINK\_1} + \eta_{LINK\_22} = \sum_{k=1}^{21} \eta_{NODE\_k} \tag{5.10}$$

During each busy event, a random number of cells with a variable range of water ages arrives at the active demand nodes. These ages represent water travel times and are stored for subsequent analysis. To discount the effect of initial conditions, cells inside the dead-end loop at the start of the travel time analysis are tagged and then discarded when withdrawn at a busy node.

Travel Time Statistics

Suppose that during the simulation period, a home receives N discrete cells with water ages τ_k (k = 1,2,...,N). Estimates of the mean, standard deviation and coefficient of variation of the travel time to this home are given by

$$\overline{T} = \frac{1}{N}\sum_{k=1}^{N} \tau_k \tag{5.11}$$

$$\hat{S}_T = \sqrt{\frac{1}{N}\sum_{k=1}^{N} \tau_k^2 - \left(\frac{1}{N}\sum_{k=1}^{N} \tau_k\right)^2} \tag{5.12}$$

$$\Theta_T = \frac{\hat{S}_T}{\overline{T}} \tag{5.13}$$

Estimates of several travel time statistics for all homes during the months of February and March 1996 are summarized in Tables 5.5 and 5.6, respectively. The mean travel time ranged from just over 8 hours near the entrance (homes 1 and 2) to about 96 hours near the apex (home 13, the most distant point from the loop entrance). Travel time trends for the dead-end loop are clearly illustrated in Figures 5.6 and 5.7 which show mean travel time versus home number and travel distance, respectively. Here travel distance is defined as the *minimum* length along either branch from the entrance of the loop to a home. Branches for each home are designated in column 2 of Tables 5.5 and 5.6.

Three features stand out, especially noticeable in Figure 5.7. First, mean travel time increases in a nonlinear manner with distance from the entrance. Second, the increase in travel time grows dramatically for homes in the apex region of the loop (in this case, homes 9-16 all more than 600 feet from the entrance). Third, there is a mild hysteresis in mean travel time to homes in the apex region of the loop. It takes about 1 day longer on average for water to reach the apex via the left branch than the right branch. This occurs because demand is not evenly balanced (i.e., uniformly distributed) with distance around the loop.

Across the entire loop the average flow-weighted travel times were 33.9 hours in February and 32.4 hours in March. Both values compare reasonably well with their nominal mean travel times of 31.8 hours and 29.0 hours, computed as the ratio of total loop volume to average loop demand [(9)]. Due to a slightly higher rate of demand (inflow) in March, travel times in March are less than those in February.

At any given home, there is relatively large variation in travel times. Apart from one exception occurring at home 14 in March, the coefficient of variation of travel times is quite stable with values generally in a range from 0.20 to 0.40, averaging about 0.30. Hence, the standard deviation of travel time increases in proportion to the mean travel time, both of which grow with distance from the loop entrance.

---

(9)

$$\text{Feb 1996: } T = \frac{V_{LOOP}}{Q_{LOOP}} = \frac{2706 \text{ gallons}}{59{,}300 \text{ gal} / 41{,}760 \text{ min}} = 1906 \text{ minutes} = 31.8 \text{ hrs}$$

$$\text{Mar 1996: } T = \frac{V_{LOOP}}{Q_{LOOP}} = \frac{2706 \text{ gallons}}{69{,}500 \text{ gal} / 44{,}640 \text{ min}} = 1738 \text{ minutes} = 29.0 \text{ hrs}$$

Table 5.5 Travel Time Statistics
From Entrance of Dead-End Loop to Homes in February 1996

Home Number	Distance (ft)	Min (hr)	Max (hr)	Avg (hr)	Stan Dev (hr)	Coef of Variation
1	208 (l)	2.3	17.8	10.5	3.0	0.28
2	235 (l)	2.2	19.1	8.9	3.1	0.35
3	316 (l)	3.7	18.8	11.4	3.7	0.33
4	383 (l)	5.4	25.5	14.5	4.5	0.31
5	400 (l)	5.8	30.3	16.0	4.9	0.31
6	473 (l)					
7	513 (l)	7.9	54.5	24.4	7.7	0.32
8	583 (l)	9.4	111.8	33.2	14.4	0.43
9	737 (l)	25.4	150.9	59.0	19.1	0.32
10	777 (l)	8.2	193.1	77.0	24.0	0.31
11	778 (l)	35.4	186.7	74.0	27.5	0.37
12	805 (l)					
13	884 (l)	13.4	208.8	81.4	27.6	0.34
14	850 (r)	6.9	114.4	60.2	14.6	0.24
15	796 (r)	6.0	83.2	51.2	10.3	0.20
16	748 (r)	20.5	66.5	41.9	7.6	0.18
17	609 (r)	6.6	139.8	31.3	5.8	0.18
18	492 (r)	6.1	75.0	22.3	4.5	0.20
19	491 (r)					
20	451 (r)	6.4	70.8	21.2	5.3	0.28
21	300 (r)	3.8	80.3	11.0	3.9	0.36
Simple Arithmetic Mean(*)				36.1(*)	(*) Not a valid measure of mean travel time.	
Flow-Weighted Mean				33.9		
Nominal Mean Travel Time, V/Q				31.8		

Notes: (1)Distance from loop entrance along shortest branch (l=left branch, r=right branch).
(2) Homes 6 & 12 were not occupied in Feb 1996; home 19 was under construction.

Table 5.6 Travel Time Statistics
From Entrance of Dead-End Loop to Homes in March 1996

Home Number	Distance (ft)	Min (hr)	Max (hr)	Avg (hr)	Stan Dev (hr)	Coef of Variation
1	208 (l)	2.2	13.8	8.2	3.1	0.38
2	225 (l)	2.6	15.6	8.4	3.3	0.39
3	306 (l)	4.0	21.2	12.1	3.8	0.31
4	374 (l)	5.8	30.7	14.5	4.1	0.28
5	391 (l)	6.0	30.0	16.0	4.1	0.25
6	464 (l)					
7	513 (l)	11.0	40.1	23.3	5.2	0.22
8	573 (l)	14.6	53.5	30.8	6.6	0.21
9	728 (l)	31.3	129.5	60.5	15.8	0.26
10	768 (l)	3.9	169.9	76.2	22.7	0.30
11	769 (l)	13.5	176.7	68.6	20.0	0.29
12	795 (l)					
13	874 (l)	27.0	287.7	95.5	36.4	0.38
14	850 (r)	8.1	256.5	84.0	50.8	0.60
15	796 (r)	22.3	174.0	61.3	16.2	0.26
16	748 (r)	15.9	76.4	51.7	10.1	0.20
17	609 (r)	7.6	107.9	35.9	10.1	0.28
18	492 (r)	5.4	93.5	23.3	4.8	0.21
19	491 (r)					
20	451 (r)	10.3	132.22	19.8	4.3	0.22
21	300 (r)	4.3	38.4	11.5	3.8	0.33
	Simple Arithmetic Mean(*)			39.0(*)	(*) Not a valid measure of mean travel time.	
	Flow-Weighted Mean			32.4		
	Nominal Mean Travel Time, V/Q			29.0		

Notes: (1)distance from loop entrance along shortest branch (l=left branch, r=right branch).
(2) homes 6 & 12 were not occupied in Mar 1996; home 19 was under construction.

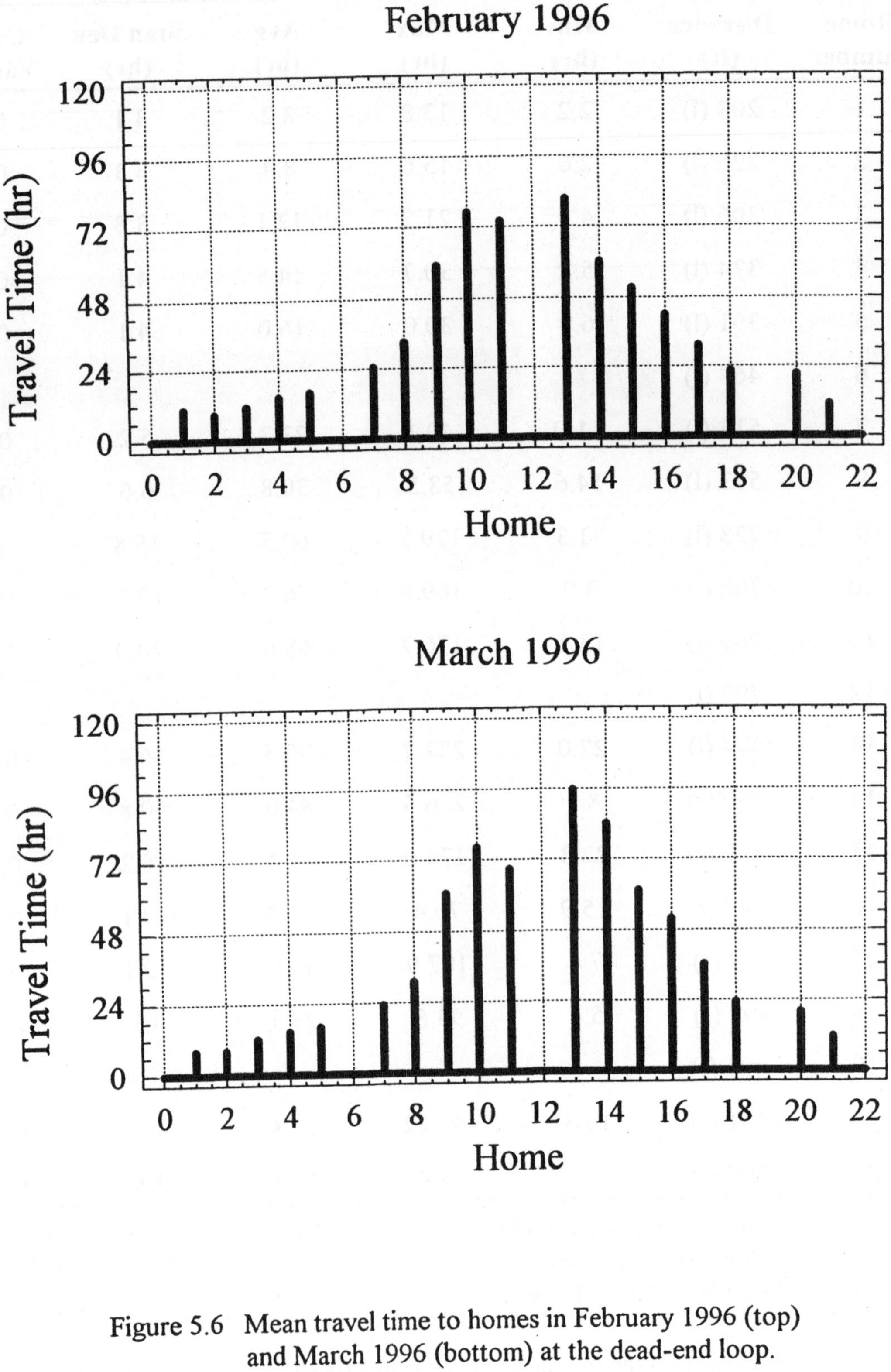

Figure 5.6 Mean travel time to homes in February 1996 (top) and March 1996 (bottom) at the dead-end loop.

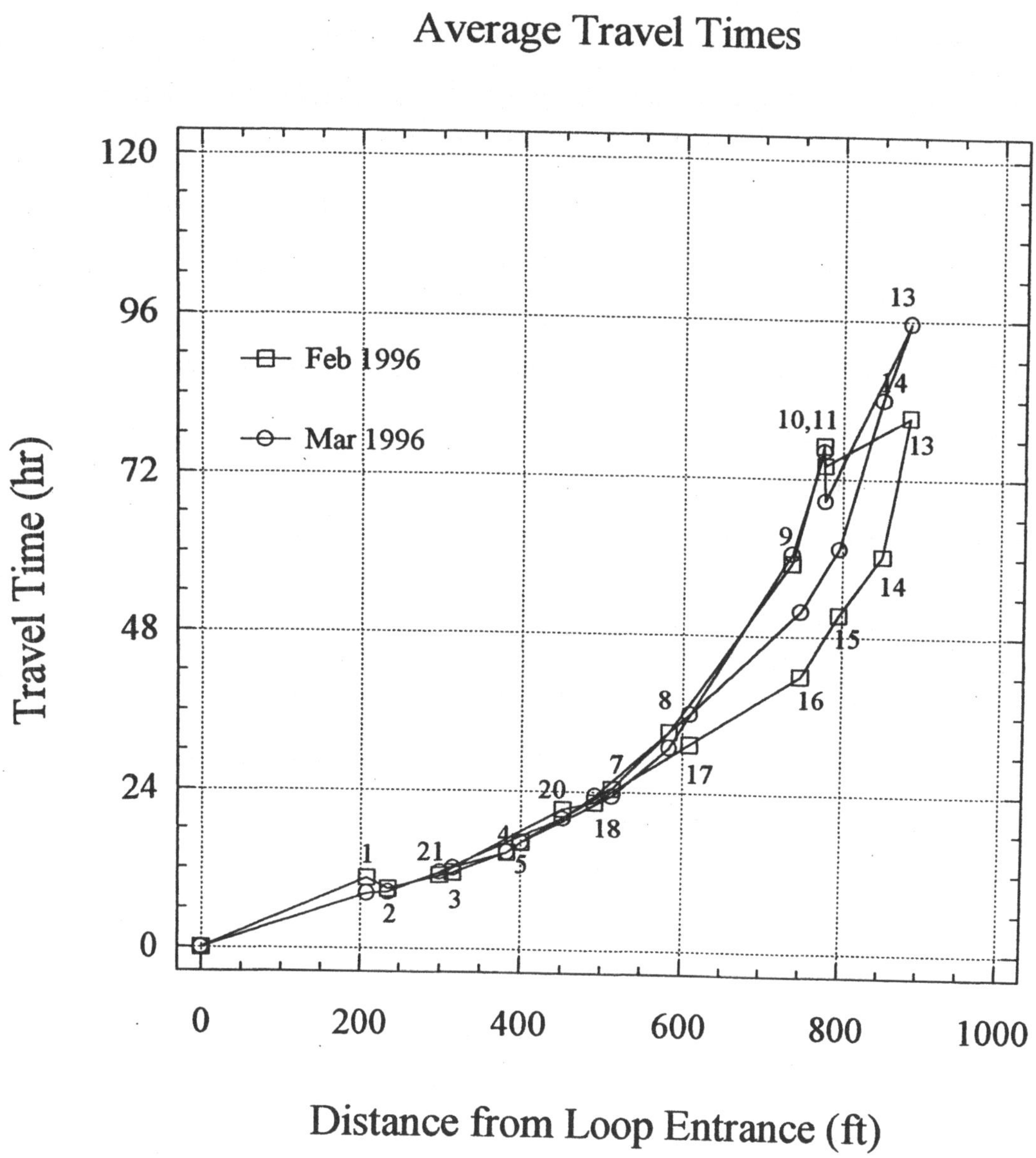

Figure 5.7 Mean travel time versus distance from loop entrance.

Travel Time Distributions

Figure 5.8 shows the relative frequency histogram of cell travel times during the month of March 1996 to homes 1 and 13, the nodes with the shortest and longest travel distances along the loop. Travel times to home 1 are clustered in the narrow interval from 2 to 14 hours; by contrast, travel times to home 13 are spread over the range from 27 to 288 hours. Note, however, that the coefficient of variation of travel times is the same for both homes. Thumbnail plots of the relative frequency and cumulative frequency histograms for all homes during February and March 1996 are presented in Appendix G.

There was a wide range in the maximum travel time. At home 1, the maximum never exceeded 14 hours, while at home 13 the maximum travel time in March approached 288 hours (12 days). This spread seems rather remarkable considering that the total loop volume was displaced on average about every 30 hours during Winter 1996. Besides having a longer travel distance, part of the increase in the travel times to home 13 (and others near the apex) can be attributed to bi-directional water movement caused by flow reversals. Once inside the loop, a discrete water cell is not withdrawn until it happens to reach an active node. As a consequence, it is possible for some cells to become temporarily trapped in the loop, especially near the apex where flow reversals are quite common (see Figure 5.5). A trapped cell will "wander" back and forth through pipe links between active demands and continue to age until eventually the right combination of residential water demands draws the cell to an active node [10].

---

[10] From a mathematical standpoint, the problem of estimating travel times in a dead-end loop is analogous to finding the "mean first passage time" in the presence of an absorbing barrier. There is a rich body of elegant and theoretical work in the scientific literature which could be applied to this problem.

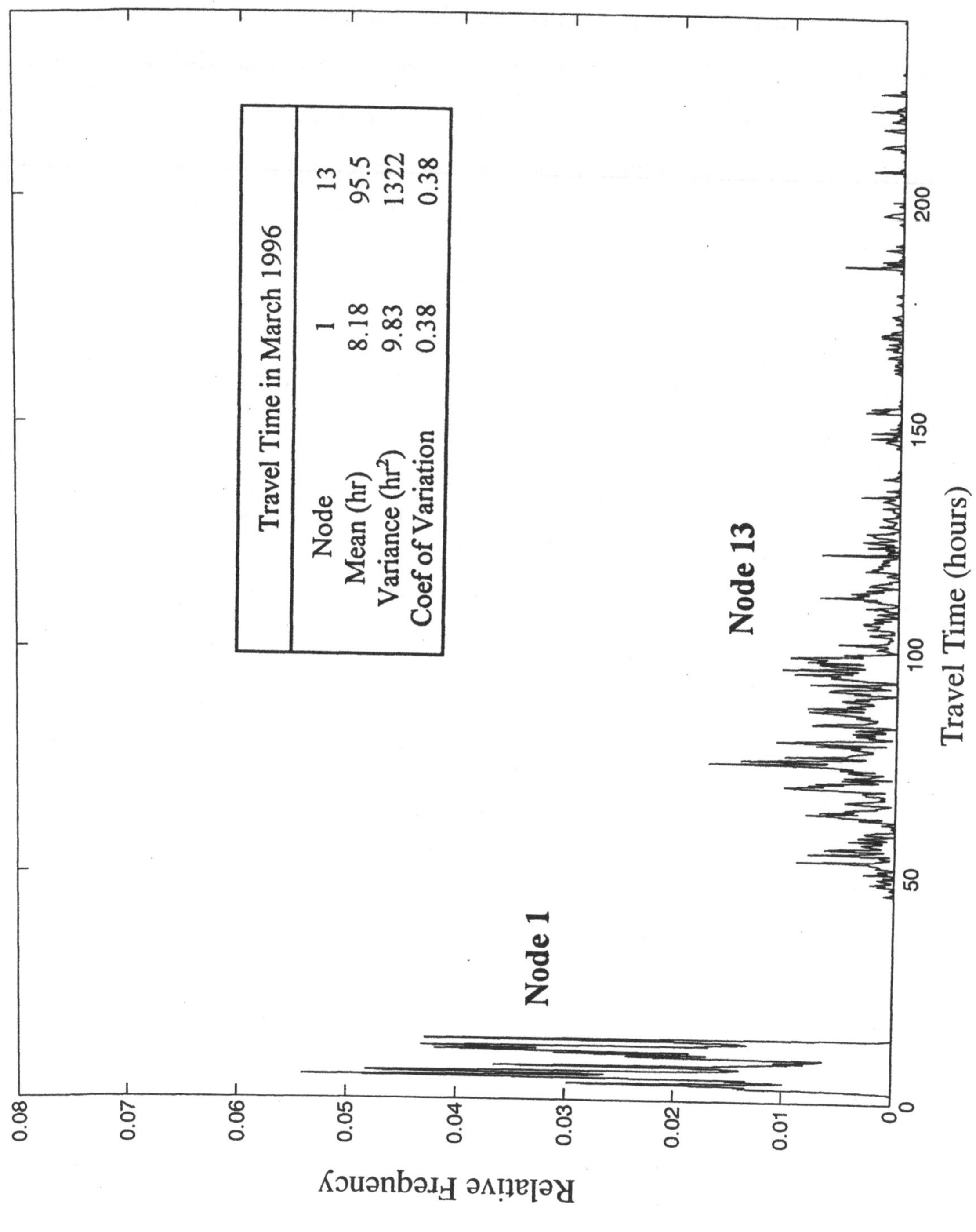

Figure 5.8 Relative frequency histogram of computed travel times during March 1996 to home 1 and home 13 along the dead-end loop.

Equivalent Dead-End Stem

Consider one-way flow through an idealized dead-end stem (see Figure 2.2) where N homes are spaced equally along the pipe length. Suppose the volume of the pipe link between each home is V and the average water demand at any residence is Q. The mean travel time through all N links of the dead-end stem is given by (see Appendix 5.2)

$$T_N^* = \frac{V}{Q}\sum_{k=1}^{N}\frac{1}{k} \tag{5.14}$$

Similarly, mean travel times along the first $m \le N$ links of the dead-end stem are given by

$$T_{m,N}^* = \frac{V}{Q}\sum_{k=N+1-m}^{N}\frac{1}{k} \tag{5.15}$$

Equation 5.14 applies only at the downstream terminal point of the dead-end stem whereas Eq 5.15 applies to all links along the dead-end stem. Hence, Eq. 5.14 is a special case of 5.15; if m=N, then Eqs 5.14 and 5.15 are identical. Both expression are the product of two terms: (i) the ratio V/Q accounts for storage and flow effects and (ii) the summation accounts for location along the dead-end stem. These expressions will be helpful when evaluating the effect of a continuous bleed on mean travel time (in Chapter 7).

Dimensionless forms of both expressions are plotted in Figure 5.9 [11]. The dashed line for Eq 5.14 indicates that the mean travel time from the entrance to the last residence on the dead-end stem grows as the number of homes increases [12]. The two solid lines for Eq 5.15 (for N=4 and N=9 homes) illustrate that the largest incremental increases in mean travel time occur in the links near the dead-end terminus, where consumers are few and demand is small. As expected, the dashed and solid lines intersect at the terminal link of the dead-end stem.

---

(11) To render Eqs 5.14 and 5.15 dimensionless, multiply by Q/V.

(12) In fact, as $N \to \infty$, the sum of reciprocals in Eq 5.14 does not converge. This makes good physical sense. It implies that the mean travel time through the dead-end keeps growing if there is no limit to the number of homes (and hence no limit to the length of the pipe).

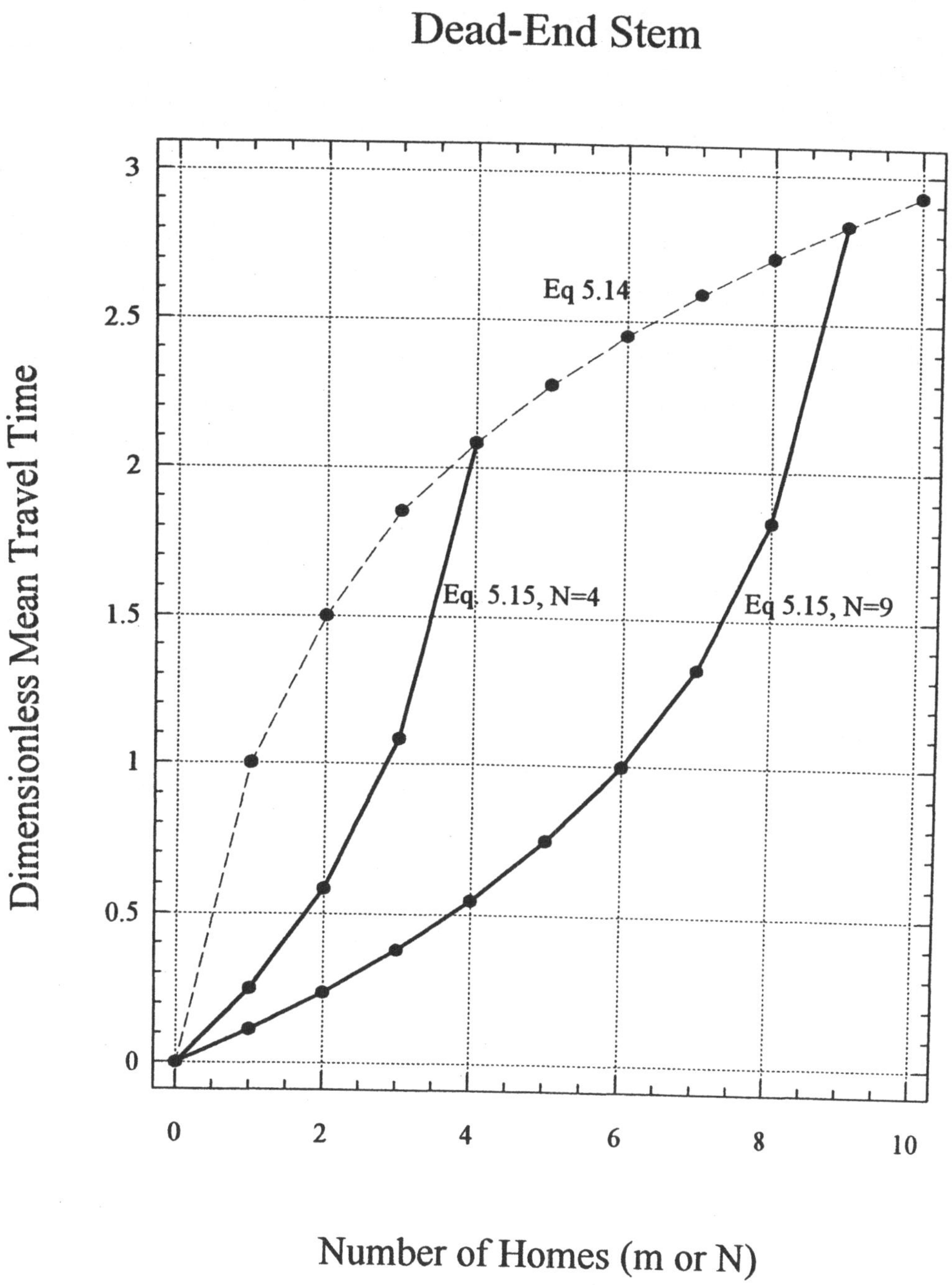

Figure 5.9 Predicted dimensionless mean travel time along a dead-end stem in a homogeneous neighborhood.

These notions can be applied to the dead-end loop at the study site. The left branch of the dead-end loop has 11 active homes along 884 feet (homes 1 through 13, with homes 6 and 12 unoccupied); the right branch has 7 active homes along 850 feet (homes 14 through 21 with home 19 under construction). In February and March 1996, the average demand generated by all 18 homes was 1.49 gpm [(13)]. The corresponding average demand at each active home is Q=1.49/18=0.0828 gpm.

Consider now a single "equivalent" dead-end stem representing the *average* of conditions along both branches. The equivalent dead-end stem supplies water to N=9 homes uniformly dispersed along a pipe with a 6 inch diameter and a total length of 867 feet. The pipe has 9 equal pipe links, each with a length, L=867/9=96.33 feet and with a volume, V=141.5 gallons. The nominal mean travel time for the dead-end stem is 28.5 hours [(14)].

The equivalent dead-end stem (described above) corresponds to the solid line in Figure 5.9 for the case N=9. Estimated travel times for all 9 links of the equivalent dead-end stem are presented in Table 5.7. Owing to the diminishing number of downstream users, the mean travel time through last link alone (~28.5 hr) exceeds the mean travel time through the first six links (~28.4 hr). The cumulative travel time through the entire dead-end (links 1-9) approaches 81 hours. The overall (flow-weighted) average, however, is 28.5 hours, consistent with the nominal mean travel time for the dead-end stem.

Predicted travel times for the equivalent dead-end stem (given in Table 5.7) are superimposed in Figure 5.10 on the average February and March travel times for the actual dead-end loop (listed in Tables 5.5 and 5.6). The agreement between travel times from the actual loop and times from the equivalent stem is quite good, especially for the links away from the apex region.

---

(13)

$$\left(\frac{128{,}800 \text{ gallons}}{60 \text{ days}}\right)\cdot\left(\frac{1 \text{ day}}{1440 \text{ minutes}}\right) = 1.49 \text{ gpm}$$

(14)

$$\left(\frac{141.5 \text{ gallons / link}}{0.0828 \text{ gpm / home}}\right)\cdot\left(\frac{9 \text{ links}}{9 \text{ homes}}\right) = 1709 \text{ min} = 28.5 \text{ hr}$$

Table 5.7 Mean Travel Time Through Links of Equivalent Dead-End Stem for N=9 Homes.

			Dimensionless Mean Travel Time		Mean Travel Time (hr)	
(1) Link Number (m)	(2) Homes Downstream (k=N+1-m)	(3) Cumulative Length from Entrance (ft)	(4) Incremental $\frac{1}{k}=\frac{1}{N+1-m}$	(5) Cumulative $\sum\frac{1}{k}$	(6) Incremental $\frac{V}{Q}\cdot\frac{1}{k}$	(7) Cumulative $\frac{V}{Q}\sum\frac{1}{k}$
1	9	96.33	0.111	0.111	3.16	3.16
2	8	192.66	0.125	0.236	3.56	6.72
3	7	289.00	0.143	0.379	4.08	10.80
4	6	385.33	0.167	0.546	4.76	15.56
5	5	481.66	0.200	0.746	5.70	21.26
6	4	578.00	0.250	0.996	7.13	28.39
7	3	674.33	0.333	1.329	9.50	37.89
8	2	770.66	0.500	1.829	14.25	52.14
9	1	867.00	1.000	2.829	28.50	80.64
	Total		2.829	9.00	80.64	256.50
	Average		0.314	1.00	8.96	28.50

Col (1) is link number starting with m=1 at the entrance.
Col (2) is number of homes downstream of link m.
Col (3) is cumulative distance from the entrance of dead-end stem, in feet.
Col (4) is dimensionless mean travel time through link m only.
Col (5) is cumulative dimensionless mean travel time through links 1,2,..,m (see Figure 5.7).
Col (6) is incremental mean travel time through link m only; V/Q = 28.50 hours at the study site
Col (7) is cumulative mean travel time through links 1,2,..,m (Eq. 5.15).

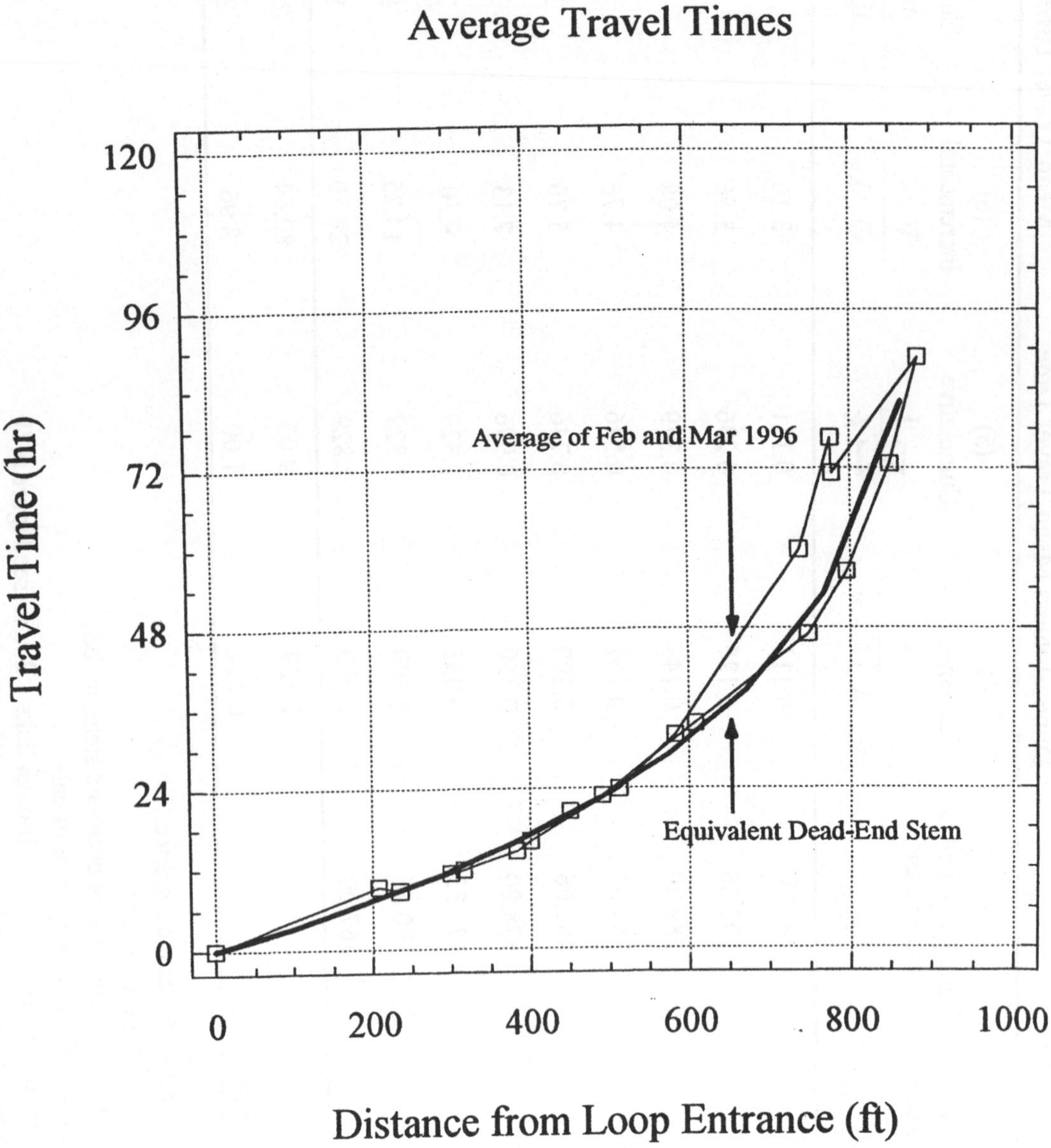

Figure 5.10 Predicted mean travel time in dead-end loop based on equivalent dead-end stem concept.

SUMMARY

A simple network hydraulic model was developed and run using water demands measured during the Winter 1996 at 18 residences along the dead-end loop. Results of the hydraulic simulation provided the rate and direction of flow on a second-by-second basis through each link of the loop. The flow rates and directions were used to characterize the prevailing flow regime and the number of flow reversals in each link and to estimate the mean travel time from the loop entrance to each residence.

The flow-weighted mean travel time for the entire neighborhood is given by the ratio of total pipe volume to average water demand, T=V/Q. Based on residential demands from Winter 1996, the mean travel time for the entire neighborhood was about 30 hours. On an individual household basis, home 1 near the entrance had the shortest travel times, ranging from 2 to 14 hours (with a mean of 8 hours). By contrast, home 13 near the apex had the longest travel times, ranging from 27 to 288 hours (with a mean of 96 hours). The coefficient of variation for travel times stayed relatively constant at about 0.30 for most homes implying that the mean and the standard deviation of travel times grew at nearly the same rate.

Besides their relatively long travel distance, part of the increase in the travel times to homes near the apex can be attributed to bi-directional water movement caused by flow reversals. Once inside the loop, a discrete water cell is not withdrawn until it happens to reach an active node. As a consequence, it is possible for some cells to become temporarily trapped in the loop, especially near the apex where flow reversals are quite common (see Figure 5.5). A trapped cell will "wander" back and forth through pipe links between active demands and continue to age until eventually the right combination of residential water demands draws the cell to an active node.

During winter season, water in the loop is stagnant over 50 percent of the time. When a demand occurs, the resulting pipe flow is almost always in the laminar regime. Hence, it is clear that water movement through the dead-end loop is not a continuous, turbulent or uni-directional flow. Instead the flow of water in the dead-end loop is characterized by random intermittent undeveloped laminar flow with frequent direction reversals. These conditions contrast sharply to assumptions invoked in most current water quality models.

Appendix 5.1: Changes in Flow Direction

Consider a small parcel of water at some location inside the pipe loop. The velocity of the parcel will change whenever a pulse begins or ends at any node on the loop. Hence, the expected number of velocity changes is twice the expected number of pulse arrivals. During a time interval of duration T with average loop-wide arrival rate λ, this gives,

$$E[\text{velocity changes}] = 2\lambda T \tag{5.1.1}$$

There are three mutually exclusive outcomes resulting from a change in pipe velocity:

(i) flow will be stagnant.
(ii) flow will be moving clockwise.
(iii) flow will be moving counter-clockwise.

Define:
p = probability of stagnant water after a velocity change
$q = 1-p$ = probability of moving water after a velocity change
rq = probability water is moving clockwise after a velocity change
sq = probability water is moving counter-clockwise after a velocity change

Note: $r+s=1$ and $p+q=1$

The event tree shown below gives probabilities of the three mutually exclusive outcomes caused by a change in pipe velocity.

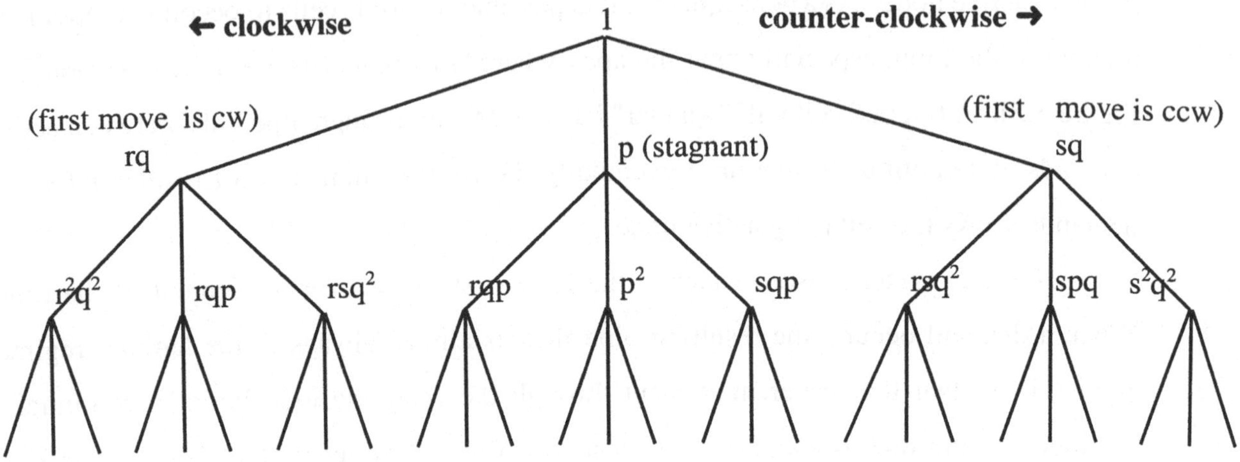

Figure 5.11 Flow directions resulting from a change in pipe velocity.

A *flow reversal* is defined as an event where the velocity of the parcel changes direction from clockwise to counter-clockwise or vice versa. The expected number of flow reversals. N_{FR}, is:

$$E[\ N_{FR}\ ] = (2\lambda T)\ \text{Prob[flow reversal | velocity change]} \tag{5.1.2}$$

Using the event trce in Figure 5.11 for velocity changes, it can be shown

$$\text{Prob[flow reversal | velocity change]} = 2rsq^2 \tag{5.1.3}$$

Putting Equation (5.1.3) into (5.1.2) gives

$$E[\ N_{FR}\ ] = 4\lambda Trsq^2 \tag{5.1.4}$$

The probabilities that flow in the loop is moving clockwise or counter-clockwise after a velocity change are, respectively:

$$r = \text{Prob[moving clockwise]} = 1 - L^*$$
$$s = \text{Prob[moving counter-clockwise]} = L^*$$

where L* is defined in Equation 5.5. These estimates for r and s assume that laminar flow prevails, that residential water demands are uniformly distributed along the length of the dead-end loop, and that no single demand node dominates the flow pattern. Finally, noting $q = 1 - P_0$ where P_0 is the stagnation probability, Equation 5.1.4 can be written,

$$E[\ N_{FR}\ ] = 4\lambda T(L^*)(1 - L^*)(1-P_0)^2 \tag{5.1.5}$$

and this appears as Equation 5.7 in the report. With $\lambda T \approx 1700$ daily arrivals at the study site, $P_0 \approx 0.50$ and considering the apex region (L*=0.50), Equation 5.1.5 gives

$$E[\ N_{FR}\ ] = 4(1700)(0.50)(0.50)(0.50)^2 = 425 \text{ daily flow reversals}$$

Since the arrival process is Poisson, the flow reversals are too. This implies that the mean and the variance of the flow reversals should be identical. The Poisson parameter for flow reversals changes with location on the loop, going from a maximum at the apex to nearly zero at the loop entrance.

Appendix 5.2: Travel Time Along a Dead-End Stem

Consider the dead-end stem with N homes shown in Figure 2.2. Let V_k be the volume of link k and Q_k be the *mean* demand at home k. Define

$$V^*_N = \Sigma\, V_k \quad k=1,2,...,N \quad \text{(total volume of the entire dead-end pipe),}$$
$$Q^*_N = \Sigma\, Q_k \quad k=1,2,...,N \quad \text{(mean demand of the entire neighborhood).}$$

The flow-weighted average travel time for the entire dead-end neighborhood is

$$\overline{T}_N = \frac{V^*_N}{Q^*_N} \tag{5.2.1}$$

To see this, note that the mean travel time through link k is

$$T_k = \frac{V_k}{Q^*_k} \tag{5.2.2}$$

For example, the mean travel time through link 3 depends only on the downstream demands,

$$T_3 = \frac{V_3}{Q^*_3} = \frac{V_3}{Q_1 + Q_2 + Q_3}$$

Noting that travel times are cumulative in the downstream direction, the flow-weighted mean travel time in the neighborhood is

$$\overline{T}_N = \frac{Q_N T_N + Q_{N-1}(T_N + T_{N-1}) + \cdot\cdot + Q_1(T_N + T_{N-1} + \cdot\cdot + T_1)}{Q^*_N} \tag{5.2.3}$$

$$= \frac{T_N Q^*_N + T_{N-1} Q^*_{N-1} + \cdot\cdot + T_1 Q^*_1}{Q^*_N}$$

Using Eq 5.2.2, this becomes,

$$\overline{T}_N = \frac{V_N + V_{N-1} + \cdot\cdot + V_1}{Q^*_N} = \frac{V^*_N}{Q^*_N} \tag{5.2.4}$$

which agrees with Eq 5.2.1. The mean travel time through an individual link (given by Eq 5.2.2) can vary dramatically from the mean travel time through the neighborhood. This is considered below.

Some insights are obtained if the dead-end stem is assumed to be homogeneous (i.e., each pipe length holds the same volume, V, and each home has the same mean demand, Q). In this case,

$$V_N^* = \sum_{k=1}^{N} V_k = NV \qquad (5.2.5a)$$

$$Q_N^* = \sum_{k=1}^{N} Q_k = NQ \qquad (5.2.5b)$$

Putting Equation 5.2.5 into 5.2.4, the flow-weighted neighborhood average travel time for the entire homogeneous block of N homes is

$$\overline{T}_N = \frac{V}{Q} \qquad (5.2.6)$$

This is identical to the mean travel time through a single link supplying a single home. Using Equation 5.2.2, the cumulative mean travel time through all N links of the dead-end stem (moving from the upstream entrance to the downstream terminus) is

$$T_N^* = \sum_{k=1}^{N} T_k = \sum_{k=1}^{N} \frac{V_k}{Q_k^*} \qquad (5.2.7)$$

For the homogeneous case, using Equations 5.2.5 and 5.2.6, this can be written

$$T_N^* = \frac{V}{Q} \sum_{k=1}^{N} \frac{1}{k} = \overline{T}_N \sum_{k=1}^{N} \frac{1}{k} \qquad (5.2.8)$$

In a similar manner, the cumulative mean travel times through the first $m \leq N$ links of the homogeneous dead-end stem are

$$T_{m,N}^* = \sum_{k=N-m+1}^{N} T_k = \overline{T}_N \sum_{k=N-m+1}^{N} \frac{1}{k} \qquad (5.2.9)$$

Equations 5.2.8 and 5.2.9 are identical to 5.14 and 5.15 presented in the text. Both expressions are the product of two terms. The term, $\overline{T}_N$, accounts for storage and flow effects on the neighborhood average travel time. The summation accounts for location along the dead-end stem.

Location is important. For example, consider a dead-end stem with N=10 homes. Using Equation 5.2.9, the home closest to the entrance has a mean travel time, $T^*_{1,10} = 0.10\,\overline{T}_N$. The home furthest from the entrance has a cumulative mean travel time, $T^*_{10,10} = 2.93\,\overline{T}_N$, nearly three time greater than the neighborhood average and 30 times greater than the mean travel time to a home at the entrance.

CHAPTER 6
THE POISSON RECTANGULAR PULSE HYPOTHESIS

OVERVIEW

This chapter tests the hypothesis that residential water use can be represented as a non-homogeneous Poisson rectangular pulse (PRP) process. The theoretical basis for the PRP model was presented by Buchberger and Wu (1995). Preliminary field evidence supporting the PRP premise has been presented by Buchberger and co-workers Greg Wells (single equivalent rectangular pulses, 1996), Trent Schade (probability distribution of busy homes, 1997) and Yeongho Lee (statistical properties of flow rates, 1999).

Derivations for the PRP model will not be repeated here, although some new findings are presented in this chapter and in the appendices. Instead, the main objective here is to compare various PRP predictions for busy homes and pipe flows against field observations from the dead-end loop study site. In addition, the versatility of the PRP model for residential water use is demonstrated with several novel applications, including:

❖ **Application 1**–The PRP model is used to explain and improve existing methods for estimating water demands in branching pipe networks.

❖ **Application 2**–A new method is outlined to screen pipe networks for leaks based on statistical analyses of measured neighborhood demands.

❖ **Application 3**–A simple algorithm is provided to simulate residential water demands that follow a non-homogeneous PRP process.

❖ **Application 4**–The PRP model for indoor use is extended to incorporate joint indoor and outdoor residential water demands.

TEST PERIOD

To test the PRP hypothesis, a representative subset of residential water demands was selected from 214 days of archived data described in Chapter 4. The test period covered 31 days from Sunday May 11 through Tuesday June 10, 1997 (see Figure 6.1) and included 52,126 residential water demands (52,090 indoor and 36 outdoor). This period was chosen because it had the highest full-day recovery rate (~95%, see Table 4.2), though any monthly period from the archived data set would be suitable.

Sun	Mon	Tues	Wed	Thur	Fri	Sat
May 1997						
4	5	6	7	8	9	10
11	12	13	14	15	16	17
18	19	20	21	22	23	24
25	26 Memorial Day	27	28	29	30	31
June 1997						
1	2	3	4	5	6	7
8	9	10	11	12	13	14

Figure 6.1 May 11 Through June 10, 1997 is the 31-Day Calendar Period Used to Test the PRP Hypothesis for Residential Water Use.

Appendix H contains 31 "quad plots", one for each day of the PRP test period. An example for Sunday May 11, 1997 appears in Figure 6.2. Each quad plot displays four snapshots of daily residential water use (both indoor and outdoor) integrated across all 21 homes at the dead-end loop:

Plot 1: 24-hr time series of instantaneous flow into the dead-end loop (upper left).
Plot 2: Cumulative frequency of instantaneous flow into the loop (upper right).
Plot 3: 24-hr time series of busy homes on the dead-end loop (lower left).
Plot 4: Relative frequency of observed and predicted busy homes (lower right).

Each of these four plots is discussed in more detail in a subsequent section of this Chapter.

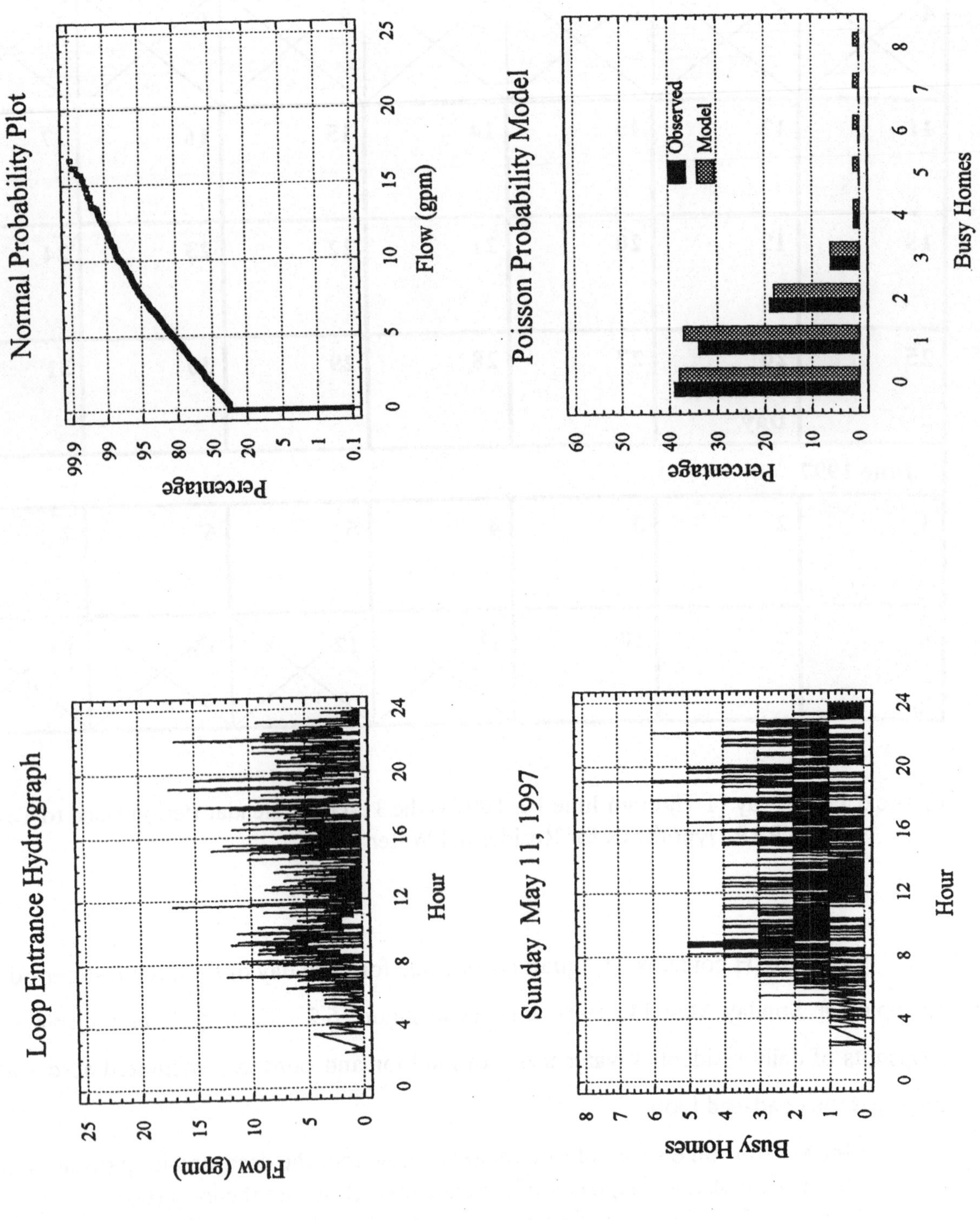

Figure 6.2 Sample quad plot for Sunday May 11, 1997.

KEY CONCEPTS

Terms and Definitions

The PRP hypothesis contains terms which are not part of the jargon typically used in the water industry. The meaning of several important terms is clarified here.

***Nonhomogeneous*:** Residential water use changes over time (during the course of a day) and varies with location (from home to home).

***Rectangular Pulse*:** Residential water use is assumed to occur as an abrupt burst of uniform *intensity*, lasting for a finite *duration*.

***Poisson Process*:** The *frequency* of residential water use during some period of time follows a random (Poisson) arrival process.

Three basic concepts reside at the core of the PRP model for residential water use:

Pulse Intensity What is the rate of a water use (e.g., gpm)?
Pulse Duration What is the length of a water use (e.g., seconds)?
Pulse Frequency How often does a water use occur (e.g., events per hour)?

Some symbols used in the PRP model have appeared in previous chapters. As a convenience, key symbols and definitions are collected in Table 6.1 (see also the Appendix).

Table 6.1 Key Symbols and Definitions for PRP Model

Symbol	Dimensions	Units	Meaning
λ	1/T	per min	customer arrival rate
μ	1/T	per min	customer service rate (at a *busy* server)
τ	T	min	average duration of water demand
ω^2	T^2	min^2	variance of water demand duration
ρ	--	--	utilization factor, $\rho=\lambda\tau=\lambda/\mu$
η	--	--	amplification factor, $\eta = \rho / [1-\exp(-\rho)]$
P_0	--	--	$Pr[Q=0] = \exp[-\rho]$ (stagnation probability)
α	L^3/T	gpm	average flow rate at a *busy* server
β^2	$(L^3/T)^2$	gpm^2	variance of flows at a *busy* server
ϕ	L^3	gallon	average volume of water pulse, $\phi=\alpha\tau$
ψ^2	$(L^3)^2$	$gallon^2$	variance of water pulse volumes
(*)	--	--	quantity summed over a block of N homes

Queueing Concepts

The PRP model for residential water use draws heavily on ideas from queueing theory[(1)], a field which deals with mathematical techniques for analyzing the flow of objects. Queueing theory has been applied extensively to solve real-world problems in diverse areas, including: transmission of calls through a telephone exchange[(2)], movement of traffic along a highway network, passage of customers through a checkout line, and flow of computer programs through a time-sharing server. A few common terms from queueing theory, in the context of municipal water distribution systems, are presented in Table 6.2.

Table 6.2 Queueing Terms Applied to Water Distribution Systems.

Queueing Term	**Point-of-Use Perspective (Hierarchy Level: H-1)**	**Point-of-Entry Perspective (Hierarchy Level: H-2)**
Customer	Any resident using water.	Any home using water.
Server	Any fixture or faucet that can extract water from the network.	Service line from main supply pipe to the home.
Arrival Rate	Number of customers engaging a server per unit time.	
Service Rate	Number of customers disengaging from a server per unit *busy* time.	
Busy	Server is active (Q>0, flow at server exceeds zero).	
Idle	Server is not active (Q=0, flow at server is zero).	

(1) A queue is a waiting line. From the **Point-of-Entry** perspective, all homes have direct access to the water distribution system and, hence, no queues form (that is, homes do not line-up to use water). In queueing theory, this corresponds to the "infinite server" scenario and often leads to a tractable analysis. From the **Point-of-Use** perspective, queues can form inside a home if the residents were to line-up while waiting to use water at a certain fixture. The PRP model presented here does not simulate queue formation inside of individual homes; instead it is aimed primarily at describing the collective behavior of busy homes in the network (i.e., **Point-of-Entry** perspective, see Figures 2.3 and 6.4). Kleinrock (1975, 1976), Newell (1982) and Gross and Harris (1985) provide excellent introductions to queueing principles.

(2) Communication utilities pioneered application of queueing theory to solve problems dealing with residential caller demands on telephone exchanges. The PRP model switches "phones with faucets" to provide a rigorous, robust framework for analyzing residential water demands on pipe networks.

Utilization Factor

The dimensionless utilization factor (ρ) listed in Table 6.1 plays a central role in the PRP hypothesis for residential water use. In the context of water demands, the utilization factor represents the mean number of busy fixtures in the home. Several other useful (and equivalent) interpretations of the utilization factor based on ratios include:

- Ratio 1: Mean arrival rate to mean service rate (λ/μ).
- Ratio 2: Mean service duration to mean inter-arrival time.
- Ratio 3: Total busy-fixture time to total elapsed time.

As summarized in Table 6.3, the utilization factor incorporates the joint effects of customer arrival rates (λ) and fixture service rates (μ). Perhaps most importantly, the utilization factor is additive across homes in the neighborhood[3]. This implies that the PRP approach can be applied to arbitrarily large networks (including entire distribution systems) simply by aggregating the utilization factors from all individual residences.

Any of the ratios listed above can be used to estimate the utilization factor from field data. To demonstrate, consider the hypothetical water demands monitored during a 60 minute period at Homes A, B and C as depicted in Figure 2.3 [4]. Estimates of the utilization factor for each home and for the entire dead-end stem are summarized in Table 6.4 (based on Ratio 3, a method of moments approach). At each home the number of busy fixtures ranges from zero to two. Owing to coincidental water use among the three homes, the number of busy fixtures across the dead-end ranges from zero to four (see Figure 2.3). On average, each of the three homes has roughly one third of a busy fixture. Taken collectively this gives an average of about one busy fixture (ρ=0.983) along the dead-end during the one hour monitoring period. The utilization factors listed in Tables 4.3 and 4.4 were computed in this manner.

---

[3] The Poisson probability distribution is closed under convolution. That is, suppose X and Y are Poisson random variables with parameters ρ_1 and ρ_2, respectively. Let Z=X+Y; then Z is Poisson with parameter $\rho_1+\rho_2$.

[4] The one-hour period used in this example is for illustrative purposes only. In practice, a much longer monitoring period would be needed to assure a representative sample.

Table 6.3 Expressions for Arrival Rate, Service Measures and Utilization Factors over a Neighborhood of N Homes.

Location	Average Arrival Rate [1/T]	Average Measures of Service: Rate [1/T]	Average Measures of Service: Duration [T]	Utilization Factor
Any Single Family Home	λ_j	$\mu_j = \frac{1}{\tau_j}$	$\tau_j = \frac{1}{\mu_j}$	$\rho_j = \frac{\lambda_j}{\mu_j} = \lambda_j \tau_j$
Block of N Homes	$\lambda^*_N = \sum_{j=1}^{N} w_j \lambda_j$	$\mu^*_N = \sum_{j=1}^{N} w_j \mu_j$	$\tau^*_N = \sum_{j=1}^{N} w_j \tau_j$	$\rho^*_N = \sum_{j=1}^{N} w_j \rho_j = \frac{\lambda^*_N}{\mu^*_N} = \lambda^*_N \tau^*_N$
Weight, w_j	$w_j = 1$	$w_j = \frac{\rho_j}{\rho^*_N}$	$w_j = \frac{\lambda_j}{\lambda^*_N}$	$w_j = 1$
Average of N Homes	$\bar{\lambda} = \frac{1}{N}\lambda^*_N$	$\bar{\mu} = \mu^*_N$	$\bar{\tau} = \tau^*_N$	$\bar{\rho} = \frac{1}{N}\rho^*_N = \frac{\bar{\lambda}}{\bar{\mu}} = \bar{\lambda} \cdot \bar{\tau}$

Notes: (1) The mean service rate is the reciprocal of the mean pulse duration, shown as the center graph in Appendix I.
(2) The arrival rate and the utilization factor are additive across the neighborhood.
(3) Service rates and durations must be expressed as a weighted average of individual homes.

Table 6.4 Estimation of Utilization Factor for 3 Homes in Figure 2.3

Time (min)	Home A	Home B	Home C	Dead-End
Total Time	60	60	60	60
Idle Time	44	42	41	24
Busy Time	16	18	19	36
Busy-Fixture Time	17	21	21	59
Utilization Factor, ρ	17/60=0.283	21/60=0.350	21/60=0.350	59/60=0.983
Amplification Factor, η	17/16=1.063	21/18=1.167	21/19=1.105	59/36=1.639
Percent Stagnant	44/60=0.733	42/60=0.700	41/60=0.683	24/60=0.400
Prob [Stagnant], P_0	0.753	0.705	0.705	0.374

Notes:

Row 1: **Total Time** is the length of the monitoring period (60 minutes in this example).
Row 2: **Idle Time** is the length of time that there is no water demand (Q=0).
Row 3: **Busy Time** is the length of time that there is at least one demand (Q>0, busy periods are denoted by black horizontal bars beneath each hydrograph in Figure 2.3).

Total Time = Idle Time + Busy Time

Row 4: **Busy-Fixture Time** is the cumulative amount of time that water is drawn from all individual fixtures. An example calculation for the dead-end stem is presented below.
Row 5: **Utilization Factor** is the ratio of busy-fixture time to total time.
Row 6: **Amplification (Amp) Factor** is the ratio of busy-fixture time to busy time.
Row 7: **Percent Stagnant** is the ratio of idle time to total time.
Row 8: **Prob[Stagnant]** is prediction of stagnation from PRP model, $P_0=\exp[-\rho]$.

Example Calculation of Busy-Fixture Time for the Dead-End Main in Figure 2.3

Number of Busy-Fixtures	Number of Minutes	Busy-Fixture Time
0	24	$0\times24 = 0$
1	18	$1\times18 = 18$
2	14	$2\times14 = 28$
3	3	$3\times3 = 9$
4	1	$4\times1 = 4$
Total	60 minutes	59 minutes

Figure 6.3 shows hourly averaged values of the arrival rate $\overline{\lambda}(t)$, pulse duration $\overline{\tau}(t)$, and utilization factor $\overline{\rho}(t)$ at the study site for the full 31-day test period. The hourly values were estimated with the expressions in Table 6.3 using indoor Point-of-Use data for each home. The Point-of-Use data (presented in Appendix I on 21 graphs, one for each home) exhibit considerable variability from home to home and during the course of a day. In contrast, the graphs in Figure 6.3 (averaged hour-by-hour across 21 homes over a 31 day period) display a more subdued diurnal pattern. Each represents the co-mingled effects of weekdays and weekends. No attempt was made to quantify day-of-the-week effects.

The most important result in Figure 6.3 is the hourly sequence for $\overline{\rho}(t)$, referred to as the "**Rho Profile**". The rho profile is the characteristic signature of residential water use at an average household. It provides essential information for making PRP predictions of indoor water use activity at the dead-end loop. In Figure 6.3, the overall single family residence average utilization factor is 0.0454 [(5)]. However, during a 24-hour period the average utilization factor for a single home ranges from an minimum of 0.0042 (at 3 AM) to a maximum of 0.0676 (at 7 PM), a 16-fold variation.

Not surprisingly, the rho profile mimics closely the diurnal pattern for customer arrivals. Minimum values correspond to low water demand activity in the early morning hours; conversely, maximum values indicate highest residential water use during the breakfast and dinner hours. The overall average arrival rate is 0.055 customers per minute (~79 per day per home), but the hourly arrival rates vary by a factor of ten during the course of the day.

Compared to customer arrivals, pulse durations are less variable. The overall average pulse duration is 0.816 minutes (quite close to the average pulse duration of 0.769 minutes obtained for the full 7-month period, as summarized in Table 4.3). The average pulse duration hits its minimum (~0.5 min) in the early morning hours, quickly rises to a maximum (~1.4 min) around 6 AM and then gradually tapers from a duration of about 1.0 minute down to 0.6 minutes over the remainder of the day.

---

(5) The rho profile is the product of customer arrivals and pulse duration. As a check, note: $\rho = \lambda\tau$ = (0.055 per minute)(0.816 minutes) = 0.0449≈0.0454 for the 31day test period. This is very close to the 7-month average value for ρ=0.0442 listed in Table 4.3.

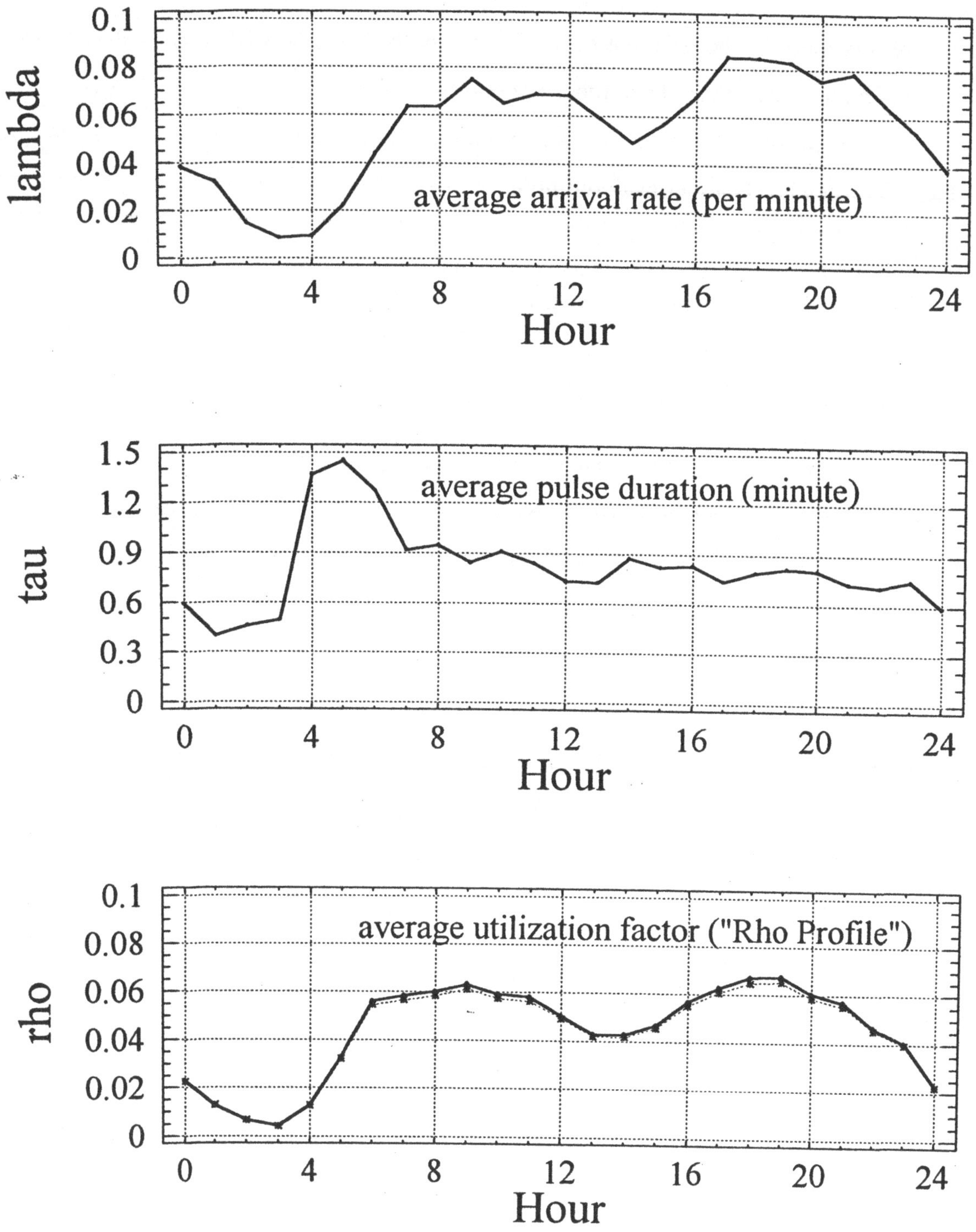

Figure 6.3 Average hourly values for arrival rate (λ), pulse duration (τ) and utilization factor (ρ) for 21 homes on dead-end loop over 31-day test period. Included with the **Rho Profile** is the binomial probability for a busy home (dotted line, $p=1-\exp[-\rho]$, see Eq 6.4).

BUSY SERVERS

Busy servers offer the first test of the PRP hypothesis. The PRP model predicts the distribution of busy servers (i.e., the number of servers drawing water from the mainline). The busy server problem can be formulated at several hierarchy levels in the network (see Figure 6.4). Two hierarchy levels are considered here:

- **Point-of-Use (H-1)**: Servers are individual fixtures inside the home.
- **Point-of-Entry (H-2)**: Servers are individual homes in the neighborhood.

Point of Use (H-1): Busy Fixtures

Suppose customers arrive at residence j according to a time homogeneous Poisson process with parameter, λ_j and receive service at a rate μ_j [(5)]. Let K_j be the number of busy fixtures in residence j. Then the total number of busy fixtures in a neighborhood of N residences is

$$K_N^* = \sum_{j=1}^{N} K_j \tag{6.1}$$

Under very mild conditions (i.e., assuming that queues do not form at fixtures), the time series of K_N^* values forms a stationary stochastic process having a Poisson probability distribution (Buchberger and Wu, 1995)

$$P_k = \text{Prob}[K_N^* = k] = \frac{(\rho_N^*)^k \exp[-\rho_N^*]}{k!} \qquad k = 0,1,2,\ldots \tag{6.2}$$

with mean and variance given by the utilization factor ρ_N^* (see Table 6.3). At a typical single family residence, the average utilization factor for indoor water use is quite low ($\rho \approx 0.05$, see Table 4.3). This implies that indoor water demands tend to be short and infrequent.

---

(5) For instance, customers at a home may engage water fixtures at an average rate, λ=3/hr and disengage from busy servers at an average rate, μ=1/min. This gives a utilization factor $\rho=\lambda/\mu$=(3/hr)/(60/hr)=0.05, typical for a single family residence. Both λ and μ could vary with the type of household fixture; however, such a distinction is not made in this report.

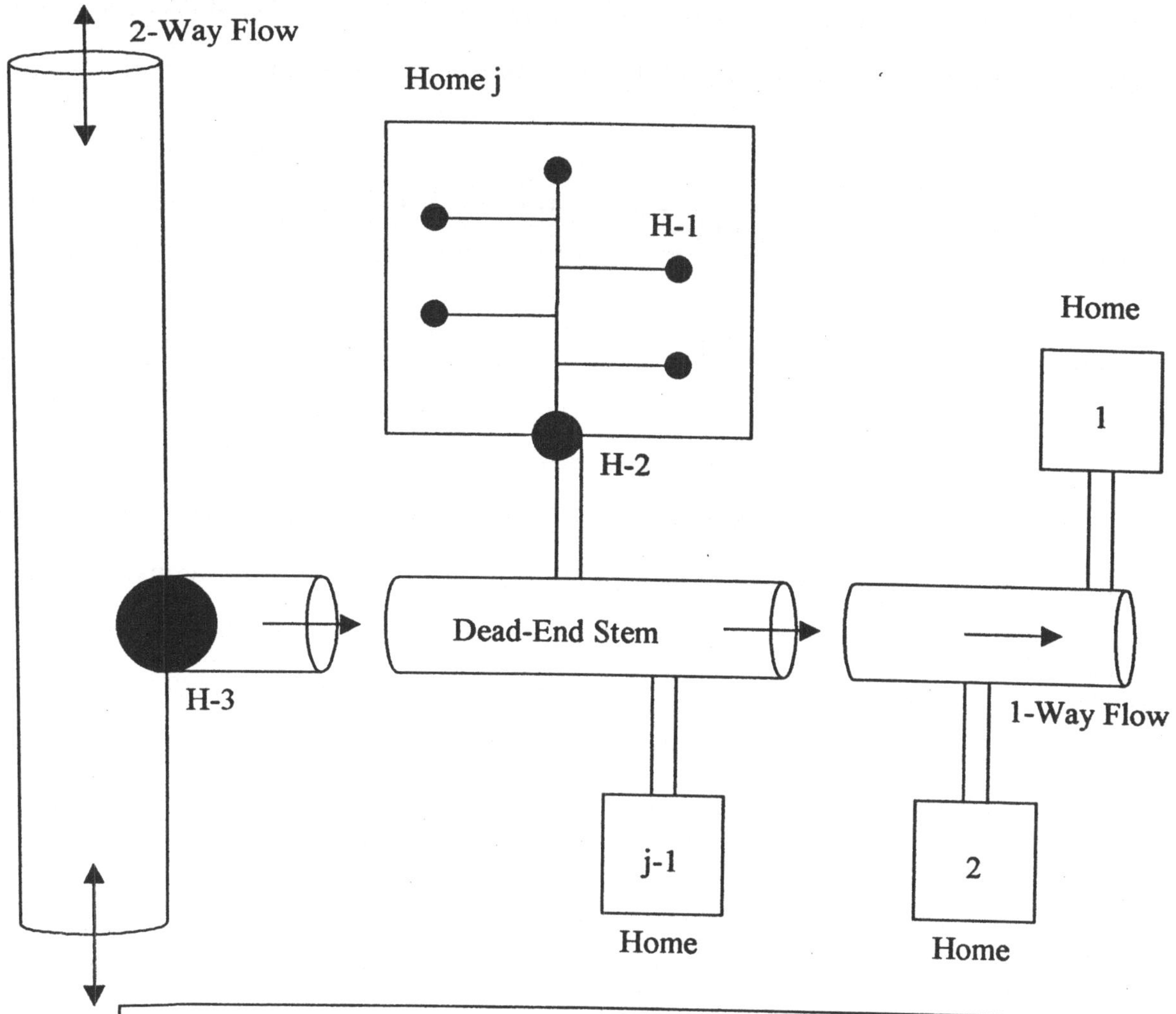

Hierarchy Level		Reference Point	Elemental Unit
●	H-1	Point of Use	Fixture
●	H-2	Point of Entry	Single Family Residence
●	H-3	Point of No Return	Neighborhood
Not Shown	H-4	Point of Treatment	Entire Network

Figure 6.4 Definition sketch for hierarchy levels in pipe network.

One way to test the PRP hypothesis is to estimate the relative frequency of busy fixtures from the archived data base and then compare results against the predicted Poisson distribution given in Eq 6.2. For example, the probability P_0 of no busy fixtures in a home (i.e., idle home) is

$$P_0 = \text{Prob}[K_1^* = 0] = \frac{(\rho_1^*)^0 \exp[-\rho_1^*]}{0!} = \exp[-\rho_1^*]$$

Hence, with $\rho \approx 0.05$, the probability of an idle home (i.e., stagnation) is $P_0 \approx 1-\rho \approx 0.95$. Conversely, the probability of a busy home (i.e., water is flowing) is $1-P_0 \approx \rho \approx 0.05$.

Two problems arise, however, with the time homogeneous busy fixture approach. First, residential water use does not follow a time-homogeneous demand process. Most days have prominent diurnal cycles with alternating periods of high and low water consumption. This problem is solved by treating the actual time dependent non-homogeneous water demands as a piecewise homogeneous process[(6)]. To do this, the day is divided into 24 one-hour intervals, each of which is considered to be time-homogeneous (for instance, see rho profile in Figure 6.3). The predicted Poisson probability distribution is then checked hour by hour against the observed relative frequency of busy servers.

The second problem is a practical consideration. Use of Eq 6.2 requires detailed Point-of-Use information on busy fixtures in each home. Presently, there is no convenient affordable way to monitor all fixtures in a home[(7)]. Fortunately, for purposes of checking the PRP hypothesis, it is not necessary to secure detailed data at the fixture level.

[(6)] This piecewise approach is analogous to the extended period simulations often adopted in current network hydraulic models to account for hourly variation in nodal demands.

[(7)] It is possible to analyze the signal of total flow into a home and from it extract the unique demand signatures of individual water fixtures. This strategy was employed in the 1999 REUW study with a sampling frequency of 1 reading per 10 seconds. Results of the current investigation (with a sampling frequency of 1 reading every second) suggest that up to 35 percent or more of indoor residential demands have a duration under 10 seconds (see Appendix D). With further refinement, the signal processing approach can provide an alternate way to generate useful data for checking the PRP hypothesis.

Instead, the second dilemma is solved by shifting the focus from the Point-of-Use (hierarchy level H-1) to the Point-of-Entry (hierarchy level H-2). In this manner, the problem changes from finding the distribution of *busy fixtures* to finding the distribution of *busy homes.* In a sense, each home is viewed as a single super server and effectively treated like a Bernoulli variable with two states: either on (busy) or off (idle).

Point of Entry (H-2): Busy Homes

Now let K(t) be the number of busy homes at time t in a neighborhood of N residences. Because the status of each home (either busy or idle) can be viewed as the outcome of an independent Bernoulli trial, the total number of busy homes in the neighborhood has a binomial distribution (see Appendix 6.1)

$$P_k = \text{Prob}[K(t) = k] = \binom{N}{k}[p(t)]^k[1 - p(t)]^{N-k} \qquad k = 0,1,2,\ldots,N \qquad (6.3)$$

where p(t) is the *average* probability that a home will be busy at time t

$$p(t) = 1 - \exp[-\overline{\rho}(t)] \qquad (6.4)$$

and $\overline{\rho}(t) = (\rho_N{}^*/N)$ is the average utilization factor from N homes at time t, as defined in Table 6.3. When $\overline{\rho} < 0.10$, Eq 6.4 reduces to

$$p = 1 - \exp[-\overline{\rho}] = 1 - \left[1 - \overline{\rho} + \tfrac{1}{2}\overline{\rho}^2 - \tfrac{1}{6}\overline{\rho}^3 + \cdots\right] \approx \overline{\rho}$$

and hence the binomial and Poisson parameters are nearly the same for small $\overline{\rho}$.

The rho profile in the bottom graph of Figure 6.3 also shows (as a dotted line) the corresponding binomial parameter p(t) computed with Eq 6.4. Owing to small average utilization factors, $\overline{\rho}(t)$ and p(t) are nearly the same over the course of a day (the average absolute difference between them is 0.0012 at the study site). Hence, the busy server predictions from Eqs 6.2 and 6.3 will be very similar.

Table 6.5 compares the predicted stagnation probability, mean, variance, and coefficient of variation of busy fixtures (Poisson model) against busy homes (binomial model) for a neighborhood of N single family residences. As required, both models provide identical estimates of stagnation, $P_0 = \exp[-N\overline{\rho}]$. This is reasonable (and necessary) because no busy fixtures imply no busy homes and vice versa. The mean and the variance of the busy fixtures exceed the mean and variance of the busy homes. This too is expected because each busy home allows more than one busy fixture.

Busy server moment ratios are plotted in Figure 6.5 as a function of the utilization factor. The rho profile in Figure 6.3 suggests $\overline{\rho}(t) < 0.10$ for indoor residential water use. In this range, the difference between the mean number of busy fixtures and the mean number of busy homes is very small (less than 5% in Figure 6.5). Hence, as noted above, it may be difficult to distinguish between results from the two probability models.

This is illustrated in Table 6.6 which compares observed and predicted busy servers (fixtures and homes) and stagnation probabilities averaged over the full 31-day test period. Three key points emerge from this broad-brush overview. First, sample statistics for busy fixtures and busy homes are very similar, though (as expected) the mean and variance of busy fixtures is slightly greater than the mean and variance of busy homes. Second, there is quite good agreement between observed and predicted mean values for busy fixtures and busy homes. Third, observed stagnation probabilities and busy server variances exceed predicted values. These discrepancies arise from the non-homogeneous (time dependent) nature of the daily water use pattern which (by design) is not reflected in Table 6.6 (see Appendix 6.5).

Poisson (Eq 6.2) and binomial (Eq 6.3) predictions for the distribution of busy servers are plotted against observed *hourly relative* frequencies in Figure 6.6 and observed *hourly cumulative* frequencies in Figure 6.7. Hourly ρ values were estimated with the method of moments procedure demonstrated in Table 6.4. The non-homogeneous nature of residential water use is clear with peak activity at breakfast and dinner and a stagnant period in the predawn hours. In all cases, there is outstanding agreement between observed and predicted hourly probabilities for busy homes during the 31-day test period. As summarized in Table 6.7, the overall average absolute value of the difference between observed and predicted hourly probabilities is around 1% for both the Poisson and binomial models.

Table 6.5 Comparison of Busy Fixtures and Busy Homes in a Neighborhood of N Single Family Residences.

Feature	Busy Fixtures	Busy Homes	Ratio BF:BH
Perspective	Point-of-Use	Point-of-Entry	--
Hierarchy Level	H-1	H-2	--
Model	Poisson	Binomial	--
Prob [Q=0]	$\exp(-N\overline{\rho})$	$\exp(-N\overline{\rho})$	1
Mean	$N\overline{\rho}$	$\dfrac{N\overline{\rho}}{\eta}$	η
Variance	$N\overline{\rho}$	$\dfrac{N\overline{\rho}}{\eta\,\mathbf{exp}(\overline{\rho})}$	$\eta\exp(\overline{\rho})$
Coefficient of Variation	$\dfrac{1}{\sqrt{N\overline{\rho}}}$	$\dfrac{1}{\sqrt{N\overline{\rho}\left[\mathbf{exp}(\overline{\rho})/\eta\right]}}$	$\sqrt{\dfrac{\mathbf{exp}(\overline{\rho})}{\eta}}$

Notes: (1) BF:BH is busy fixtures to busy homes.
(2) $\eta = \overline{\rho}\,/\,[1-\exp(-\overline{\rho})]$, the amplification factor, see Table 6.1.
(3) Hierarchy Level, see Table 2.3 and Figure 6.4

Table 6.6 Comparison of Observed and Predicted Busy Fixtures and Busy Homes at the Dead-End Loop for 31-Day Test Period.

Network Feature	Busy Fixtures	Busy Homes	Ratio BF:BH
Observed, P[Q=0]	0.423	0.423	1
Predicted, P[Q=0]	0.386	0.386	1
Observed Mean	0.952	0.907	1.050
Predicted Mean	0.952	0.931	1.023
Observed Variance	1.196	1.041	1.149
Predicted Variance	0.952	0.890	1.070
Obs Coef Variation	1.149	1.125	1.021
Pred Coef Variation	1.025	1.104	1.011

Notes: (1) BF:BH is ratio of busy fixtures to busy homes.
(2) "Observed" values are based on data from the 31-day test period.
(3) "Predicted" values are computed using expressions given in Table 6.5, with N=21 homes, $\overline{\rho}$=0.0454 (from Fig 6.3) and $\eta=\overline{\rho}/[1-\exp(-\overline{\rho})]=1.023$.

Average hourly stagnation probabilities range from a maximum of P_0=0.92 at 3 AM (when ρ is lowest) to a minimum of P_0=0.24 at 7 PM (when ρ is highest). The overall daily average probability of stagnation during the test period is 42.3 %. This is lower than the frequency of stagnation noted during Winter 1996 when only 18 homes on the dead-end loop were occupied (see Figures 5.1 and 5.2). The effect of adding three more homes is to reduce the occurrence of stagnation along the dead-end loop.

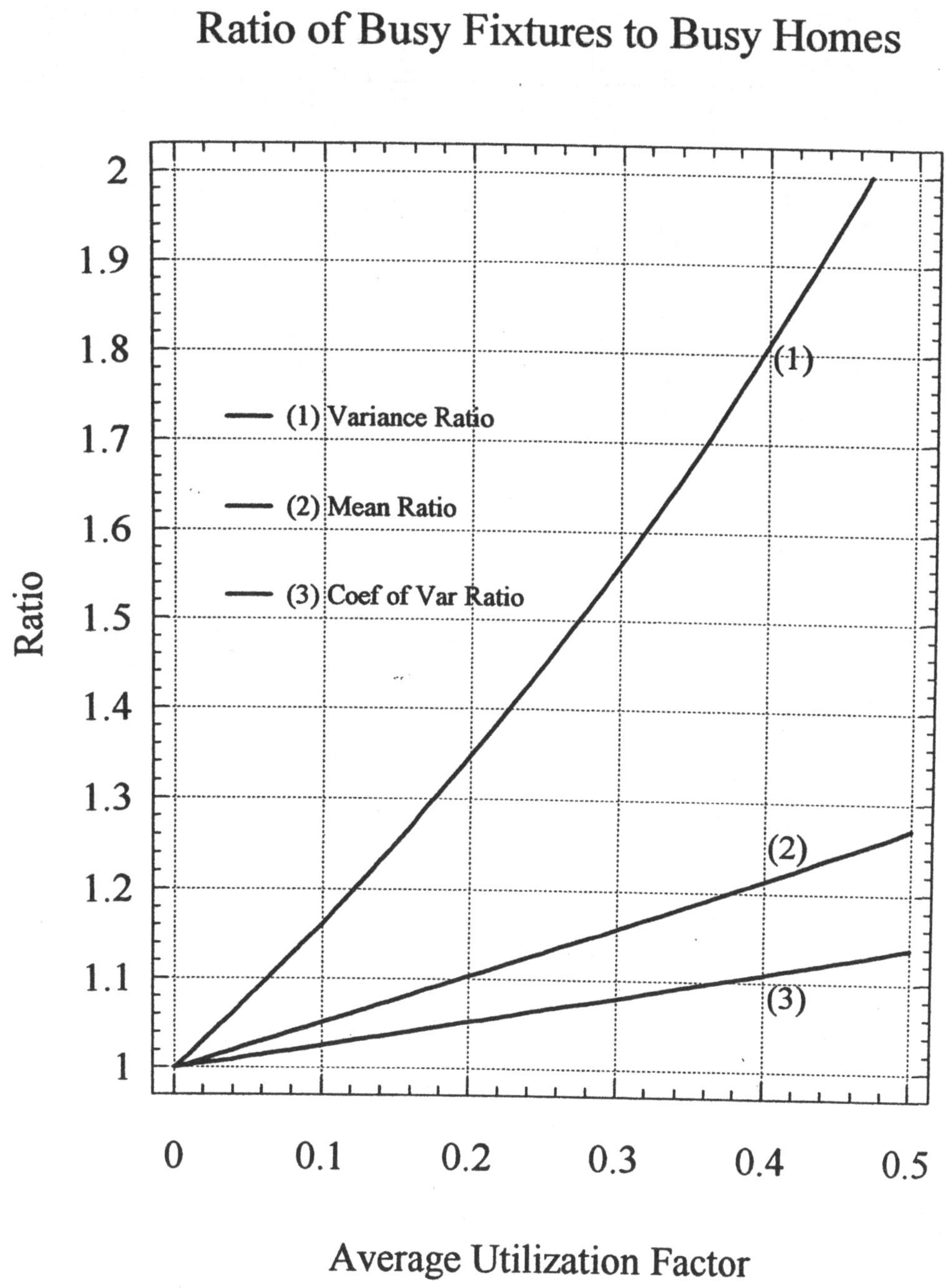

Figure 6.5 Theoretical moment ratio of busy fixtures to busy homes versus utilization factor. Based on expression in Table 6.5.

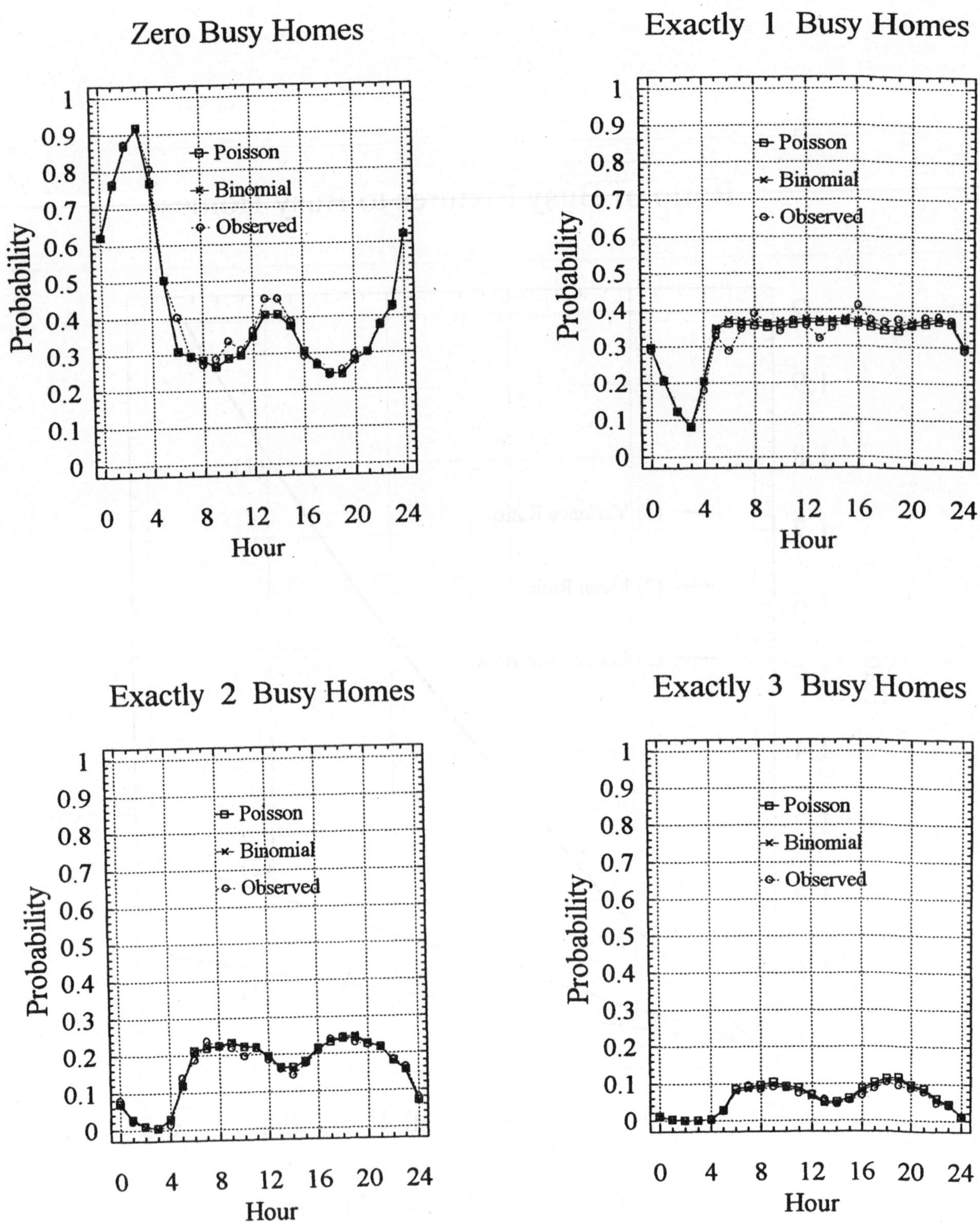

Figure 6.6 Predicted and observed hourly probability distribution of busy homes during the 31-day test period at the dead-end loop (N=21 total homes).

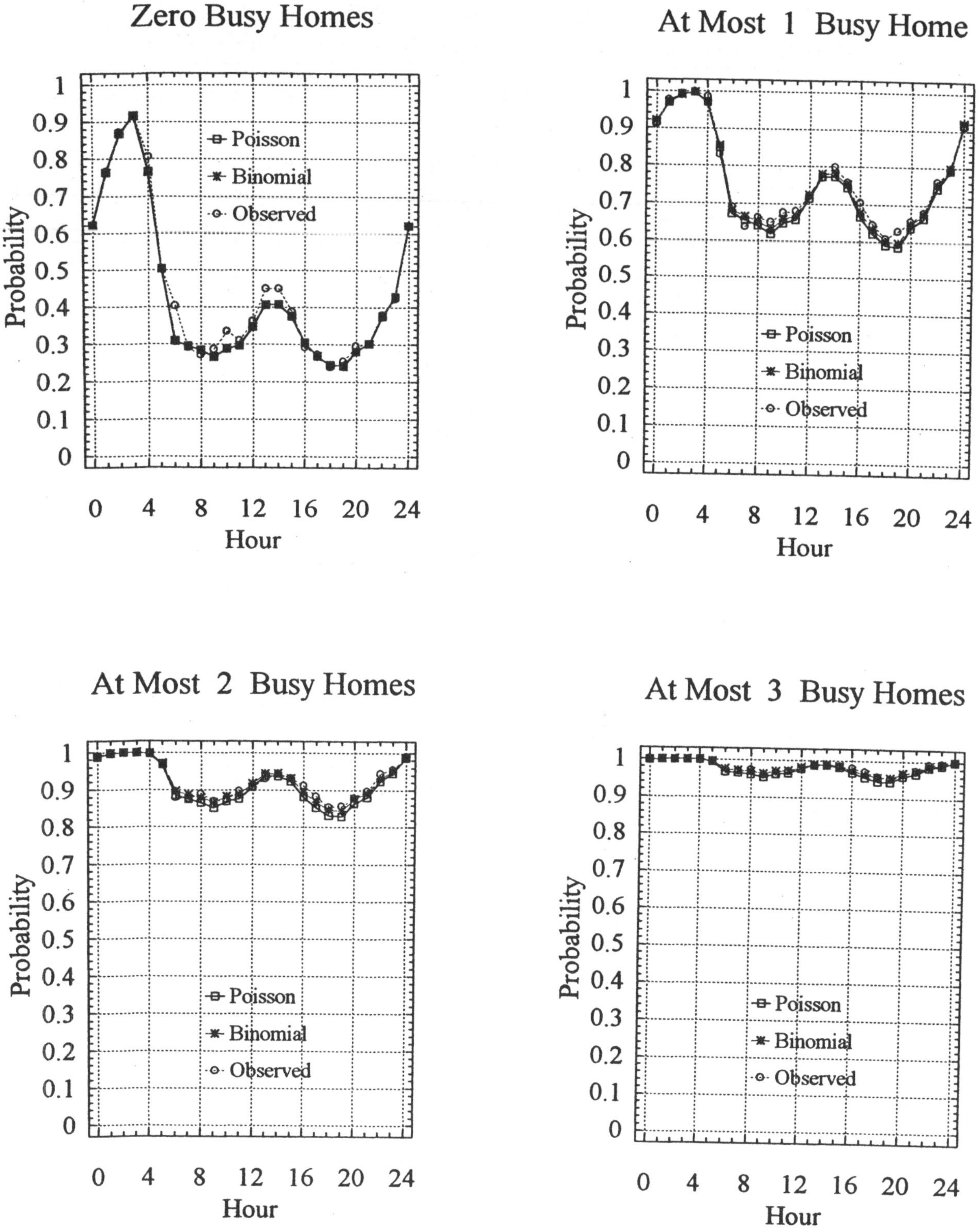

Figure 6.7 Predicted and observed hourly cumulative distribution of busy homes during the 31-day test period at the dead-end loop (N=21 total homes).

Table 6.7 Average Absolute Differences Between Hourly Observed Frequencies and Predicted Probabilities for Busy Homes.

Number of Busy Homes	Poisson Model Relative Frequency (Fig 6.6)	Poisson Model Cumulative Frequency (Fig 6.7)	Binomial Model Relative Frequency (Fig 6.6)	Binomial Model Cumulative Frequency (Fig 6.7)
0	0.0173	0.0173	0.0173	0.0173
1	0.0190	0.0174	0.0180	0.0123
2	0.0094	0.0117	0.0098	0.0069
3	0.0072	0.0050	0.0050	0.0031
4	0.0031	0.0021	0.0021	0.0013
Total	0.0560	0.0535	0.0522	0.0409
Average	0.0112	0.0107	0.0104	0.0082

Table 6.8 Probability of Exactly One Busy Server (with N=21 homes).

$\overline{\rho}(t)$	$N\overline{\rho}(t)$	Poisson Model	Binomial Model	Percent Difference
0.00	0.00	0.000	0.000	0.00
0.01	0.21	0.170	0.171	0.58
0.02	0.42	0.276	0.279	1.07
0.03	0.63	0.336	0.341	1.47
0.04	0.84	0.363	0.370	1.89
0.05	1.05	0.367	0.377	2.65
0.06	1.26	0.357	0.368	2.99
0.07	1.47	0.338	0.350	3.42
0.08	1.68	0.313	0.326	3.99
0.09	1.89	0.286	0.299	4.35
0.10	2.10	0.257	0.270	4.81

$$\text{Poisson Model: } \mathrm{Prob}[K=1] = \left(N\overline{\rho}\right)\exp[-N\overline{\rho}]$$

$$\text{Binomial Model: } \mathrm{Prob}[K=1] = \binom{N}{1}\left[1-\exp(-\overline{\rho})\right]\left[\exp(-\overline{\rho})\right]^{N-1}$$

Consider, for example, a few outcomes for the hour starting at 7 AM. Field observations show the relative frequency of stagnation (0 busy homes) is 29.0%; both probability models give 29.4%. Field data indicate the chance of having exactly 1 busy home is 34.7%; the PRP models predict 36.0% and 37.0% (Poisson and binomial, respectively in Figure 6.6). The frequency of having at most 2 busy homes at 7 AM is 87.6%; the PRP models predict 87.5% and 88.7% (Poisson and binomial, respectively in Figure 6.7). Glancing now at the hour of peak water use activity (around 7 PM), it seems remarkable to note that there is less than a 15% chance that more than 2 out of 21 homes will be busy. This underscores the sporadic random nature of residential indoor water use.

Curiously, the probability of sustaining exactly 1 busy home out of 21 total homes on the loop is relatively constant (~36%) from 5 AM to 11 PM (see Figure 6.6). During these hours, the average utilization factor varies from 0.03 to 0.07 (see Figure 6.3). Results in Table 6.8 for a neighborhood of N=21 homes (with $0.03 < \overline{\rho}(t) < 0.07$) show that the predicted probability of exactly 1 busy server is confined to the narrow range of 34 to 37 percent for both the Poisson and binomial models. This agrees well with observed relative frequencies for the 31-day test period. The percent difference between the two probability models grows as $\overline{\rho}(t)$ increases; however, in the range expected for residential indoor water use, the differences are not important.

Due to the relatively low hourly residential utilization factors, the difference between the Poisson and binomial models is almost indistinguishable (see Equation 6.4 and Table 6.7). A comparison between both options (based on predictions for 1 busy server) is presented in Table 6.8. Assuming that these results are typical, it appears the Poisson model for busy fixtures can be used without much error in place of the binomial model to approximate the probability distribution of busy homes in a water distribution system[(8)]. Given a choice between the two, the Poisson model would be the preferred alternative owing to ease of use (it avoids the binomial coefficient). In the next section, the Poisson model is used to examine the distribution of busy homes on a daily time scale.

[(8)] The PRP predictions for busy homes can be applied *anywhere* in the distribution system, not only to dead-end zones.

Busy Homes and the Daily Quad Plots

Each of the 31 daily quad plots in Appendix H includes a time series and a frequency histogram of busy homes compiled from *indoor and outdoor* water use (Data Set 3, Table 3.2). Occasionally the time series plots suggest a nearly uniform busy home activity throughout much of the day (see Friday May 16 or Wednesday June 4). In most cases, however, the non-homogeneous nature of daily water use is vividly displayed (see Thursday May 15, Friday May 23, Tuesday May 27, Sunday June 1) with distinct periods of high and low activity at the loop. The actual number of busy homes ranged from 0 (the mode) up to a maximum of 10 busy homes, which occurred once (Sunday morning May 18).

The busy home frequency histograms include a Poisson probability model fitted with a daily utilization factor. Even though the 1-day time scale is rather coarse (in contrast to the 1-hr increment used in Figures 6.6 and 6.7), most of the daily distributions of busy homes achieve a surprisingly good fit to the Poisson model. Exceptions occur when outdoor use is dominant on the loop (for example, see Monday May 19, Saturday May 24, Monday May 26, Saturday June 7) (9). Despite providing a reasonably good fit to busy homes, the Poisson model consistently under-estimates the daily stagnation probability (this discrepancy was first noted in Table 6.6) (10).

---

(9) Outdoor use is treated later in this Chapter.

(10) The under-estimation occurs because the exponential function is nonlinear and the utilization factor is non-homogeneous (time dependent). To see this, note the Poisson model in the quad plots is based on first finding a daily average utilization factor and then estimating the probability of stagnation (see Case 1). A more accurate option is first to estimate the hourly probability of stagnation and then take the average (see Case 2). It can be shown that the prediction from Case 1 never exceeds the prediction from Case 2 (see Appendix 6.5).

$$\textit{Case 1:}\ \overline{P}_0 = \exp[-\overline{\rho}] = \exp\left[-\frac{1}{N}\sum_{j=1}^{N}\rho_j\right]$$

$$\textit{Case 2:}\ \overline{P}_0 = \frac{1}{N}\sum_{j=1}^{N}(P_0)_j = \frac{1}{N}\sum_{j=1}^{N}\exp[-\rho_j]$$

Busy Server Transitions

The number of busy servers changes in response to customer arrivals and departures. The average number of customer arrivals in a block of N homes during time interval ΔT is $\lambda_N{}^*\Delta T$ (where $\lambda_N{}^*$ is the cumulative arrival rate, Table 6.3). Under steady-state conditions, the expected number of jumps upward from **k→ (k+1)** busy servers and jumps downward from **(k+1)→ k** busy servers are equal (see Appendix 6.2). The busy server state transitions can be formulated for Point-of-Use (H-1) and Point-of-Entry (H-2) perspectives.

Point-of-Use: Busy Fixtures–The expected number of jumps[11] is $E[J_{k,k+1}] = \lambda_N{}^*\Delta T\, P_k$, where P_k is the Poisson model for k busy fixtures (Eq 6.2). In dimensionless form this is

$$\frac{E[J_{k,k+1}]}{\lambda_N^* \Delta T} = \frac{\left(\rho_N^*\right)^k \exp\left[-\rho_N^*\right]}{k!} \tag{6.5}$$

Point-of-Entry: Busy Homes–Here $E[J_{k,k+1}] = [(N-k)/N]\, \lambda_N{}^*\Delta T\, P_k$, where P_k is the binomial probability model for k busy homes (Eq 6.3). In dimensionless form this can be written

$$\frac{E[J_{k,k+1}]}{\lambda_N^* \Delta T} = \binom{N-1}{k}\left(1-\exp\left[-\frac{\rho_N^*}{N}\right]\right)^k\left(\exp\left[-\frac{\rho_N^*}{N}\right]\right)^{N-k} \tag{6.6}$$

Figure 6.8 shows hourly busy home transitions observed during the 31-day test period versus the neighborhood hourly utilization factor, $\rho_N{}^*$. Also included are dimensionless busy server transitions for k = 0 to 3 fixtures (dotted lines computed with Eq 6.5) and k = 0 to 3 homes (solid lines based on Eq 6.6). As expected, Eqs 6.5 and 6.6 agree when k=0; however, when k>0 there are fewer jumps with Eq 6.6 than 6.5 because the arrival (or departure) of a customer does not necessarily change the number of busy homes (each home can have more than one busy fixture). The fit between observed and predicted busy server transitions is quite good, bolstering support for the PRP model of residential water use.

---

[11] This is the number of one-way jumps (either up or down). Multiply by 2 to get the total number of jumps *up and down* between states k and k+1.

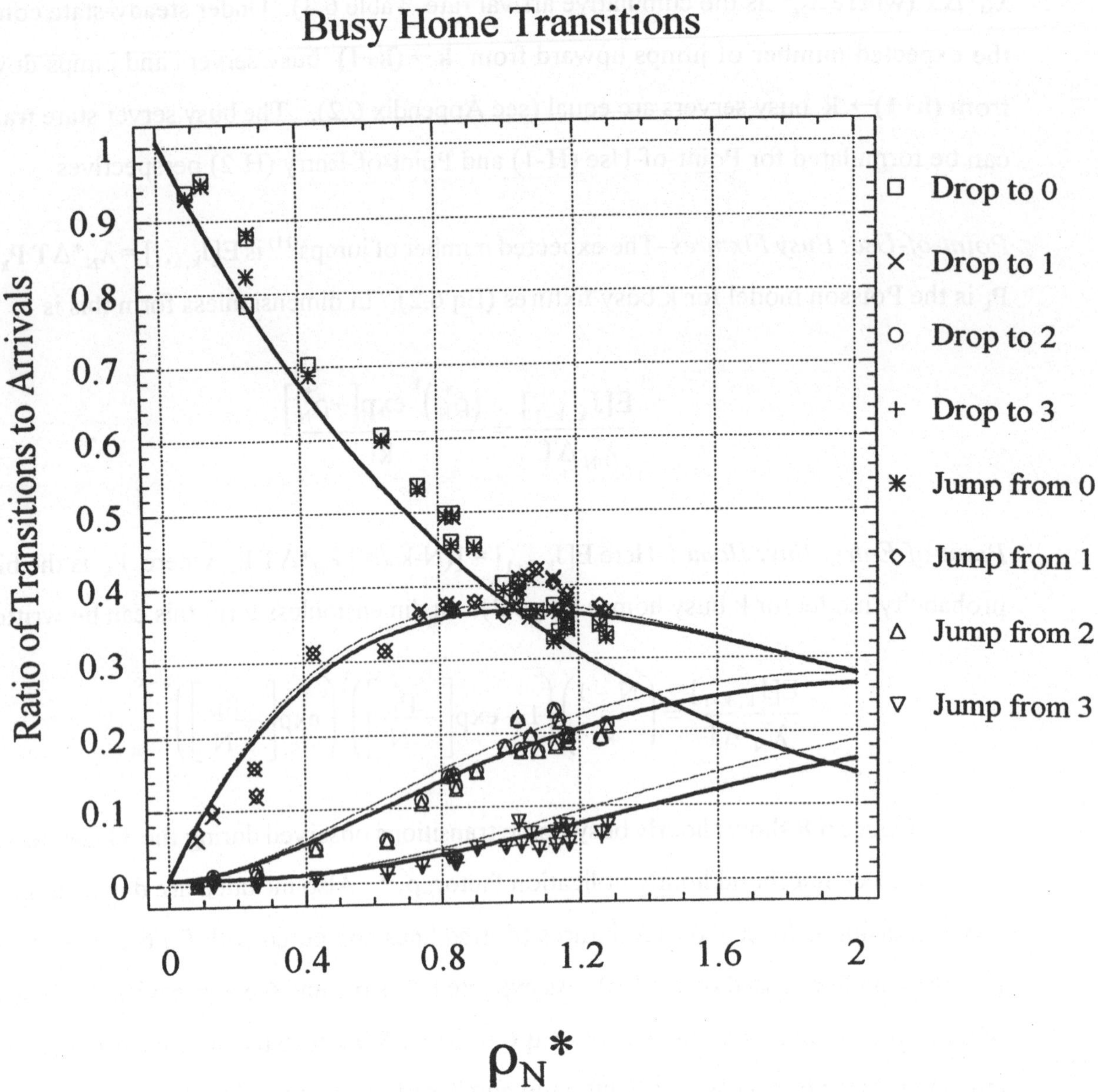

Figure 6.8 Predicted and observed busy server transitions during the 31-day test period at the dead-end loop (solid lines are homes; dotted lines are fixtures).

Busy Server Autocorrelation Function

Consider a neighborhood of N homes subject to a homogeneous Poisson arrival process with parameter, ρ_N*. If the service times are exponentially distributed with mean duration τ_N* (or mean rate, μ_N*), then the number of busy fixtures, K_N*(t), has an exponential covariance function (Parzen, 1962, p 149):

$$\mathrm{CoVar}[K_N^*(t)] = \rho_N^* \exp\left[-\frac{|t|}{\tau_N^*}\right] = \rho_N^* \exp\left[-\mu_N^* |t|\right] \qquad (6.7)$$

When t=0, the covariance function recovers the busy fixture variance, ρ_N*. Dividing by ρ_N* gives the dimensionless autocorrelation function (ACF: exp[$-\mu_N$* |t|]). Figure 6.9 shows two examples of the hourly ACF for busy fixtures at the study site based on sample estimates and on theoretical predictions from Eq 6.7. Appendix J has 144 additional hourly ACFs, computed for six consecutive days (May 12 through May 17, 1997). Generally speaking, the agreement between the observed and predicted busy server ACF is very good.

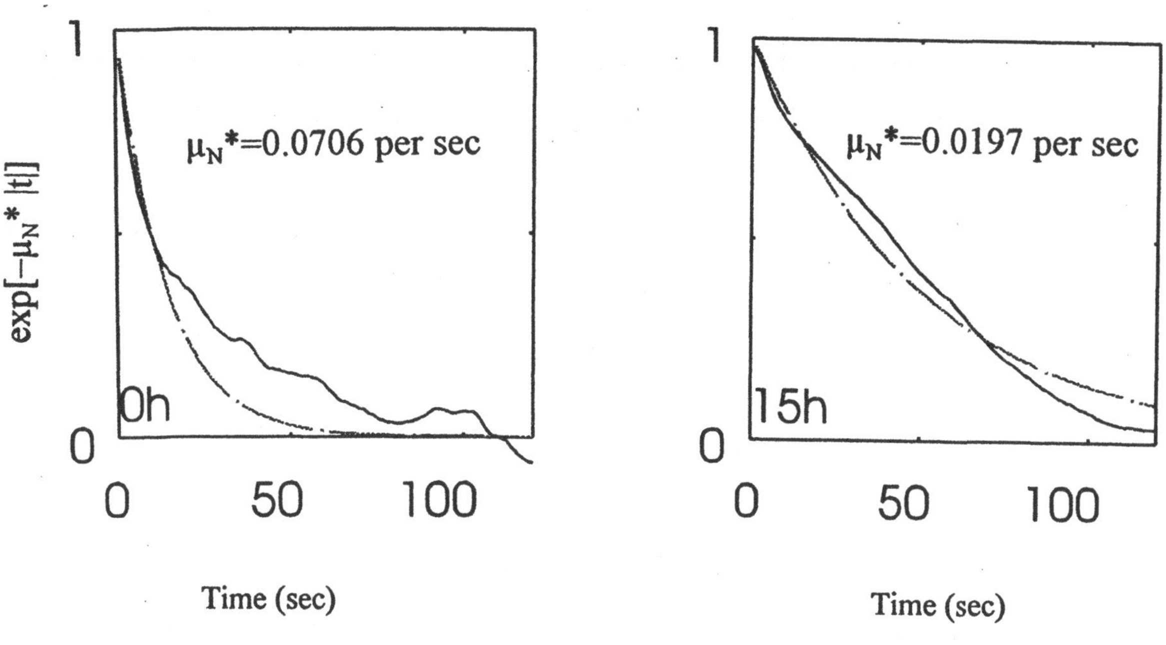

Figure 6.9 Predicted and observed busy server ACF, exp[$-\mu_N$* |t|] at hours 0:00 and 15:00 on May 13, 1997 at the dead-end loop.

FLOW RATES

Flow rates provide a second test of the PRP hypothesis for residential water use. Flow rates represent the cumulative instantaneous water demand resulting from downstream users. The PRP model predicts the moments of the flow rate at various locations throughout the dead-end loop. Similar to busy servers, the analysis of flow rates can be formulated at several hierarchy levels. Three hierarchy levels are considered (see Figure 6.4):

❖ **Point-of-Use (H-1)**: Flow rates at individual fixtures inside the home.
❖ **Point-of-Entry (H-2)**: Flow rates into an individual single family home.
❖ **Point-of-No-Return (H-3)**: Flow rates into dead-end stem supplying N homes.

In addition, each of the three hierarchy levels can be viewed from two different time horizons, namely, "busy-time" and "total-time". Both of these time frames have appeared already in the example calculation for the utilization factor given in Table 6.4. Hence, up to six potential cases can be considered when dealing with flow rates in dead-end regions.

Mean and Variance of Flows

Table 6.9 summarizes expressions for the mean and variance of water demands for each of the six cases mentioned above (three hierarchy levels each with two time horizons). The results are based on the assumption of a homogeneous Poisson arrival process (see Appendix 6.4). If the arrivals are non-homogeneous (and almost surely they will be), then the variance expressions in Table 6.9 will under-estimate the actual flow variance, particularly at the H-3 level. Non-homogeneous arrivals, however, have no effect on the mean flow (see Appendix 6.5).

Two key SERP parameters appear in Table 6.9, namely, α (mean) and β^2 (variance) of the water demand intensity at a busy fixture. Under ideal experimental conditions, α and β^2 would be measured directly at the H-1 Point-of-Use location and then all other flow quantities in Table 6.9 would be computed from them. In this study, however, demand monitoring was performed at the H-2 Point-of-Entry (service line into each home). As explained in Chapter 3, coincident pulses were separated from the Point-of-Entry input signal (see flowchart in Figure 3.6) and subsequently used to estimate α and β^2.

Table 6.9 Moments of Residential Water Demands Depend on Time Horizon and Location in Water Distribution System.

Busy Time (only when water flows)		Total Time (busy *and* idle periods)	
Hierarchy Level: H-1 Single Fixture...SERPs (Point-of-Use)			
$E[Q] = \alpha$	(6.8a)	$E[Q] = \rho\alpha/\eta$	(6.9a)
$Var[Q] = \beta^2$	(6.8b)	$Var[Q] = (\rho/\eta)[\beta^2 + \alpha^2 \exp(-\rho)]$	(6.9b)
$E[Q^2] = \alpha^2 + \beta^2$	(6.8c)	$E[Q^2] = (\rho/\eta)(\alpha^2 + \beta^2)$	(6.9c)
Hierarchy Level: H-2 Single Home...SORBs (Point-of-Entry)			
$E[Q] = \eta\alpha$	(6.10a)	$E[Q] = \rho\alpha$	(6.11a)
$Var[Q] = \eta(\alpha^2 + \beta^2) - \exp(-\rho)\eta^2\alpha^2$	(6.10b)	$Var[Q] = \rho(\alpha^2 + \beta^2)$	(6.11b)
$E[Q^2] = \eta(\alpha^2 + \beta^2) + \eta\rho\alpha^2$	(6.10c)	$E[Q^2] = \rho(\alpha^2 + \beta^2) + (\rho\alpha)^2$	(6.11c)
Hierarchy Level: H-3 Single Neighborhood...Dead-End Entrance (Point-of-No-Return)			
$E[Q_N^*] = \Sigma\, \eta_j\, \alpha_j$	(6.12a)	$E[Q_N^*] = \Sigma\, \rho_j\, \alpha_j$	(6.13a)
$Var[Q_N^*] = \Sigma\, \eta_i\, (\alpha_j^2 + \beta_j^2) - \exp(-\rho_N^*)(\Sigma\, \eta_j\, \alpha_j)^2$	(6.12b)	$Var[Q_N^*] = \Sigma\, \rho_j\, (\alpha_j^2 + \beta_j^2)$	(6.13b)
$E[(Q_N^*)^2] = \Sigma\, \eta_j\, (\alpha_j^2 + \beta_j^2) + [1 - \exp(-\rho_N^*)](\Sigma\, \eta_j\, \alpha_j)^2$	(6.12c)	$E[(Q_N^*)^2] = \Sigma\rho_j\, (\alpha_j^2 + \beta_j^2) + (\Sigma\rho_j\, \alpha_j)^2$	(6.13c)

Notes: (1) Results assume time homogeneous Poisson arrival process.
(2) Summations in level H-3 are taken over all N homes (j=1,2,...,N).

$$\eta = \frac{\rho}{1 - \exp(-\rho)} \qquad \eta_j = \frac{\rho_j}{1 - \exp\left(-\rho_N^*\right)} \qquad \rho_N^* = \sum_{j=1}^{N} \rho_j \qquad Q_N^* = \sum_{j=1}^{N} Q_j$$

Moving from H-1 to H-2 to H-3 in Table 6.9 (down the busy-time or total-time column) leads to an *increase* in the mean and the variance of the flow rate. These increases occur because each step of this progression (from fixture to home to neighborhood) adds more potential downstream users. Moving left to right across rows (from a busy-time to a total-time perspective at a fixed hierarchy level) leads to a *decrease* in the mean and variance of the flow rate. The drop occurs because total-time calculations include idle periods when the flow is stagnant and demands are zero.

This behavior is evident in Figure 6.10 which shows predicted mean flow ratios versus average household utilization factor (recall, $\overline{\rho} \approx 0.05$). There are three lines in Figure 6.10 and each can be visualized as a transfer out of the H-1 busy-time cell in Table 6.9. Line (1) is the amplification factor (η) and illustrates the effect of movement from H-1 to H-2 in the busy-time column. Line (2) is the utilization factor (ρ) and shows the transfer of flow rate information from the H-1 busy-time cell to the H-2 total-time cell (diagonally in Table 6.9). Line (3) corresponds to the dimensionless ratio $\rho/\eta = 1-\exp(-\rho) = \text{Prob}[Q>0]$ and indicates the effect of movement from a busy-time to a total-time perspective for a given hierarchy level (left to right across a row in Table 6.9).

In moving from level H-2 to H-3, it is assumed that homes act independently of each other so that residential flow variances are additive. The expression for the total-time mean flow in level H-3 (Eq 6.13a) is a generalized version of the result by Linaweaver *et al* (1966, pg 14) for the expected demand on a water system serving N homes. When the number of homes gets large, the H-3 busy-time and total-time results in Table 6.9 converge[(12)]. Hence, as neighborhood size increases, stagnation gradually disappears and flow through the supply main evolves from a random *intermittent* signal to a random *continuous* pattern. In practice, this transition from intermittent to continuous flow depends on $N\overline{\rho}$. Assuming average daily conditions ($\overline{\rho} \approx 0.050$), about N=100 homes are needed to achieve nearly continuous flow. During peak hours ($\overline{\rho} \approx 0.067$) about 75 homes may be adequate while in the early morning ($\overline{\rho} \approx 0.005$) up to 1000 homes or more may be needed to prevent stagnation.

---

(12) To see this, note that as $N \rightarrow \infty$, the amplification factor approaches the utilization factor ($\eta_j \rightarrow \rho_j$) and the neighborhood stagnation probability $P_0 = \exp(-\rho_N^*) \rightarrow 0$. See also Eq 6.15.

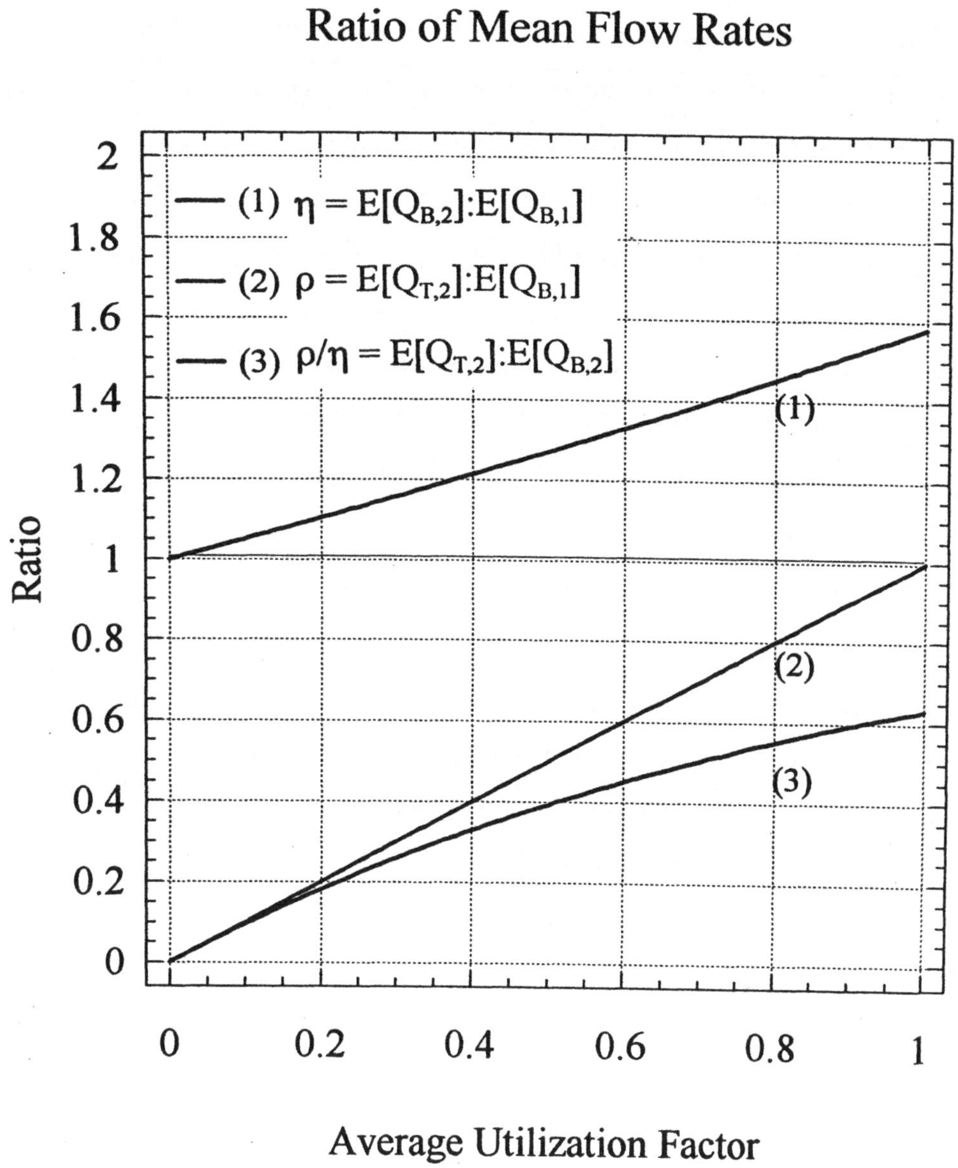

Figure 6.10 Theoretical ratio of mean flow rates versus utilization factor, based on expressions in Table 6.9. (note: $Q_{B,2}$ denotes busy-time flow at level H-2 similarly, $Q_{T,2}$ denotes total-time flow at hierarchy level H-2).

Residential Point-of-Use (SERP) water demand statistics for the 31-day test period are summarized in Table 6.10. The information in columns 1 through 11 is extracted from the May11_June10 subset of the monthly statistical summaries of residential indoor water use presented in Appendix C. Probability distributions and other graphics for the SERP intensity, duration and volume for each home are presented in Appendix D. Columns 8 and 9 give estimates of α and β^2 (mean and variance). Both are key seed parameters in the H-1 busy-time cell of Table 6.9. Column 12 shows the household average utilization factor (estimated using the method of moments approach demonstrated in Table 6.4) and column 13 gives the corresponding amplification factor, $\eta=\rho/[1-\exp(-\rho)]$. A separate legend accompanies Table 6.10 with exact details on how values in each column were obtained.

Table 6.11 has comparisons between observed and predicted means and variances for busy-time flows and total-time flows for all homes at the H-2 Point-of-Entry. The *observed* flow values (even-numbered columns) are based on field measurements archived in Data Set No 2, the SORBs. The *predicted* flow values (odd-numbered columns) are based on equations from level H-2 in Table 6.9 using household SERP values for α, β, ρ, and η from Table 6.10. Computation details are summarized in a separate legend. Figure 6.11 shows graphs of these tabulated household mean and variance data, plus the coefficient of variation for the flows. The graphs reveal very good agreement between observed and predicted trends (correlation coefficients exceed 97% in all cases), though the busy-time results exhibit a slight but systematic bias with low PRP predicted means and high PRP predicted variances.

In Table 6.12, the comparison between observed and predicted moments of the flow is extended to hierarchy level H-3, the neighborhood of 21 homes. The total flow into the study site was not measured directly. Instead, the inflow hydrograph to the dead-end loop was reconstructed by superimposing second-by-second the indoor water demands measured at all homes along the dead-end loop. Taken over the entire 31-day test period (May 11 to June 10, 1997), this produced a time series of 2,678,400 readings. Daily snapshots of the total inflow hydrograph (with both indoor and outdoor water demands) are included in the quad plots of Appendix H. These inflows were analyzed to estimate the mean and variance of busy-time and total-time demands into the neighborhood. Results, summarized in Table 6.12, exhibit four expected trends:

Table 6.10 SERP Water Use Statistics from Indoor Residential Demand Data Hierarchy Level: H-1 Point-of-Use, Collected May 11 to June 10, 1997

	Quantities Recorded at the Dead-End Loop Study Site			Arrival Rate	SERP Duration		SERP Intensity		SERP Volume		Utilization Factor	Amp Factor
(1)	(2)	(3)	(4)	(5)	(6)	(7)	(8)	(9)	(10)	(11)	(12)	(13)
home	total SERP count	total volume (gal)	busy-fixture time (min)	average λ (min^{-1})	average τ (min)	variance ω^2 $(min)^2$	average α (gpm)	variance β^2 $(gpm)^2$	average ϕ (gal)	variance ψ^2 $(gal)^2$	ρ	η
1	2421	3502	1580	0.0542	0.652	1.791	2.217	1.160	1.447	10.76	0.0354	1.0178
2	1361	3693	1339	0.0305	0.984	3.693	2.758	2.943	2.713	39.82	0.0300	1.0151
3	1553	3331	1550	0.0348	0.998	4.615	2.149	2.165	2.145	27.88	0.0347	1.0175
4	8649	5473	2700	0.1944	0.312	0.210	2.027	1.323	0.633	2.16	0.0605	1.0306
5	1897	5417	3166	0.0425	1.669	11.41	1.711	1.696	2.856	35.40	0.0709	1.0359
6	2389	2393	1089	0.0535	0.456	0.414	2.197	1.638	1.002	4.67	0.0244	1.0123
7	2831	4867	1795	0.0634	0.634	1.720	2.712	1.901	1.719	23.23	0.0402	1.0202
8	2240	5709	1902	0.0502	0.849	1.927	3.002	2.856	2.549	32.26	0.0426	1.0215
9	1590	6588	2162	0.0356	1.360	5.153	3.047	2.478	4.144	76.56	0.0484	1.0244
10	1815	4516	1982	0.0407	1.092	2.913	2.279	1.877	2.488	31.92	0.0444	1.0224
11	1483	1912	903	0.0332	0.609	0.786	2.118	0.763	1.289	5.34	0.0202	1.0101
12	3062	5236	2309	0.0686	0.754	1.951	2.268	1.588	1.710	17.31	0.0517	1.0261
13	1555	4491	1926	0.0348	1.239	3.204	2.332	0.476	2.888	23.81	0.0431	1.0217
14	2633	2761	1331	0.0590	0.506	1.823	2.075	1.106	1.049	12.82	0.0298	1.0150
15	2053	4506	2292	0.0460	1.117	3.435	1.966	1.368	2.195	12.96	0.0513	1.0259
16	2436	2818	1244	0.0546	0.511	0.734	2.265	0.976	1.157	6.05	0.0279	1.0140
17	1441	3191	1390	0.0323	0.965	1.647	2.296	1.277	2.214	16.73	0.0311	1.0156
18	3328	6087	3921	0.0746	1.178	5.476	1.552	0.761	1.829	26.83	0.0878	1.0445
19	837	3242	1824	0.0188	2.179	5.300	1.777	0.827	3.873	44.49	0.0409	1.0206
20	3051	9373	2950	0.0683	0.967	3.033	3.177	1.439	3.072	36.85	0.0661	1.0334
21	3447	7732	3184	0.0772	0.924	6.717	2.428	1.796	2.243	30.03	0.0713	1.0361
Total	52,072	96,838	42,539	1.1672	19.955	67.952	48.353	32.414	45.215	517.88	0.9527	21.4807
GA-1	2,480	4,611	2,026	0.0556	0.950	3.236	2.303	1.544	2.153	24.66	0.0454	1.0229
GA-2					0.817	2.820	2.277	1.530	1.860	20.70		

Legend is on next page; Additional statistics are in Appendix C; GA=global average (see Table 4.5); SERP=Single Equivalent Rectangular Pulse.

Legend for Tables 6.10 and 6.11
Water Use Statistics from Indoor Residential Demand Data
Collected May 11 to June 10, 1997

TABLE 6.10 Hierarchy Level H-1, Point-of-Use, SERP Statistics
Columns (2) through (4) are measured values obtained from Data Set 1 (The SERPs).
Columns (5) through (13) are computed values.

Column (1)	is the assigned home ID number.	
Column (2)	is the total number of individual water demand pulses recorded at the home; (coincident water uses have been separated).	
Column (3)	is the total volume of water used at the home.	
Column (4)	is the total busy-fixture time recorded at the home (see Table 6.4 for example)	
Column (5)	arrival rate,	λ = Col (2) ÷ 44,640
Column (6)	mean duration,	τ = Col (4) ÷ Col (2)
Column (7)	variance duration,	$\omega^2 = [\Sigma(d_i)^2/\text{Col }(2)] - [\text{Col }(6)]^2$
Column (8)	mean intensity,	α = Col (3) ÷ Col (4) = Col (10) ÷ Col (6)
Column (9)	variance intensity,	$\beta^2 = [\Sigma d_i q_i^2/\text{Col }(4)] - [\text{Col }(8)]^2$
Column (10)	mean volume,	ϕ = Col (3) ÷ Col (2)
Column (11)	variance volume,	$\psi^2 = [\Sigma(d_i q_i)^2/\text{Col }(2)] - [\text{Col }(10)]^2$
Column (12)	utilization factor,	ρ = Col (4) ÷ 44,640 = Col (5) × Col (6)
Column (13)	amplification factor,	η = Col(12) ÷ {1-exp[-Col(12)]}

TABLE 6.11 Hierarchy Level H-2, Point-of-Entry, SORB Statistics
Odd columns are predictions from PRP theory computed with expressions in Table 6.9.
Even columns are computed from observed PoE flows in Data Set 2 (the SORBs).
(α, β, ρ, and η are household **SERP** parameters obtained from Table 6.10)

Column (1)	is the assigned home ID number.	
Column (2)	$\Sigma(D_iQ_i)/\Sigma D_i$	
Column (3)	$E[Q] = \eta\alpha$	(Eq 6.10a in Table 6.9)
Column (4)	$\Sigma(D_iQ_i^2)/\Sigma D_i - [\Sigma(D_iQ_i)/\Sigma D_i]^2$	
Column (5)	$Var[Q] = \eta(\alpha^2+\beta^2) - \exp(-\rho)(\eta\alpha)^2$	(Eq 6.10b in Table 6.9)
Column (6)	$\Sigma(D_iQ_i)/44{,}640$	
Column (7)	$E[Q] = \rho\alpha$	(Eq 6.11a in Table 6.9)
Column (8)	$\Sigma(D_iQ_i^2)/44{,}640 - [\Sigma(D_iQ_i)/44{,}640]^2$	
Column (9)	$Var[Q] = \rho(\alpha^2+\beta^2)$	(Eq. 6.11b in Table 6.9)

Notes: (1) There are 44,640 minutes in the 31 day period from May 11 to June 10.
(2) d_i is the duration and q_i is the intensity of ith **SERP** at the home.
(3) D_i is the duration and Q_i is the intensity of the ith **SORB** at the home.

Table 6.11 Comparison of Observed and Predicted Flows into Homes Along Dead-End Loop Hierarchy Level: H-2 Point-of-Entry, Data Collected May 11 to June 10, 1997.

Home Number	Busy-Time Perspective: Mean Flow (gpm) (2) Observed	(3) Predicted	Variance of Flow $(gpm)^2$ (4) Observed	(5) Predicted	Total-Time Perspective: Mean Flow (gpm) (6) Observed	(7) Predicted	Variance of Flow $(gpm)^2$ (8) Observed	(9) Predicted
1	2.2614	2.2565	1.1287	1.2687	0.0785	0.0784	0.2045	0.2150
2	2.9626	2.7996	2.9059	3.1026	0.0827	0.0827	0.3127	0.3164
3	2.3437	2.1865	2.2546	2.2838	0.0746	0.0746	0.2357	0.2355
4	2.1025	2.0889	1.3667	1.4902	0.1226	0.1226	0.3084	0.3293
5	1.7887	1.7740	1.8014	1.8631	0.1214	0.1213	0.3108	0.3280
6	2.3663	2.2239	1.7059	1.7174	0.0536	0.0536	0.1598	0.1577
7	2.7851	2.7669	1.8669	2.0893	0.1090	0.1090	0.3534	0.3721
8	3.4135	3.0664	2.5411	3.1120	0.1279	0.1279	0.4995	0.5058
9	3.4054	3.1213	2.1698	2.7667	0.1476	0.1476	0.5540	0.5698
10	2.3535	2.3300	1.9825	2.0360	0.1012	0.1012	0.3033	0.3140
11	2.1296	2.1395	0.7596	0.8164	0.0428	0.0428	0.1029	0.1062
12	2.3908	2.3271	1.6637	1.7647	0.1173	0.1173	0.3352	0.3483
13	2.4445	2.3826	0.3121	0.6052	0.1006	0.1006	0.2390	0.2551
14	2.1071	2.1061	1.1019	1.1874	0.0619	0.0619	0.1552	0.1613
15	2.0542	2.0169	1.3795	1.5042	0.1010	0.1009	0.2553	0.2688
16	2.3328	2.2967	0.9696	1.0619	0.0631	0.0631	0.1657	0.1702
17	2.4955	2.3319	1.1315	1.3798	0.0715	0.0715	0.2007	0.2037
18	1.6433	1.6211	0.7317	0.9037	0.1364	0.1316	0.2491	0.2785
19	1.7909	1.8136	0.8164	0.9095	0.0726	0.0726	0.1528	0.1629
20	3.2575	3.2831	1.3415	1.8280	0.2100	0.2100	0.6851	0.7622
21	2.5370	2.5156	1.8537	2.0760	0.1732	0.1732	0.5080	0.5485
Total	50.9659	49.4466	31.7847	35.7667	2.1695	2.1693	6.2911	6.6094
GA-1	2.427	2.355	1.514	1.703	0.103	0.103	0.300	0.315
GA-2	2.399	2.335	1.504	1.707				

Notes: (1) Observed values are computed from H-2 Point-of-Entry **SORBs** (Data Set 2); See legend on preceding page.
(2) Predicted means and variances are computed from expressions in Table 6.9 using H-1 Point-of-Use **SERP** parameters from Table 6.10.
(3) Correlation coefficients: busy-time comparisons have R>97.4%; total-time comparisons have R>99.8%

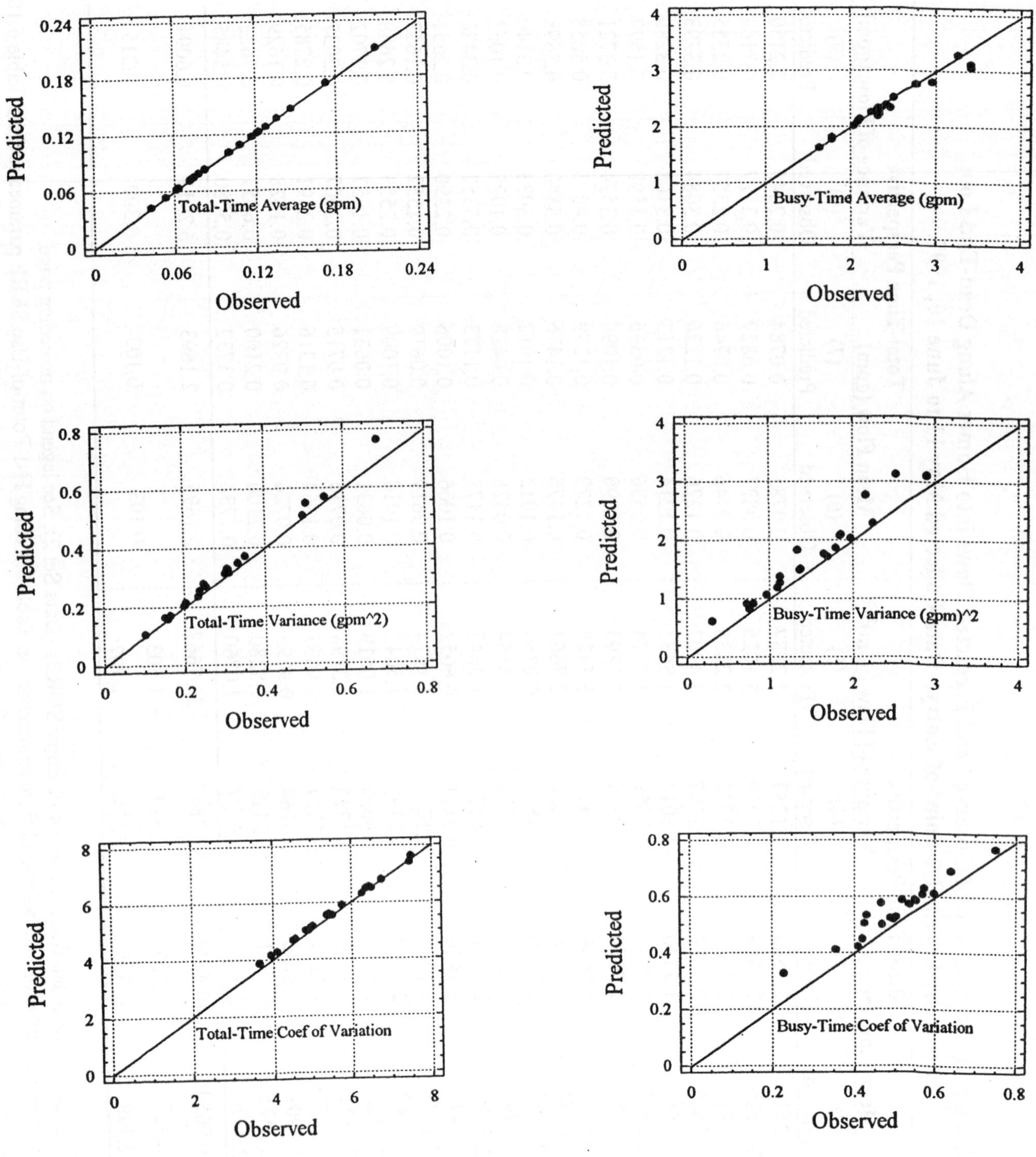

Figure 6.11 Predicted and observed total-time and busy-time flow statistics during the 31-day test period at the dead-end loop.

Table 6.12 Observed and Predicted Moments of Indoor Residential Water Demands Data Collected May 11 to June 10, 1997 at the Dead-End Loop.

Busy-Time Perspective (only when water flows)				Total-Time Perspective (busy *and* idle periods)			
Hierarchy Level: H-1 Single Fixture (Point-of-Use)							
Method	E[Q] (gpm)	Var[Q] $(gpm)^2$	$\Theta_B[Q]$	Method	E[Q] (gpm)	Var[Q] $(gpm)^2$	$\Theta_T[Q]$
(a)	α=2.28	β^2=1.53	0.54	(b)	0.101	0.288	5.31
Hierarchy Level: H-2 Single Home (Point-of-Entry)							
Method	E[Q] (gpm)	Var[Q] $(gpm)^2$	$\Theta_B[Q]$	Method	E[Q] (gpm)	Var[Q] $(gpm)^2$	$\Theta_T[Q]$
(c)	2.399	1.514	0.51	(c)	0.102	0.300	5.36
(d)	2.335	1.703	0.56	(d)	0.102	0.315	5.50
(e)	2.336	1.702	0.56	(f)	0.103	0.311	5.41
Hierarchy Level: H-3 Single Neighborhood (N=21 homes) (Point-of-No-Return)							
Method	$E[Q_N^*]$ (gpm)	$Var[Q_N^*]$ $(gpm)^2$	$\Theta_B[Q_N^*]$	Method	$E[Q_N^*]$ (gpm)	$Var[Q_N^*]$ $(gpm)^2$	$\Theta_T[Q_N^*]$
(g)	3.853	6.996	0.69	(g)	2.169	7.587	1.27
(h)	3.800	5.702	0.63	(i)	2.169	6.291	1.16
(j)	3.743	5.947	0.65	(k)	2.169	6.607	1.19

Legend for Methods:

(a) Calculated from Eqs 6.8a,b in Table 6.9 with loop-averaged SERP parameters from Table 6.10.
(b) Calculated from Eqs 6.9a,b in Table 6.9 with loop-averaged SERP parameters from Table 6.10(*).
(c) Loop-averaged *observations* from even columns of Table 6.11 based on household SORB data.
(d) Loop-averaged *predictions* from odd columns of Table 6.11 based on SERP parameters in Table 6.10.
(e) Calculated from Eqs 6.10a,b in Table 6.9 with loop-averaged SERP parameters from Table 6.10(*).
(f) Calculated from Eqs 6.11a,b in Table 6.9 with loop-averaged SERP parameters from Table 6.10(*).
(g) Calculated from 31-day loop inflow hydrograph generated w/ SERP data aggregated across 21 homes.
(h) Calculated from Eqs 6.12a,b in Table 6.9 with loop-averaged SERP parameters from Table 6.10(*).
(i) Loop-wide sum of predictions from column 8 of Table 6.11 based on household SORB data.
(j) Calculated from Eqs 6.12a,b in Table 6.9 with household SERP parameters from Table 6.10(*).
(k) Calculated from Eqs 6.13a,b in Table 6.9 with household SERP parameters from Table 6.10(*).
(*) Equations 6.9 through 6.13 use: ρ=0.0454, η=1.0229, ρ_N^*=0.9529, N=21 homes (see Table 6.10)

- ❖ Moments of the flow increase from levels H-1→H-2→H-3.
- ❖ Moments of the flow decrease from busy-time to total-time periods.
- ❖ Busy-time and total time results show signs of convergence at the H-3 level.
- ❖ Agreement between observed and predicted results is quite good.

Predicted and observed mean total-time flows at hierarchy level H-3 agree perfectly ($E[Q_N^*] = 2.169$ gpm). As required by mass conservation, the sum of the mean flows at 21 individual homes matches the mean flow of the hydrograph into the loop (e.g., compare totals from columns 6 and 7 of Table 6.11 against results from methods (g), (i), (k) in Table 6.12). There is a minor discrepancy (<2%) between the observed and predicted mean busy-time flows at level H-3 (3.85 gpm vs. 3.80 gpm).

Effect of Nonhomogeneous Arrivals

The predicted statistics in Table 6.9 are based on a homogeneous arrival pattern. Consequently, some of the discrepancies between observed and predicted flow variances in Table 6.12 can be attributed to diurnal trends in the demand data. To illustrate, consider Figure 6.12 which shows two rho profiles for the 31-day test period at the study site:

Case (1): Daily average rho value ($\overline{\rho} = 0.0454$) for a home on the loop.
Case (2): Hourly average rho profile (values are from Figure 6.3).

The area under each rho profile is the same; hence, both cases preserve the daily average utilization factor at the study site. The flow statistics listed in Table 6.9 correspond to Case (1) where the utilization factor is uniform throughout the day.

It can be shown that the mean and the variance of the total-time flow at level H-3 for Case (2) are (see Appendix 6.5):

$$E[Q_N^*] = E\left[\rho_N^*\right]\alpha \tag{6.14.a}$$

$$Var[Q_N^*] = E\left[\rho_N^*\right]\left(\alpha^2 + \beta^2\right) + \alpha^2 Var\left[\rho_N^*\right] \tag{6.14b}$$

Here α and β^2 represent the mean and variance of flow at a busy server while $E[\rho_N^*]$ and $Var[\rho_N^*]$ are the *temporal* (hourly) mean and variance of the neighborhood utilization factor.

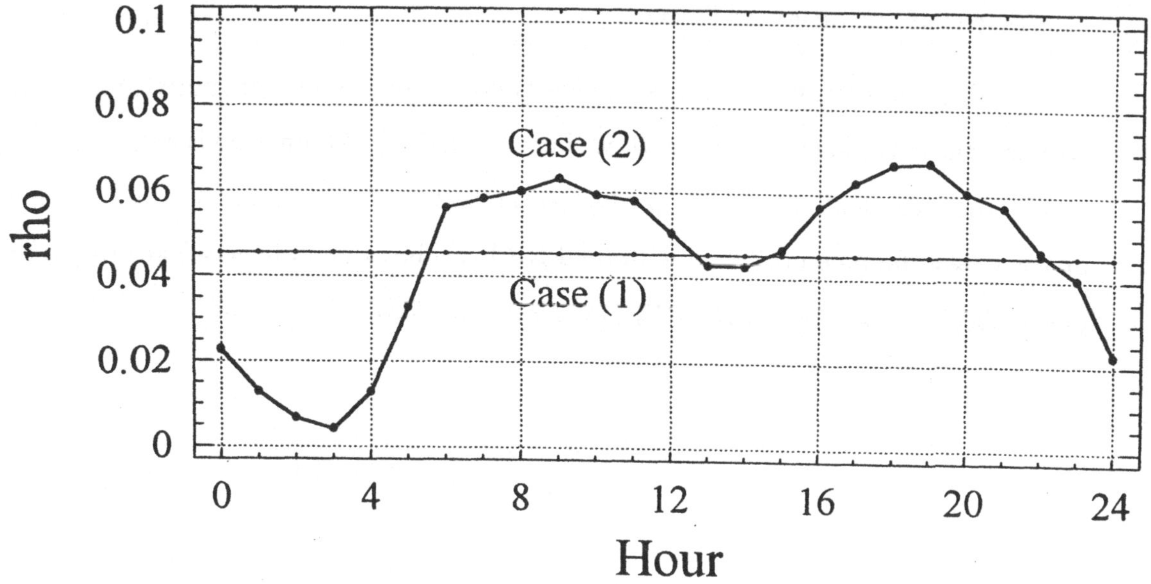

Figure 6.12 Homogeneous and nonhomogeneous rho profiles with identical total time mean flow rates.

Equation 6.14a is identical to Equation 6.13a in Table 6.9. Owing to conservation of mass, the total-time mean flow for the nonhomogeneous case is the same as the mean for the homogeneous case. However, in the presence of nonhomogeneous arrivals, the variance of the total-time flow increases by an amount α^2 Var[ρ_N*] over the homogeneous case. Estimates from Equation 6.14a,b, summarized in Table 6.13, show excellent agreement with the observed mean total-time flow. The flow variance for the nonhomogeneous case is within 4% of the variance computed from the loop inflow hydrograph (see Table 6.9 method g)

Table 6.13 Comparison of Homogeneous and Nonhomogeneous Estimates of Total-time Flow Statistics for May 11 to June 10, 1997 at Dead-end Loop.

Arrival Conditions	E[ρ_N*]	Var[ρ_N*]	E[Q_N*] (gpm)	Var[Q_N*] (gpm)^2
Homogeneous	0.953	0	2.169	6.41
Nonhomogeneous	0.953	0.174	2.169	7.32
Observed	–	–	2.169	7.59

Note: From Table 6.11, α=2.28 gpm and β^2=1.53 gpm^2.

Coefficient of Variation of Flows

Flow rates and busy servers are closely connected. The probability of zero busy servers corresponds to the likelihood of no flow (i.e., stagnation). Stagnation is linked in an interesting way to the mean and variance of the busy-time and total-time flow. For an intermittent flow process (like the type predicted with the PRP hypothesis and observed at the study site), the probability of stagnation is given by (see Appendix 6.6),

$$\mathrm{Prob}[Q=0]=P_0=1-\frac{\Theta_B^2[Q]+1}{\Theta_T^2[Q]+1} \tag{6.15}$$

where $\Theta_T[Q]$ and $\Theta_B[Q]$ are the coefficient of variation of flows for total-time and busy-time periods, respectively.

Using means and variances of the flow at the Point-of-Entry (see Table 6.11), stagnation probabilities were computed with Equation 6.15. Results, summarized in Table 6.14 and plotted in Figure 6.13, show excellent agreement between observed and predicted stagnation probabilities at each residence on the dead-end loop and across the entire neighborhood.

The numerator in Equation 6.15 goes to zero as total-time and busy-time flow statistics converge. Hence, stagnation in the supply lines disappears ($P_0 \rightarrow 0$) when there is no difference between total-time and busy-time flow rates. This observation is consistent with earlier findings (and with intuition) that the probability of stagnation at hierarchy level H-3 diminishes as the size of the neighborhood increases.

Table 6.14
Comparison of Observed and Predicted Stagnation Probabilities
Data Collected May 11 to June 10, 1997 at the Dead-End Loop

Home on Loop	Observed Coefficient of Variation		Predicted Coefficient of Variation		(P_0) Probability of Stagnation	
	(2) $\Theta_T[Q]$	(3) $\Theta_B[Q]$	(4) $\Theta_T[Q]$	(5) $\Theta_B[Q]$	(6) Observed	(7) Predicted
1	5.761	0.470	5.907	0.499	0.9643	0.9652
2	6.762	0.575	6.802	0.629	0.9715	0.9705
3	6.508	0.641	6.505	0.691	0.9675	0.9659
4	4.530	0.556	4.681	0.584	0.9392	0.9414
5	4.592	0.750	4.722	0.770	0.9292	0.9316
6	7.458	0.552	7.409	0.589	0.9770	0.9759
7	5.454	0.491	5.596	0.522	0.9596	0.9606
8	5.526	0.467	5.561	0.575	0.9614	0.9583
9	5.043	0.433	5.156	0.533	0.9551	0.9535
10	5.442	0.598	5.537	0.612	0.9556	0.9566
11	7.495	0.409	7.614	0.422	0.9796	0.9800
12	4.936	0.540	5.031	0.571	0.9491	0.9496
13	4.860	0.229	5.021	0.327	0.9573	0.9578
14	6.364	0.498	6.488	0.517	0.9699	0.9706
15	5.003	0.572	5.133	0.608	0.9490	0.9499
16	6.451	0.422	6.542	0.449	0.9724	0.9726
17	6.266	0.426	6.312	0.504	0.9706	0.9693
18	3.659	0.521	3.869	0.586	0.9117	0.9158
19	5.384	0.505	5.559	0.526	0.9582	0.9600
20	3.942	0.356	4.157	0.412	0.9319	0.9360
21	4.115	0.537	4.277	0.573	0.9282	0.9311
Loop	1.27	0.69	1.18	0.67	0.435	0.394

Notes:

(1) The coefficient of variation is based on the mean and variance of flows in Table 6.11
--Observed $\Theta[Q]$ values are from data in even-numbered columns of Table 6.11
--Predicted $\Theta[Q]$ values are from data in odd-numbered columns of Table 6.11

(2) Observed P_0 values are from Eq 6.15 using $\Theta[Q]$ from columns (2) and (3); also identical to P_0 values obtained directly from SORB data (Data Set #2).

(3) Predicted P_0 values are from Eq 6.15 using $\Theta[Q]$ from columns (4) and (5).

(4) Loop $\Theta[Q]$ values are from Table 6.12; loop P_0 values are from Eq 6.15; the actual P_0=0.423.

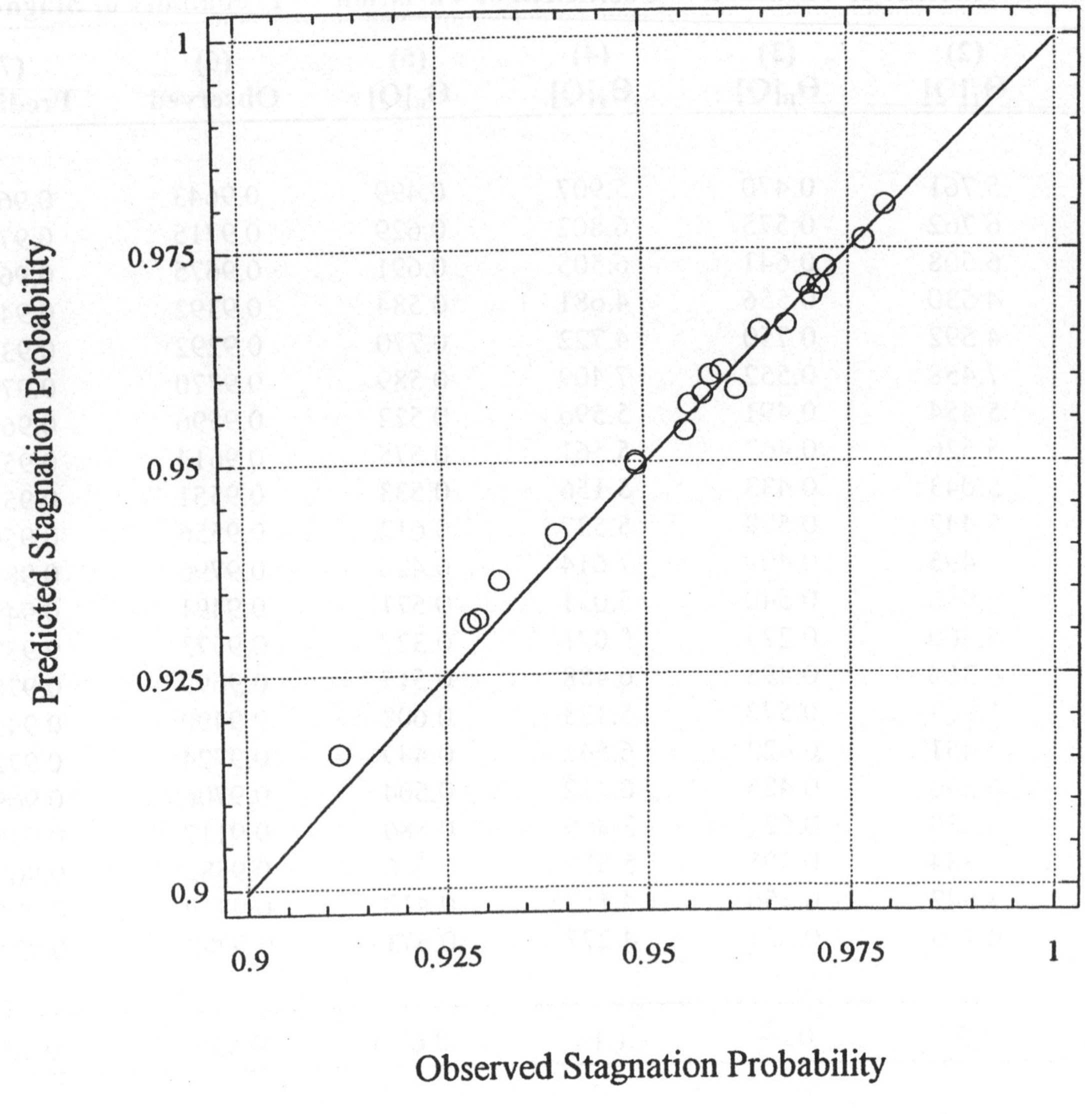

Figure 6.13 Predicted and observed stagnation probabilities during the 31-day test period at 21 homes along the dead-end loop.

Frequency Distribution of Flows

The PRP hypothesis does not specify the frequency distribution of the flow rates. However, some basic characteristics of the flow at various locations in the network can be ascertained. Flows resulting from residential water demands along dead-end lines will exhibit a mixed probability distribution having two distinct parts.

One part consists of a discrete probability mass $P_0 = \exp(-\rho_N{}^*)$ at Q=0. This mass corresponds to the chance that all servers (fixtures and homes) are idle. At the service line into a single family home (hierarchy level H-2), average daily values for the stagnation probability (P_0) generally range from 90% to 98% (see Figure 6.13). At the study site entrance to the neighborhood of 21 homes (hierarchy level H-3), P_0 was usually in the range of 35% to 50% (see any of the 31 quad plots in Appendix H).

The second component of the flow distribution is a continuous density function with a total mass $1-\exp(-\rho_N{}^*)$ for Q>0. This, of course, gives the probability that at least one server is busy. At the H-1 level under a *busy-time perspective*, the frequency distributions of SERP intensity, duration and volume over the 31-day test period at each residence show a reasonably good fit to the log-normal probability model (see Appendix D). This observation is consistent with findings from a previous study in which water consumption at four homes only was monitored (Buchberger and Wells 1996).

Moving up to the H-3 level with a *total-time perspective*, the flow into the loop generally follows a normal distribution, truncated at Q=0 (see quad plots in Appendix H). The probability mass at the truncation point corresponds to the relative frequency of idle servers (i.e., stagnant conditions when Q=0). The flow at level H-3 is the instantaneous cumulative demand of all downstream users. Hence, the appearance of the normal distribution for the loop inflows arises as a consequence of the Central Limit Theorem which states that sums of identically distributed random variables (i.e., sums of simultaneous downstream demands) quickly approach the normal distribution.

FURTHER APPLICATIONS OF THE PRP HYPOTHESIS

The Poisson Rectangular Pulse model offers a simple robust description of residential water use. PRP is a powerful tool, opening new avenues for analyzing, understanding and simulating the complex hydraulic behavior of water distribution systems. To highlight the versatility of the PRP model, a few new example applications are described here.

Application 1: Demands on Branching Systems

In rural water systems, branching (i.e., non-looping) pipes are common. Estimates of the instantaneous maximum rate of flow through a branching pipe often are obtained with a peak demand diversity (PDD) relation of the form (Lingireddy *et al* 1998)

$$Q_N = A\sqrt{N} + BN + C \tag{6.16}$$

Here N is the number of homes served by the pipe and the terms A, B and C are empirical coefficients determined from field data. When N=0, Q_N=C; hence, C measures unaccounted for water due to leaks and/or unregistered connections.

The PRP model sheds some theoretical insight on the basis of the PDD relation. Consider a period of *high* water use activity (with average household utilization factor $\overline{\rho}$) in a neighborhood of N homes. The corresponding busy-time mean and variance of residential water demands can be written (from Eqs 6.12a,b in Table 6.9 and Appendix 6.4),

$$E[Q_N^*] = \frac{N\overline{\rho}\overline{\alpha}}{1-\exp(-N\overline{\rho})} \tag{6.17}$$

$$\mathrm{Var}[Q_N^*] = \frac{N\overline{\rho}\left(\overline{\alpha}^2 + \overline{\beta}^2\right)}{1-\exp(-N\overline{\rho})} - \exp(-N\overline{\rho})\left(\frac{N\overline{\rho}\overline{\alpha}}{1-\exp(-N\overline{\rho})}\right)^2 \tag{6.18}$$

Peak design flows can be estimated using a "frequency factor" approach,

$$Q_{N,k} = E[Q_N^*] + z_k\sqrt{\mathrm{Var}[Q_N^*]} = E[Q_N^*]\left(1 + z_k\Theta_B[Q_N^*]\right) \tag{6.19}$$

where z_k is the kth percentile[13] of the underlying probability distribution and $\Theta_B[Q_N^*]$ is the coefficient of variation of the busy-time demands in the neighborhood of N homes. Putting Eqs 6.17 and 6.18 into Eq 6.19 and simplifying gives,

$$Q_{N,k} = \left(\frac{N\overline{\rho}\overline{\alpha}}{1-P_0}\right)\left(1+z_k\sqrt{\left(\frac{1-P_0}{N\overline{\rho}}\right)\left(1+\frac{\overline{\beta}^2}{\overline{\alpha}^2}\right)-P_0}\right) \tag{6.20}$$

where $P_0 = \exp(-N\overline{\rho})$ is the neighborhood stagnation probability. For a single home (N=1 and $P_0 \approx 1-\overline{\rho}$), Eq 6.20 reduces to

$$Q_{1,k} \approx \overline{\alpha} + z_k\sqrt{\overline{\beta}^2 + \overline{\rho}\overline{\alpha}^2}$$

When the number of homes is "large" (N>10 will suffice provided $P_0 \rightarrow 0$ during the peak busy hour), then Equation 6.20 can be written,

$$Q_{N,k} \approx N\overline{\rho}\overline{\alpha} + z_k\sqrt{\overline{\rho}\left(\overline{\alpha}^2 + \overline{\beta}^2\right)N} \tag{6.21}$$

This has the form of a PDD relation (see Eq 6.16) with coefficients A, B and C given by

$$A = z_k\sqrt{\overline{\rho}\left(\overline{\alpha}^2 + \overline{\beta}^2\right)} \tag{6.22a}$$

$$B = \overline{\rho}\overline{\alpha} \tag{6.22b}$$

$$C = 0 \tag{6.22c}$$

A is the product of the frequency factor (z_k) and the standard deviation of total-time demand at a typical residence. B is the mean total-time demand (see Table 6.9)[14]. Both A and B should be estimated during a period with high water use activity (to maximize $\overline{\rho}$). In Eq 6.22c, C=0 because the PRP model does not explicitly account for leaks.

---

[13] In many engineering applications, z_k is called the "frequency factor".

[14] Since $P_0 \rightarrow 0$, there is no real distinction between busy-time and total-time statistics.

Table 6.15 lists values of A, B, and C for several versions of the peak demand diversity relation. The negative values for B and C that appear in the IRDA-2 version are difficult to justify. This underscores the empirical nature of the current PDD application. Also included are coefficient values for indoor water use obtained from Eq 6.22 for the Poisson rectangular pulse interpretation[15]. Two PRP options are listed in Table 6.15, one of which includes a leak occurring at an arbitrarily assigned rate of C=4 gpm.

Figure 6.14 shows peak flow versus number of homes using the PDD parameter values given in Table 6.15 (some data are summarized in Table 6.16). Results from the PRP approach are very close to peak flow estimates from the IRDA-1 relation (R^2 = 0.999). The PRP approximation for large N agrees well with the general PRP result for N>10 homes (compare columns 5 and 6 in Table 6.16). The peak flow at the dead-end loop predicted by the PRP relation for N=21 homes is 30 gpm. Actual peak flows during the 31-day test period rarely exceeded 25 gpm (see Appendix H) and then only when outdoor use was present.

The PRP approach offers two key advantages over the empirical PDD method. First, the PRP method provides a sensible way to estimate values for A, B and C directly from the mean and variance of the SERPs and the utilization factor. Second, the frequency factor attaches an inherent measure of reliability (or risk) with the results. In this example, assuming the underlying parent distribution is Extreme Value Type I (i.e., Gumbel distribution), the frequency factor of z_k=6.73 corresponds to an exceedence probability of P_E = 0.01%. This means there is about a 1 in 10,000 chance that the actual instantaneous daily demand from indoor use will exceed the estimated peak design flow. It appears that the conventional PDD approach implicitly incorporates a very high factor of safety.

---

(15) The PRP coefficients are based the 31-day test period at the dead-end loop site:

(a) $\overline{\alpha} = 2.3$ gpm and $\overline{\beta^2} = 1.55$ gpm^2 (from Table 6.12 for indoor water use).
(b) $\overline{\rho}$=0.10 is from the average busiest minute of the day.
(c) The frequency factor for the Gumbel distribution is (Bras, 1990)

$$z_k = -\frac{\sqrt{6}}{\pi}\left\{0.5772 + \ln\left[\ln\left(\frac{1}{1-P_E}\right)\right]\right\}$$

Using an exceedence probability of $P_E = 0.0001$, this expression gives $z_k = 6.73$.

Table 6.15 Coefficients for Peak Flow Estimates in Branching Pipes

PDD Model Coefficient	Peak Demand Diversity			Poisson Rectangular Pulse	
	MSDH	IRDA-1	IRDA-2	Leak = 0	Leak = 4 gpm
A	7.20	4.00	9.78	5.57	5.57
B	0.24	0.30	-0.27	0.23	0.23
C	4.82	7.00	-0.51	0	4.00

Notes: (1)MSDH is Mississippi State Department of Health
(2) IRDA is Iowa Rural Development Association
(3) PDD coefficients are from Lingireddy *et al*, 1998
(4) PRP coefficients are from Eq 6.22 as discussed in footnote (15).

Table 6.16 Comparison of PDD and PRP Estimates for Peak Flows in Branching Pipes.

Homes (N)	MSDH (gpm)	IRDA-1 (gpm)	IRDA-2 (gpm)	Eq 6.20 (gpm)	Eq 6.21 (gpm)
0	4.8	7.0	-0.5	0	0
1	12.3	11.3	9	11.7	5.8
2	15.5	13.3	12.8	12.7	8.3
3	18.0	14.8	15.6	13.7	10.3
4	20.2	16.2	18.0	14.6	12.1
5	22.1	17.4	20.0	15.5	13.6
10	30.0	22.7	27.7	20.6	19.9
20	41.8	30.9	37.8	28.6	29.5
21	**42.9**	**31.6**	**38.6**	**29.5**	**30.4**
30	51.5	37.9	45.0	36.6	37.4
40	60.0	44.3	50.5	43.9	44.4
50	67.7	50.3	55.2	50.6	50.9
100	100.8	77.0	70.3	78.6	78.7

In some cases, it will be necessary to consider peak flows resulting from both indoor and outdoor use. A method called *coincident frequency analysis* can be used to combine indoor and outdoor demands and estimate the total maximum water demand on a branching pipe system. The details are beyond the scope of this report, but the basic ideas are discussed in Kubik (1992) and Buchberger (1995).

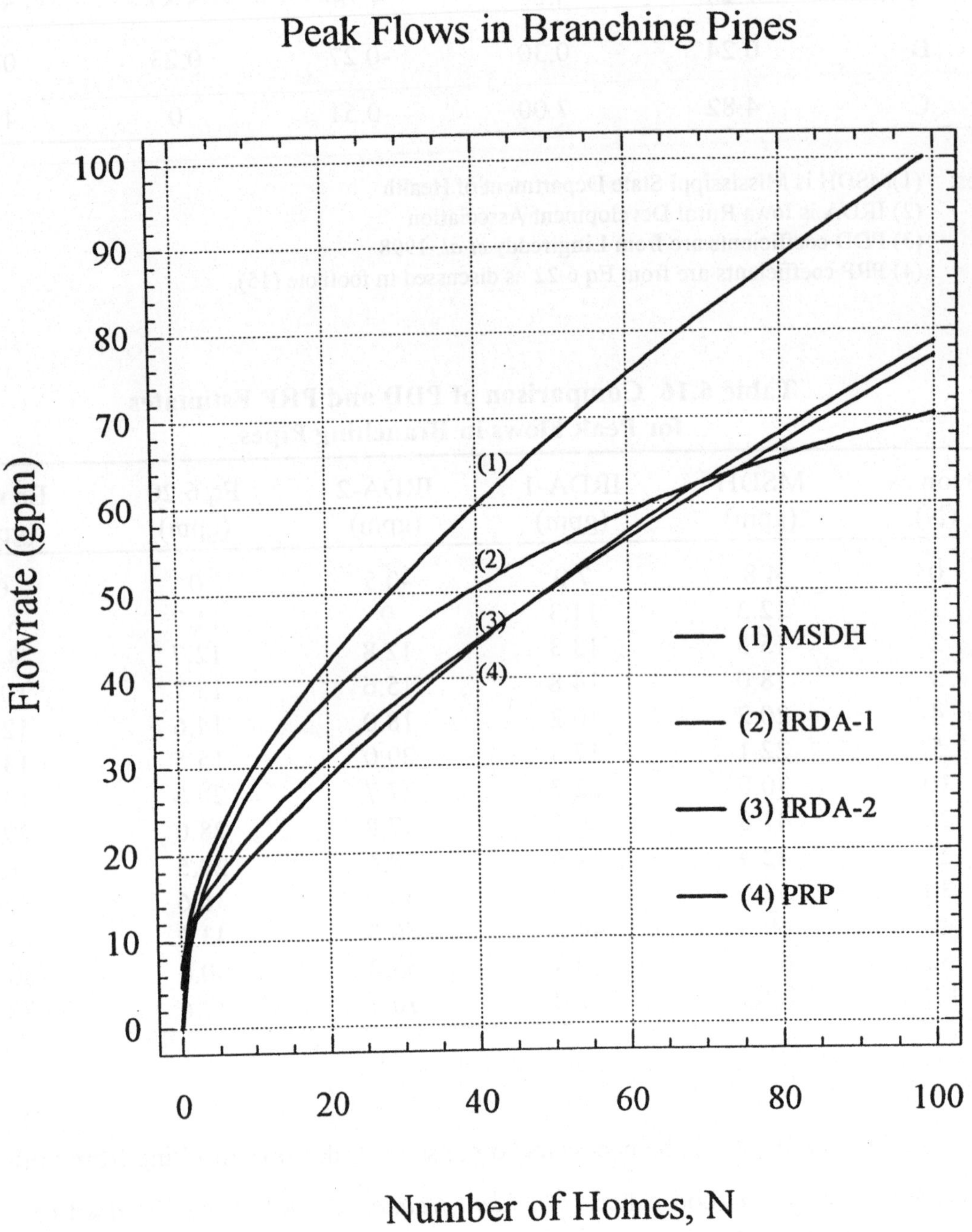

Figure 6.14 Peak flows in branching pipes.

Application 2: Screening for Leaks

Leaking pipes are a significant concern for many water utilities. Leaking fixtures can be a problem for homeowners. The statistical signature of flows that result from leaks is very different from the signature of legitimate domestic water demands.

Leaking pipes generate continuous flows with relatively minor fluctuations; stagnant periods are absent. By contrast, residential water demands are intermittent with abrupt bursts of highly variable flows between random periods of stagnation. When considered separately, it is not difficult to distinguish between a leak and a legitimate use. The challenge is that pipe flow a mixture of leaks and legitimate uses. When these contrasting water demands are superimposed, the distinction between the two signals becomes blurred.

Total-time statistics of residential water demands from the PRP model (see Table 6.9) can be used as a diagnostic tool to screen pipe flows and to quantify the magnitude of any background leakage. To illustrate the method, consider Figure 6.15 which shows a hypothetical sequence of flow rates recorded during a *low* demand period at some point along a branching pipe[(16)]. For the sake of example, assume the pipe delivers water to a large neighborhood (N=435 homes) where the busy fixtures statistics are $\alpha = 2.30$ gpm and $\beta^2 = 1.55$ gpm^2 and the early morning average household utilization factor is $\overline{\rho} = 0.005$ (these PRP parameters are based on typical values from the dead-end loop study site).

The flow sequence in Figure 6.15 is composed of two types of water demands. The unshaded base flow corresponds to a steady background discharge resulting from a constant rate of leakage (Q_L = 2.5 gpm). The shaded region represents random intermittent flow resulting from typical indoor residential water use (neighborhood mean is $N\overline{\rho}\alpha = 5.0$ gpm). Normally the breakdown between leaks and legitimate water use is not known *a-priori*.

---

(16) The method outlined here applies to any region in the distribution system where the supply of water is delivered through a single pipe. This includes all branching pipe systems (e.g., dead-end stems and dead-end loops). Leakage in looped neighborhoods can also be screened with the PRP method if the local network is properly isolated when the pipe flow readings are obtained (e.g., DMA with only one open supply line). In practice, the PRP method for screening leaks will be most effective when $\overline{\rho}$ is nearly constant and when residential water demands are low. Measuring pipe flows during an early morning hour would be ideal.

Table 6.17 compares several total-time flow statistics at hierarchy level H-3 in a pipe network with leaks and without leaks. In the presence of a steady leakage rate, four key points emerge from Table 6.17:

- Stagnation disappears.
- The total-time mean flow increases by a percentage 100S*.
- The total-time variance of the flow is not affected.
- The coefficient of variation decreases by a percentage 100[S*/(1+S*)]

Here S* is a dimensionless leakage rate given by

$$S^* = \frac{Q_L}{N\bar{\rho}\alpha} \tag{6.23}$$

where Q_L is the leakage rate and $N\bar{\rho}\alpha$ is the mean water demand from N homes. In the hypothetical problem under consideration here, S* = (2.5 gpm)/(5.0 gpm) = 0.50. The corresponding total-time flow statistics are summarized in Table 6.18. Comparing column (4) against (2) in Table 6.18, the effect of the steady leak is to eliminate stagnation, increase the mean flow (by 50%), and decrease the coefficient of variation (by 33%). The variance of the pipe flow does not change.

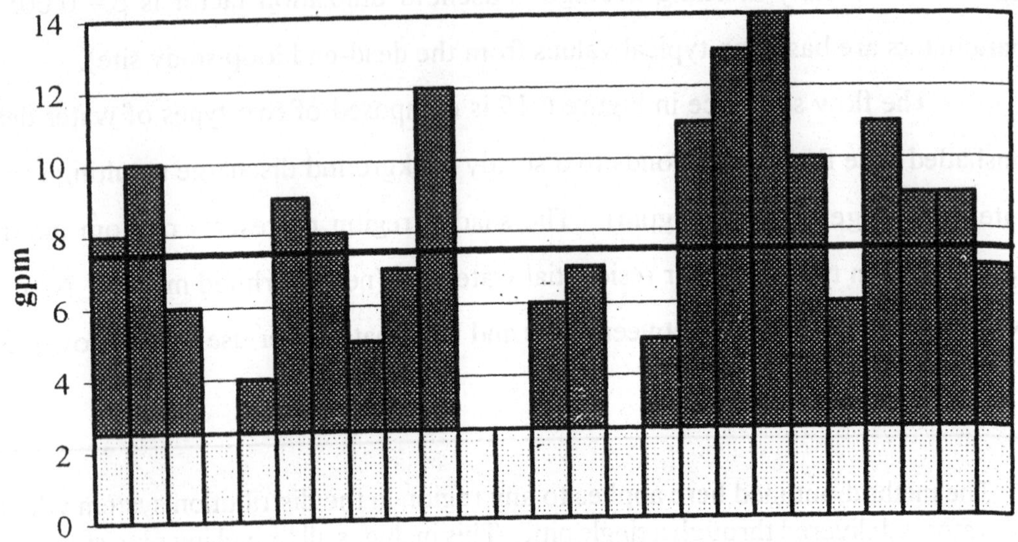

Figure 6.15 Hypothetical pipe flow sequence in the presence of significant system system leakage. Shaded bars represent random intermittent pipe flow. Unshaded bars represent constant (but unknown) rate of leakage.

Table 6.17 Total-Time Statistics for Pipe Network with Leaks
Hierarchy Level: H-3 Single Neighborhood with N homes

Total-Time Flow Statistic	Network without Leaks	Network with Leaks
Prob[Q=0]	$\exp[-N\overline{\rho}]$	0
E[Q]	$N\overline{\rho}\alpha$	$N\overline{\rho}\alpha(1+S^*)$
Var[Q]	$N\overline{\rho}(\alpha^2+\beta^2)$	$N\overline{\rho}(\alpha^2+\beta^2)$
$\Theta_T[Q]$	$\sqrt{\dfrac{1+\left(\dfrac{\beta}{\alpha}\right)^2}{N\overline{\rho}}}$	$\left(\dfrac{1}{1+S^*}\right)\sqrt{\dfrac{1+\left(\dfrac{\beta}{\alpha}\right)^2}{N\overline{\rho}}}$

Table 6.18 Effect of Constant Leak on Flow Statistics in a Branching Pipe Feeding a Hypothetical Neighborhood of N=435 Homes.

(1) Total-Time Flow Statistic	(2) Network without Leak	(3) Leakage Rate	(4) Network with Leak	(5) Change due to Leak
Prob[Q=0]	0.114	0	0	(–) 100%
E[Q]	5.00 gpm	2.50 gpm	7.50 gpm	(+) 50%
Var[Q]	14.88 gpm^2	0	14.88 gpm^2	0
Θ[Q]	0.771	0	0.514	(–) 33%

Note: Columns (2) and (4) are based on expressions in Table 6.17 with the following parameters: $\alpha = 2.30$ gpm, $\beta^2 = 1.55$ gpm^2, N = 435 homes, $\overline{\rho} = 0.005$, $S^* = 0.50$.

The proposed method for estimating the magnitude of the pipe leakage involves examining how total-time flow statistics change with the degree of truncation (censorship) applied to the sequence of recorded pipe flows. The process starts with a censor level of zero ($S=0$), ends when the censor level reaches the maximum flow, $S=Q_{MAX}(t)$. If steady leakage is present, each of the four total-time flow statistics will display a kink when plotted against censor level. The location of the kink indicates where the censor threshold matches the leakage rate. At this break point, stagnation frequency will increase dramatically. A simple seven-step algorithm for the proposed screening test is outlined in Figure 6.16.

Continuing with the example problem, Figure 6.17 shows how the total-time flow statistics for the hypothetical neighborhood will vary as a function of *dimensionless* censor level (censor values are divided by mean neighborhood demand, see Eq 6.23). The data points plotted at $S=0$ (no censor applied) correspond to the flow statistics in column 4 of Table 6.18. The data points plotted at $S=S^*=0.50$ (censor threshold matches leakage rate) correspond to values in column 2 of Table 6.18. In the region where pipe leakage occurs ($0 \le S \le S^*$), the relationships in Table 6.17 are used to define the plotted lines[17]. In the region of legitimate water demand ($S^* < S < S_{MAX}$), the plotted results are simply straight line interpolations between theoretical results at the endpoints[18].

This hypothetical example reveals obvious kinks in each of the four plotted lines at the point ($S=S^*=0.50$) where the censor threshold matches the (known) leakage rate. In practice, leakage rates are not steady and the break point will not be so crisp. Hence, some judgement will be needed to estimate the magnitude of leakage. However, the weight of all the evidence should provide a good indication of where the cut-off between legitimate water use and uncontrolled leakage may lie.

---

[17] Substitute (S^*-S) for (S^*) in the expressions for E[Q] and $\Theta_{T'}[Q]$ given in Table 6.17.

[18] At $S=S^*$, leakage is absent and the flow sequence arises from a pure homogeneous PRP demand process; hence, total-time flow statistics come directly from the theoretical PRP results given in Table 6.17 (network without leakage). At $S=S_{MAX}$ all flows are completely censored, so stagnation goes to 100% while the mean and variance of the flow go to 0. A more detailed treatment and example application of the proposed leak detection approach can be found in Buchberger and Nadimpalli (2003).

Step 1
Initialize: Set S=0; specify ΔS

↓

Step 2
Censor recorded flows: Compute $Q_S(t) = Q(t) - S$

↓

Step 3
Eliminate negative flows: If $Q_S(t) < 0$, set $Q_S(t) = 0$

↓

Step 4
Compute total-time statistics for $Q_S(t)$; plot versus S

↓

Step 5
Increment the censor threshold: $S = S + \Delta S$

↓

Step 6
If $S < Q_{MAX}(t)$ return to Step 2

↓

Step 7
Estimate leakage rate from break point on graphs

Figure 6.16 Algorithm for estimating rate of leakage from measured pipe flows.

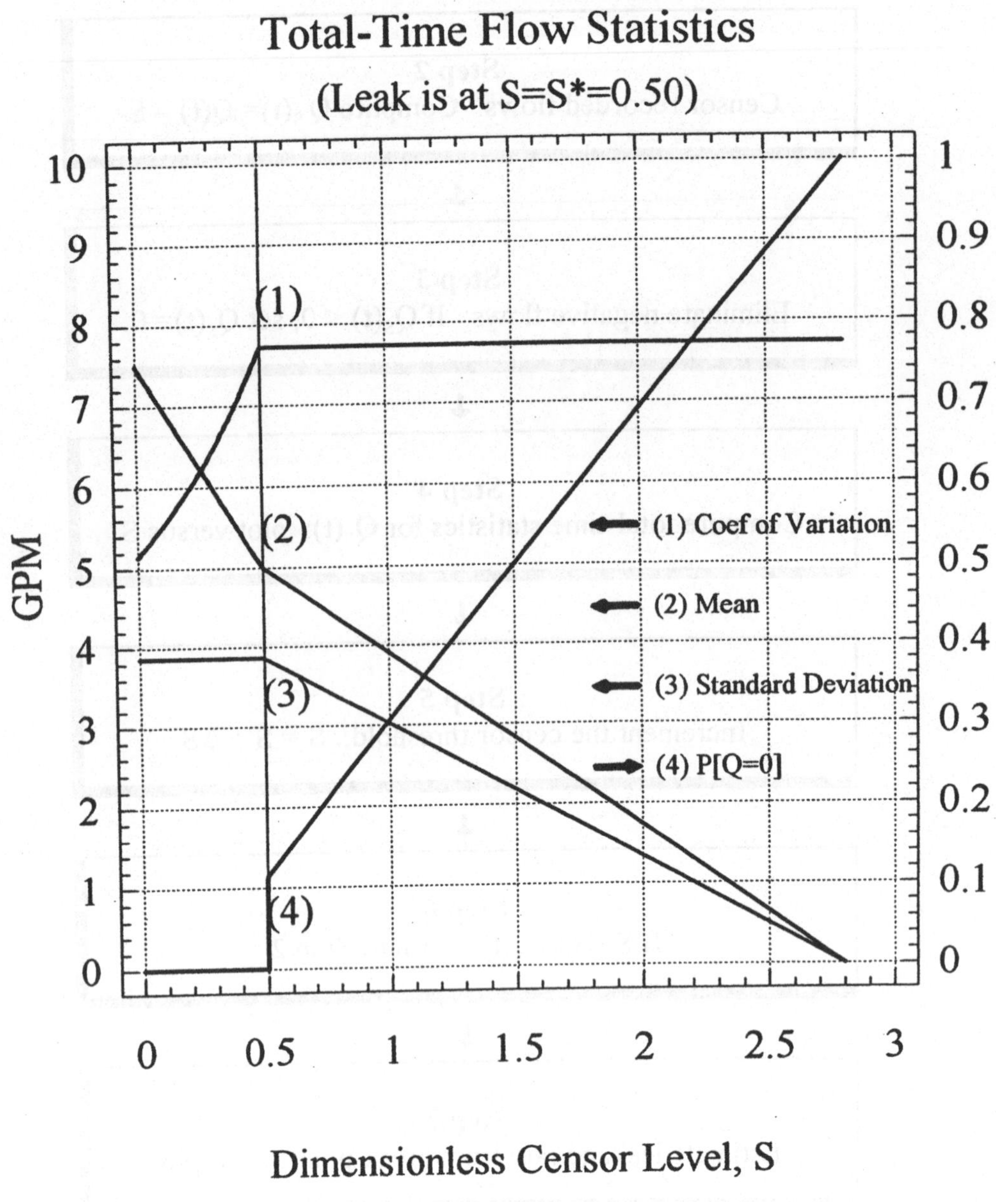

Figure 6.17 Variation of total-time flow statistics versus censor level in a network with leakage

Application 3: Simulating Residential Water Demands

Current network models generally operate with average total-time demands from hierarchy level H-3. These time-averaged flows can preserve mass; however, they can not replicate the random sporadic nature of water use nor the dynamic transitions among flow regimes. As a consequence, the ability of current network models to make accurate water quality predictions is limited by their rather coarse approximations of the prevailing hydraulic conditions. This problem is most pronounced in peripheral zones of the distribution system, where branching pipes are common and the consumer base is large.

A switch from the H-3 total-time perspective to the H-2 busy-time perspective is necessary to capture more realistically the dynamic hydraulic behavior of municipal water distribution systems. The PRP model offers an efficient and effective way to simulate residential water demands at the H-2 level. Seven simple steps to simulate nonhomogeneous PRP residential water demands at the H-2 level are outlined in the algorithm given in Figure 6.18. The five parameters needed in the simulation (λ, τ, ω^2, α, β^2) are the key statistics describing SERP duration and SERP intensity (see Tables 6.1 and 6.10). Nonhomogeneous sequences can be obtained by allowing $\lambda(t)$ to vary with each time step. Efficient schemes are available to generate the *psuedo*-random variables needed to simulate the SERPs.

A program to simulate residential water demands in any size network based on the algorithm in Figure 6.18 is available from the authors (see Appendix A). It can be incorporated as a subroutine option in current network models. This upgrade will be especially important for network models that use results from the hydraulic simulation to drive a water quality subroutine. The chief drawback with adopting the H-2 busy-time perspective for water demands is the increased computational burden.

This approach, however, appears to be the only viable way to replicate the random dynamic flow conditions that accompany water quality transformations through the pipe network from the point of treatment to the point of consumption. Advances in computational hardware and software eventually will dispel reservations about the increased computational effort. In the near future, concerns about the computational burden will fade and the primary thrust of network models will focus on generation and interpretation of realistic predictions of water quality at all points throughout the distribution system.

Step 1: Initialize

Set t=0. Specify Δt and number of time intervals, M

↓

Step 2: Generate Arrival Rate

Generate a fixed arrival rate λ for given time interval

↓

Step 3: Simulate Busy Homes

Poisson arrival process implies that waiting times (T) have an exponential distribution, $T \sim E[\lambda^{-1}]$

↓

Step 4: Simulate Pulse Duration

For each arrival, generate the SERP duration (D) from a log normal distribution, $D \sim LN[\tau, \omega^2]$

↓

Step 5: Simulate Pulse Intensity

For each arrival, generate the SERP intensity (Q) from a log normal distribution, $Q \sim LN[\alpha, \beta^2]$

↓

Step 6: Compute Pulse Volume

For each arrival, compute the SERP Volume, $V = Q \cdot D$

↓

Step 7: Check for End

$t = t + \Delta t$; If $t < M\Delta t$, then return to Step 2.

Figure 6.18 Algorithm for simulating PRP residential water demands.

Application 4: Joint Indoor and Outdoor Demands

Up to now the focus has been on indoor residential water use. Outdoor water use is also important, especially during the warm summer season. For instance, outdoor use is very prominent in the quad plots for Monday May 19, Sunday May 25 and Monday May 26 (see Appendix H). These days of heavy outdoor use also produced histograms exhibiting the poorest fits with the average daily PRP model for busy homes.

Indoor water use is a perennial high frequency signal characterized by many random bursts of short duration. By contrast, outdoor water use is a seasonal low frequency signal characterized by a few random bursts with long durations. The mean duration of an indoor pulse at the study site is about 45 seconds; the mean duration of an outdoor pulse is just under 80 minutes. These quantities differ by a factor of 100. The exceptionally long duration for outside water use is attributed to pool filling at four residences (see Table 3.1). Essentially, the question of mixing indoor and outdoor water demands is a problem of incommensurate time scales. There are two main choices:

- ❖ **Option 1**–Lump both the indoor and the outdoor demands together and model total residential water use as a single aggregate PRP process or

- ❖ **Option 2**–Separate the indoor and outdoor demands and model total residential water use as the superposition of two distinct PRP processes.

Under option one, the mean pulse duration for the lumped PRP model is

$$\tau_L = \frac{\rho_{IN} + \rho_{OUT}}{\lambda_{IN} + \lambda_{OUT}} \tag{6.24}$$

where the subscripts "IN" and "OUT" refer to indoor and outdoor use, respectively. The mean pulse duration of the lumped model (τ_L) given in Eq 6.24 can differ significantly from the mean durations of the parent indoor and outdoor water demand processes[19].

---

[19] For example, consider a home with an average of 1 outdoor use per day lasting 1 hour. This gives ρ_{OUT}=0.042 and λ_{OUT}=0.0007 per minute. Taking ρ_{IN}=0.045 and λ_{IN}=0.058 per minute (from study site statistics), Eq 6.24 gives τ_L=1.48 minutes. This lumped values is twice the indoor mean pulse duration and 40 times less than the outdoor mean duration. Although it is statistically correct, it lacks a strong physical basis.

Lee and Buchberger (1999) demonstrate that option one will almost certainly *not* be appropriate as a model for joint indoor and outdoor residential water use when it is important to preserve the pulse characteristics of the two parent processes. The presence of extended outdoor use destroys the local character of the indoor pattern. Hence, when residential water use includes indoor and outdoor demands, each should be represented as a separate water demand process with a distinct utilization factor, arrival rate and other appropriate SERP statistics.

The superposition strategy of option 2 implies that two PRP processes should emerge when the combined set of indoor and outdoor water demands are disaggregated. As a simple test, the water demands recorded on Monday May 19, 1997 (N=1,672 pulses) were separated into indoor (N_{IN} = 1,669) and outdoor (N_{OUT} = 3) uses. The original co-mingled set of water demands showed poor agreement between observed and predicted busy homes(Figure 6.19 top). The discrepancies arise from the non-homogeneous time pattern of the water demands and because the one-day period is too short to estimate ρ_L accurately for joint indoor and outdoor water uses. When the outdoor demand data are filtered (by removing 3 high volume, long duration uses), then the agreement between observed and predicted busy homes is much improved (Figure 6.19 bottom). Three data points are not sufficient to test the PRP distribution of the outdoor pulses; however, all evidence suggests that the PRP concept will remain valid in the case of outdoor water demands.

The results presented here illustrate that the versatile PRP model can provide a very good description of both indoor and outdoor residential water demands. During the summer months and other periods of significant outdoor water demand, residential water use should be represented as the superposition of two distinct (i.e., indoor and outdoor) PRP processes rather than by a single lumped PRP model.

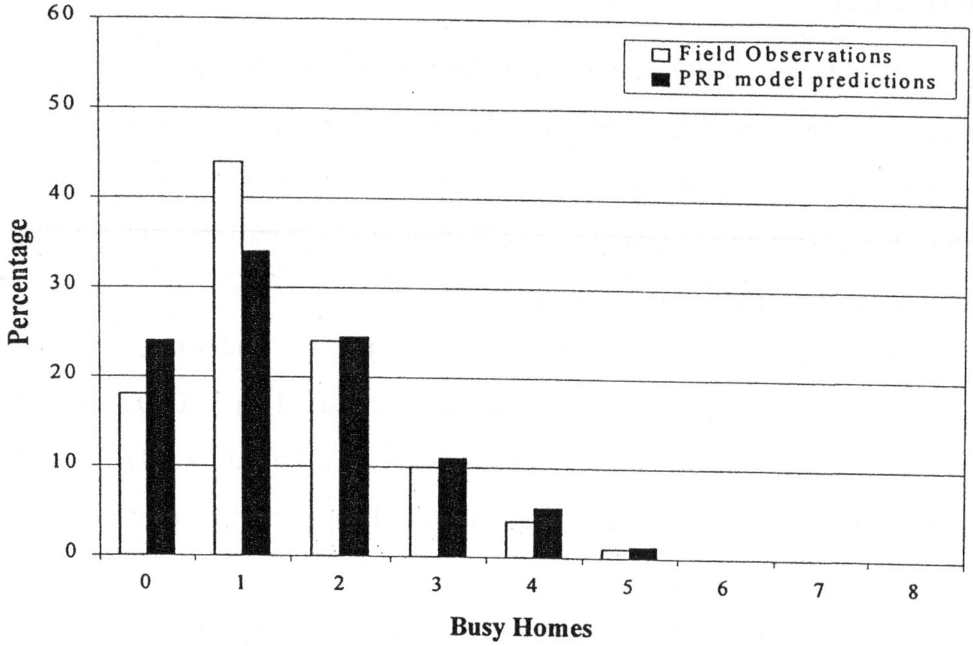

a. Lumped (indoor and outdoor water use, $\rho_L = 1.01$ and N = 1672)

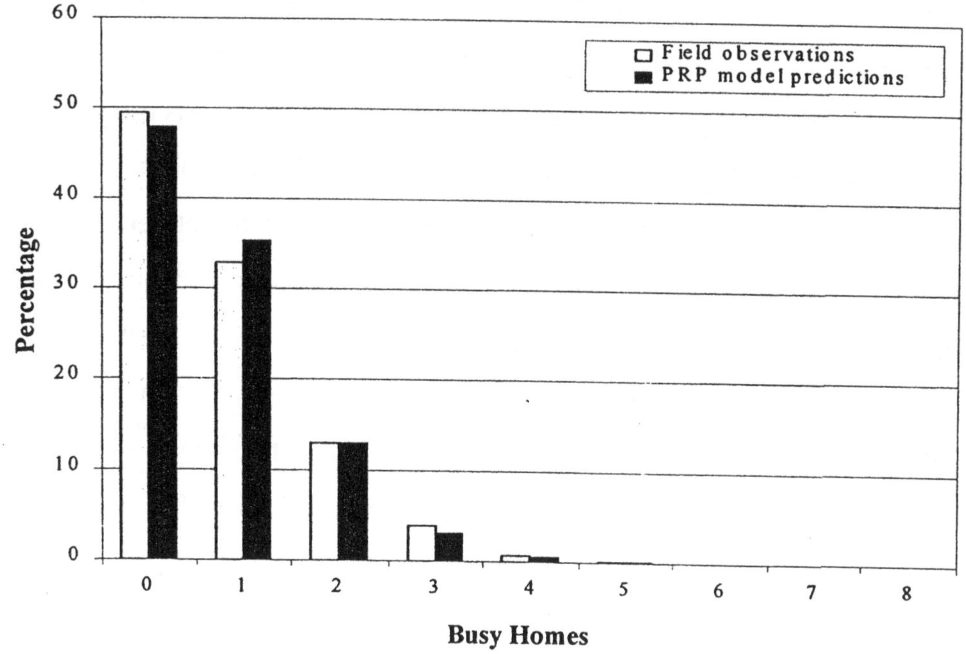

b. Indoor water use only ($\rho_I = 0.74$ and N = 1669)

Figure 6.19 Comparison of predicted and observed busy home distributions from water demand data collected May 19, 1997 at 21 homes along the dead-end loop.

SUMMARY

Residential water use is visualized as a customer-server interaction of the kind encountered in queueing theory. While this view is new to the drinking water industry, it follows approaches pioneered in the 20th century by communication utilities who investigated demands on telephone exchanges. The PRP model taps into this rich existing knowledge base and makes a transition from "phones to faucets".

Compelling evidence is presented supporting the Poisson rectangular pulse hypothesis for residential water use. Most of the analyses are based on 52,000 archived residential water demands recorded at 21 homes along the dead-end loop during a 31-day period from May 11 to June 10, 1997. The test period represents only one month out of the 7 month collection of archived water demand data.

The probability distribution of busy fixtures (inside a home) is Poisson; the probability distribution of busy homes (along the loop) is binomial. The difference between the two is so slight that, practically speaking, the more tractable Poisson model can be used without serious error to describe the probability distribution of busy homes. Diurnal variation in water use patterns is easily accommodated using a piecewise homogeneous rho profile. Results indicate that the PRP hypothesis provides an outstanding fit to the number of busy homes on an hour by hour basis (Figs 6.6 and 6.7). There is good agreement between predicted and observed busy server transitions (Fig 6.8) and predicted and observed busy server autocorrelation functions (Fig 6.9 and Appendix J).

Pipe flows at various locations in the distribution system (at hierarchy levels H-1, H-2 and H-3) and over two time horizons (busy-time and total-time) show excellent agreement between observed and predicted mean flow rates (see Table 6.12). Consistency among mean values confirms that mass is conserved across each level. There are, however, slight discrepancies between predicted and observed flow variances. This is largely attributed to the nonhomogeneous arrival pattern (see Fig 6.12). In accord with the Central Limit Theorem, the frequency distribution of instantaneous flows into the loop over the course of a day is truncated normal with a mass concentration at Q=0 equal to the probability that the flow through the loop is stagnant. Four novel applications demonstrate the utility of the PRP model for residential water use.

Appendix 6.1: Fixtures are Poisson, Homes are Binomial

A home is either busy or idle. This dual outcome is a classic "Bernoulli trial", implying that the probability distribution of busy homes in any neighborhood in any water distribution system is binomial. Let p = Pr[home is busy]. In a neighborhood with N homes, p can be found as

$$p = 1 - \mathbf{Pr}[\textit{home is idle}] = 1 - \mathbf{exp}(-\bar{\rho}) \qquad \textit{where } \bar{\rho} = \frac{\dot{\rho}_N}{N} = \frac{1}{N}\sum_{i=1}^{N} \rho_i \tag{6.1.1}$$

Here $\bar{\rho}$ represents the average utilization factor for the neighborhood. The probability of having exactly k busy homes in a neighborhood of N homes is given by

$$\mathbf{Pr}[\text{k } \textit{busy homes}|\text{N,p}] = \binom{N}{k} p^k (1-p)^{N-k}$$

$$= \frac{N!}{(N-k)!k!}\left[1 - \mathbf{exp}(-\bar{\rho})\right]^k \left[\mathbf{exp}(-\bar{\rho})\right]^{N-k} \tag{6.1.2}$$

The probability of zero busy homes must be the same as the probability of zero busy fixtures. When k=0, Equation (6.1.2) reduces to

$$\mathbf{Pr}[0 \ \textit{busy homes}] = \left[\mathbf{exp}(-\bar{\rho})\right]^N = \mathbf{exp}(-N\bar{\rho}) = \mathbf{exp}(-\dot{\rho}_N) \tag{6.1.3}$$

This matches exactly the Poisson probability of zero busy fixtures in a neighborhood of N homes.

Unfortunately, the binomial distribution is cumbersome to use. Fortunately, for most applications in water supply systems, $\bar{\rho}$ is small (i.e., < 0.10) and, hence, the binomial model given in (6.1.2) is well approximated with the Poisson distribution.

To see how the Poisson distribution arises from the binomial model (E.E. Lewis 1996 **Introduction to Reliability Engineering**, Wiley, NY, p. 24), multiply Eq. (6.1.2) by (N^k/N^k), regroup terms, and take limits,

$$\binom{N}{k} p^k (1-p)^{N-k} = \frac{N!}{(N-k)!k!}\left(\frac{N^k}{N^k}\right) p^k (1-p)^{N-k}$$

$$= \frac{N!}{(N-k)!N^k} \frac{(Np)^k}{k!} \left(1 - \frac{Np}{N}\right)^N (1-p)^{-k} \tag{6.1.4}$$

Taking limits in (6.1.4) as $N \to \infty$ and $p \to 0$ but the product $Np=\lambda$ is a constant gives.

$$\lim_{N\to\infty}\left(\frac{N!}{(N-k)!N^k}\right)=1$$

$$\lim_{N\to\infty}\left(1-\frac{Np}{N}\right)^N = \lim_{N\to\infty}\left(1-\frac{\lambda}{N}\right)^N = \exp(-\lambda)$$

$$\lim_{p\to 0}(1-p)^{-k}=1$$

Putting these results into Eq (6.1.4) gives

$$\Pr[k \textit{ busy homes}|\lambda] = \frac{\lambda^k \exp(-\lambda)}{k!} \qquad (6.1.5)$$

which is Poisson with parameter

$$\lambda = Np = N[1-\exp(-\bar{\rho})] \qquad (6.1.6)$$
$$= N[1-(1-\bar{\rho}+\tfrac{1}{2}\bar{\rho}^2-\tfrac{1}{6}\bar{\rho}^3 \pm \cdots)]$$

The Poisson model for busy homes given in (6.1.5) works well provided $\bar{\rho} \leq 0.20$ (see Figure 6.20). For small values of $\bar{\rho}$, (6.1.6) can be written,

$$\lambda \approx N\bar{\rho} = \rho_N^* \qquad (6.1.7)$$

which is identical to the Poisson parameter (i.e., the utilization factor) for busy fixtures in the neighborhood of N homes.

In summary, the following key points hold for a neighborhood of N homes:

(i) The distribution of busy fixtures is Poisson with parameter, $\rho_N^* = \Sigma\, \rho_i \quad i=1,2,\ldots,N$.

(ii) The distribution of busy homes is binomial with parameter, $p=1-\exp(-\bar{\rho})$.

(iii) The probability of stagnation is Pr[0 homes] = Pr[0 fixtures] = $\exp(-N\bar{\rho})$.

(iv) The Poisson approximation for the distribution of busy homes (with parameter, $\lambda=Np$) works exceedingly well provided $\bar{\rho} \leq 0.20$; this is almost always true for indoor (and most outdoor) residential water use.

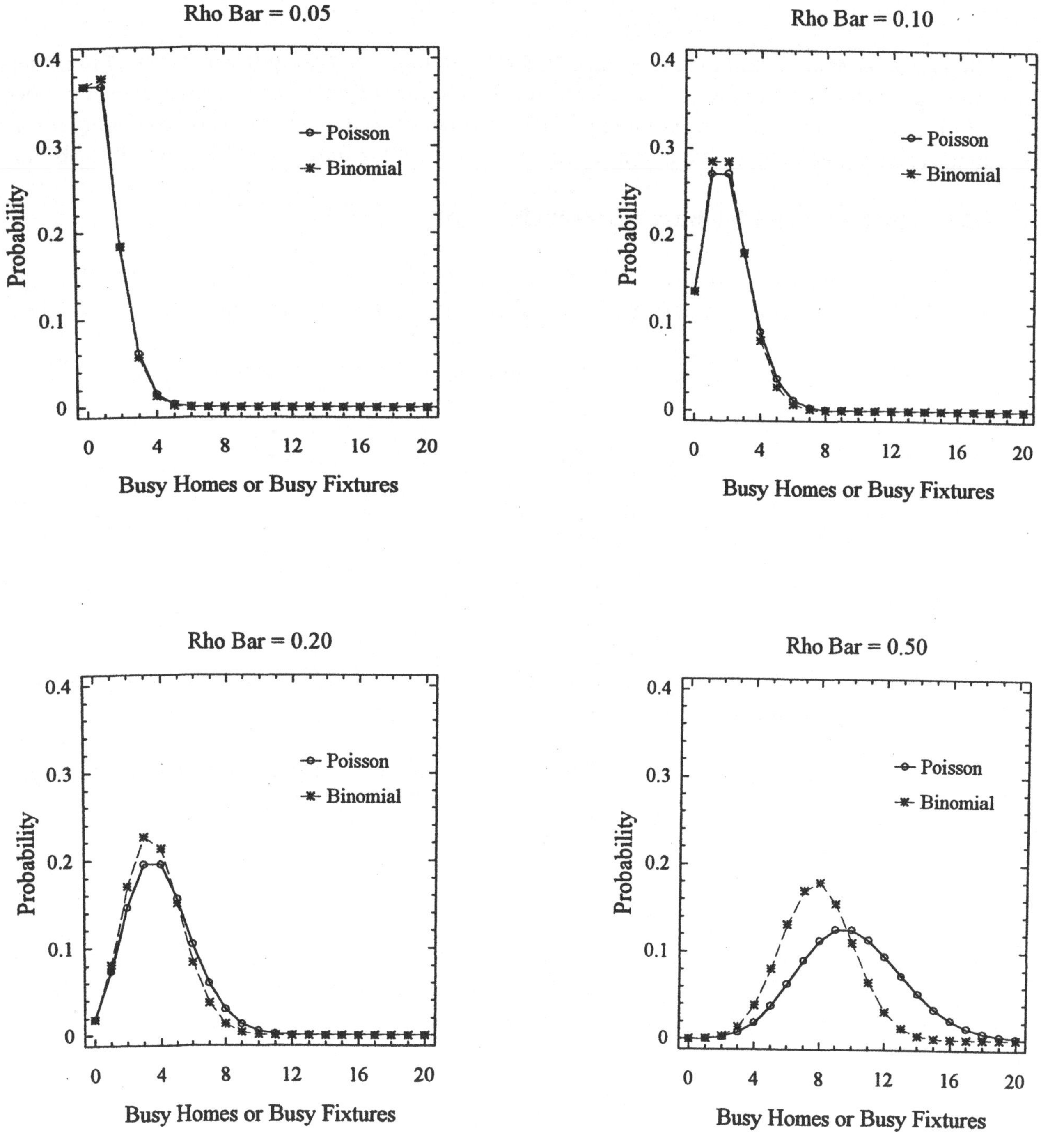

Figure 6.20 Poisson approximation of binomial distribution is acceptable when residential utilization factor is under 20% (ρ_1*<0.20).

Appendix 6.2: Busy Server Transitions

Under steady-state conditions, there is a dynamic equilibrium among the number of busy servers in the neighborhood. As new customers arrive, the number of busy servers increases; conversely, as existing customers depart, the number of busy servers decreases. The transitions among busy-server states can be predicted at two hierarchy levels, H-1 for busy fixtures and H-2 for busy homes.

Hierarchy Level: H-1 for Busy Fixtures (Poisson)

Let $J_{k+1,k}$ denote a jump down from (k+1)→ k busy fixtures (see Figure 6.21). During a time period of duration ΔT in a neighborhood of N homes, the expected number of jumps down is,

$$E[J_{k+1,k}] = E[\text{departures} \mid k+1 \text{ busy servers}] \cdot \Pr[k+1 \text{ busy fixtures in block of N homes}]$$

$$E[J_{k+1,k}] = \left[(k+1)\mu_N^* \Delta T\right] \frac{(\rho_N^*)^{k+1} \exp(-\rho_N^*)}{(k+1)!}$$

$$= \lambda_N^* \Delta T \frac{(\rho_N^*)^{k} \exp(-\rho_N^*)}{k!} \qquad (6.2.1)$$

Note that $\rho_N^* = \lambda_N^* / \mu_N^*$ as defined in Table 6.3. Similarly, the expected number of jumps up from k→ (k+1) busy fixtures is,

$$E[J_{k,k+1}] = E[\text{arrivals} \mid k \text{ busy servers}] \cdot \Pr[k \text{ busy fixtures in block of N homes}]$$

$$E[J_{k,k+1}] = \lambda_N^* \Delta T \frac{(\rho_N^*)^{k} \exp(-\rho_N^*)}{k!} \qquad (6.2.2)$$

Equations 6.2.1 and 6.2.2 show that the expected number of arrivals and departures from state "k" are equal. In dimensionless form this can be written,

$$\frac{E[J_{k\leftrightarrow k+1}]}{\lambda_N^* \Delta T} = \frac{(\rho_N^*)^{k} \exp(-\rho_N^*)}{k!} = P(k, N) \qquad (6.2.3)$$

where P(k,N) is the Poisson probability of k busy fixtures in a block of N homes. This result is given in Equation 6.5 of the text.

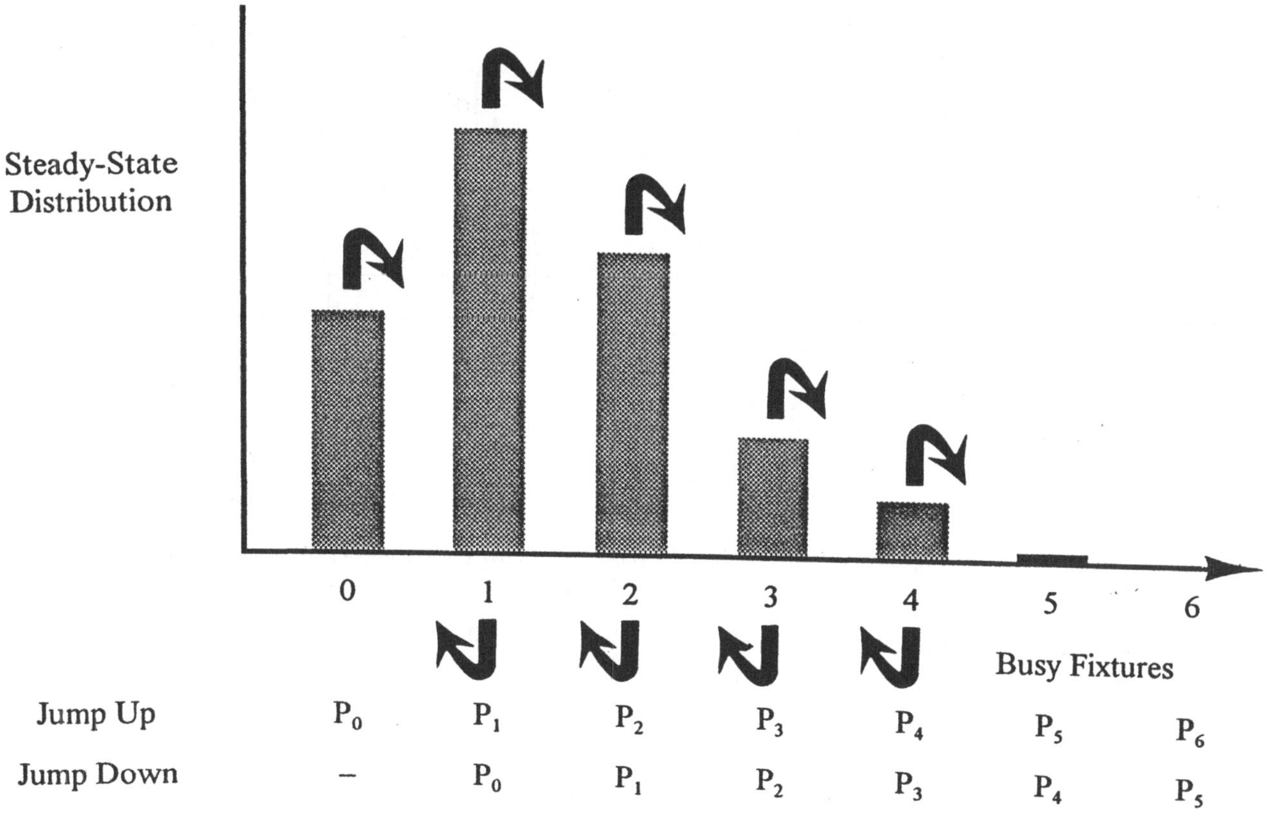

Figure 6.21 Steady State Transitions Among Busy Fixtures are Poisson.

Hierarchy Level: H-2 for Busy Homes (Binomial)

Busy homes can have more than one busy fixture. This has two consequences. First, the number of busy homes will never exceed the number of busy fixtures (see Table 6.5). Second, the expected number of transitions among busy homes will be less than the transitions among busy fixtures.

The number of busy homes decreases only if a departure occurs at a home with one busy server. Let $J_{k+1,k}$ denote a jump down from $(k+1) \rightarrow k$ busy homes. During a time period of duration ΔT in a neighborhood of N homes, the expected number of jumps down is,

$$E[\,J_{k+1,k}\,] \;=\; E[\text{departures} \mid k+1 \text{ busy homes}] \cdot \Pr\begin{bmatrix} k+1 \text{ busy homes} \\ \text{among } N \text{ homes} \end{bmatrix} \cdot \Pr\begin{bmatrix} \text{exactly 1 busy fixture} \\ \text{at a busy home} \end{bmatrix}$$

$$E[J_{k+1,k}] = \left[(k+1)\mu_N^* \Delta T\right] \cdot \left[\binom{N}{k+1} p^{k+1} q^{N-k-1}\right] \cdot \left(\frac{\rho_N^*}{N}\right) \cdot \left(\frac{q}{p}\right)$$

$$= \left[\left(\frac{k+1}{N}\right)\lambda_N^* \Delta T\right] \cdot \left[\binom{N}{k+1} p^k q^{N-k}\right] \qquad (6.2.4)$$

where

$$p = 1 - \exp\left[-\frac{\rho_N^*}{N}\right] \quad \text{and} \quad q = 1 - p$$

The definition for the binomial "p and q" follows from Equation 6.4 in the text.

The number of busy homes increases only if an arrival goes to an idle home. Following the approach above, the expected number of jumps up from k→ (k+1) busy homes is,

$$E[\, J_{k,k+1} \,] = E[\text{arrivals} \mid \text{k busy homes}] \cdot \Pr\begin{bmatrix}\text{k busy homes} \\ \text{among N homes}\end{bmatrix} \cdot \Pr\begin{bmatrix}\text{arrival occurs} \\ \text{at idle home}\end{bmatrix}$$

$$E[J_{k,k+1}] = \left[\lambda_N^* \Delta T\right] \cdot \left[\binom{N}{k} p^k q^{N-k}\right] \cdot \left(\frac{N-k}{N}\right) \qquad (6.2.5)$$

The expected number of transitions given by Equations 6.2.4 and 6.2.5 are equal[20]. In dimensionless form this can be written,

$$\frac{E[J_{k\leftrightarrow k+1}]}{\lambda_N^* \Delta T} = \left(\frac{N-k}{N}\right) P(k,N) \qquad k \le N-1 \qquad (6.2.6)$$

where P(k,N) is the binomial probability of k busy homes in a block of N homes (see Equation 6.3 in text). This result is identical to Equation 6.6 of the main report.

---

[20] To see this note,

$$(k+1)\binom{N}{k+1} = (N-k)\binom{N}{k}$$

Appendix 6.3: Duration of Idle and Busy Periods

Case 1: Idle Period

During a time interval of duration T in a neighborhood of N homes, the total amount of time that the flow is expected to be stagnant is

$$T_S = P_0 T \tag{6.3.1}$$

where $P_0 = \exp[-\rho_N^*]$ is the stagnation probability and ρ_N^* is the neighborhood utilization factor. From Appendix 6.2, the expected number of jumps from 0 to 1 busy server is

$$J_{0\to 1} = \lambda_N^* T P_0 \tag{6.3.2}$$

where λ_N^* is the cumulative neighborhood arrival rate (see Table 6.3). The expected duration of a stagnant period in a branching pipe supplying N homes is,

$$\Delta T_S = \frac{T_S}{J_{0\to 1}} = \frac{1}{\lambda_N^*} \tag{6.3.3}$$

Case 2: Busy Period

Conversely, during a time period of duration T, the total amount of time during which at least one server is expected to be busy is

$$T_B = (1 - P_0)T \tag{6.3.4}$$

The expected number of drops from 1 to 0 busy servers ("returns to zero") is also given by 6.3.2. Hence, the expected duration of a busy period in a branching pipe supplying N homes is

$$\Delta T_B = \frac{T_B}{J_{1\to 0}} = \frac{1}{\lambda_N^*}\left(\frac{1 - P_0}{P_0}\right) \tag{6.3.5}$$

Assuming a loop-wide arrival rate of λ_N^*=1,800 demands per day and an average stagnation probability of P_0 = 54% (Winter 1996), Equations 6.3.3 and 6.3.5 give:

- Mean duration of idle time: ΔT_S = 48 seconds.
- Mean duration of busy time: ΔT_B = 41 seconds.

Appendix 6.4: Mean and Variance of PRP Flows

Busy-Time Flows at Hierarchy Level, H-1: Single Fixture

Let Y be the intensity of flow at a *busy* fixture in a residence. Across a neighborhood of N homes, assume Y is an independent and identically distributed random variable with mean α and variance β^2. Let Y(k) be the sum of k simultaneous water use intensities in the neighborhood,

$$Y(k) = \sum_{i=0}^{k} Y_i \qquad k = 0,1,2,\ldots \tag{6.4.1}$$

Y(0) corresponds to stagnation. Y(1) is a single server. Y(k) has mean $k\alpha$ and variance $k\beta^2$.

Total-Time Flows at Hierarchy Level, H-3: Branching Pipe

The rth moment of the flow through a branching pipe supplying N homes is given by

$$E\left[\left(Q_N^{\bullet}\right)^r\right] = \sum_{k=0}^{\infty} E\left[Y^r(k)\right] \cdot \Pr(k \text{ busy fixtures}) \tag{6.4.2}$$

where Pr(k busy fixtures) is from the Poisson distribution in Eq 6.2. Setting r=1, gives the mean,

$$E\left[Q_N^{\bullet}\right] = \sum_{k=0}^{\infty} (k\alpha) \cdot \frac{\left(\rho_N^{\bullet}\right)^k \exp\left[-\rho_N^{\bullet}\right]}{k!} = \rho_N^{\bullet}\alpha \tag{6.4.3}$$

where ρ_n* is the cumulative utilization factor for the neighborhood. Setting r=2 gives the mean square,

$$E\left[\left(Q_N^{\bullet}\right)^2\right] = \sum_{k=0}^{\infty} \left[k^2\alpha^2 + k\beta^2\right] \cdot \frac{\left(\rho_N^{\bullet}\right)^k \exp\left[-\rho_N^{\bullet}\right]}{k!} = \rho_N^{\bullet}\left(\alpha^2+\beta^2\right) + \left(\rho_N^{\bullet}\alpha\right)^2 \tag{6.4.4}$$

Combining Equations 6.4.3 and 6.4.4 gives the variance,

$$\mathrm{Var}\left[Q_N^{\bullet}\right] = E\left[\left(Q_N^{\bullet}\right)^2\right] - E\left[Q_N^{\bullet}\right]^2 = \rho_N^{\bullet}\left(\alpha^2+\beta^2\right) \tag{6.4.5}$$

Busy-Time Flows at Hierarchy Level, H-3: Branching Pipe

Busy-time flow statistics come from conditioning on the probability of having at least one busy server,

$$\Pr[\text{at least 1 busy fixture}] = 1 - P_0 = 1 - \exp[\rho_N^{\bullet}] \tag{6.4.6}$$

Hence, the busy-time mean flow is,

$$E[Q_N^{\bullet} \mid \text{busy time}] = \frac{\rho_N^{\bullet}\alpha}{1 - \exp[\rho_N^{\bullet}]} = \eta_N^{\bullet}\alpha \tag{6.4.7}$$

likewise, the busy-time mean square of the flow is,

$$E[(Q_N^{\bullet})^2 \mid \text{busy time}] = \frac{\rho_N^{\bullet}(\alpha^2 + \beta^2) + (\rho_N^{\bullet}\alpha)^2}{1 - \exp[\rho_N^{\bullet}]} = \eta_N^{\bullet}(\alpha^2 + \beta^2) + (1 - \exp[\rho_N^{\bullet}])(\eta_N^{\bullet}\alpha)^2 \tag{6.4.8}$$

so the variance of the busy-time flow is

$$\mathrm{Var}[Q_N^{\bullet} \mid \text{busy time}] = \eta_N^{\bullet}(\alpha^2 + \beta^2) + \exp[-\rho_N^{\bullet}](\eta_N^{\bullet}\alpha)^2 \tag{6.4.9}$$

Key Observations

An alternate more detailed formulation with equivalent results for the total-time flow statistics is presented in Buchberger and Wu (1995).

To get flow statistics for hierarchy level H-2 (Point-of-Entry to a home), use hierarchy level H-3 with number of homes, N=1.

All results presented here agree with the corresponding expressions in Table 6.9, provided the mean α and variance β^2 of the flow at a busy fixture are obtained from a properly *weighted* average of the values from all residences in the neighborhood. It is imperative to use global average GA-2, as defined in Table 4.5. A simple arithmetic average (GA-1) is not appropriate here.

Expressions for the mean flow (Eqs 6.4.3 and 6.4.7) are valid for either homogeneous or nonhomogeneous Poisson arrivals. Expressions for the variance (Eqs 6.4.5 and 6.4.9) are valid only for the homogeneous case. The nonhomogeneous case is treated in Appendix 6.5.

Appendix 6.5: Nonhomogeneous Demand Patterns

All results in Table 6.9 are based on the assumption of a homogeneous Poisson arrival process. This implies that the frequency of residential water use does not change. In real distribution systems, however, demand patterns change across time scales ranging from seconds, to hours, to seasons.

This appendix examines the error that is introduced when a nonhomogeneous demand pattern is approximated with an equivalent homogeneous demand pattern.

Case 1. Daily $\overline{\rho}$ and Stagnation

When the PRP model is applied to the *daily* distribution of busy servers, model predictions of stagnation are consistently less than observed values (see quad plots in Appendix H). Predicted stagnation probabilities in the quad plots are based on a daily average neighborhood utilization factor $\overline{\rho}$ computed as follows

$$\Pr[Q=0]_1 = \exp[-\overline{\rho}] = \exp\left[-\frac{1}{M}\sum_{k=1}^{M}\rho_k\right] \tag{6.5.1}$$

Here M is the number of time intervals in a day. For example, M=24 if hourly periods are used. A more accurate alternate approach averages the stagnation probabilities for each interval,

$$\Pr[Q=0]_2 = \frac{1}{M}\sum_{k=1}^{M}(P_0)_k = \frac{1}{M}\sum_{k=1}^{M}\exp[-\rho_k] \tag{6.5.2}$$

The utilization factor for time interval k can be expressed as the sum of the daily average and a deviation, $\rho_k = \overline{\rho} + \Delta\rho_k$. Hence, Equation 6.5.2 can be written,

$$\Pr[Q=0]_2 = \frac{1}{M}\sum_{k=1}^{M}\exp[-(\overline{\rho}+\Delta\rho_k)] = \frac{\exp[-\overline{\rho}]}{M}\sum_{k=1}^{M}\exp[-\Delta\rho_k] \tag{6.5.3}$$

Combining Equations 6.5.3 and 6.5.1 leads to a ratio of stagnation probabilities,

$$\frac{\Pr[Q=0]_2}{\Pr[Q=0]_1} = \frac{1}{M}\sum_{k=1}^{M}\exp[-\Delta\rho_k] \tag{6.5.4}$$

Note that the deviations $\Delta\rho_k$ have a mean value of zero. Assuming that the deviations are evenly distributed around zero with some maximum value (say $\pm\Delta\rho_{max}$), then Equation 6.5.4 can be approximated as,

$$\frac{\Pr[Q=0]_2}{\Pr[Q=0]_1} = \left(\frac{1}{2\Delta\rho_{max}}\right)\int_{-\Delta\rho_{max}}^{\Delta\rho_{max}} \exp(x)\,dx$$

$$= \frac{\exp(\Delta\rho_{max}) - \exp(-\Delta\rho_{max})}{2\Delta\rho_{max}} = \frac{\sinh(\Delta\rho_{max})}{\Delta\rho_{max}} \qquad (6.5.5)$$

Equation 6.5.5 is plotted below in Figure 6.22. The point where $\Delta\rho_{max}$=0 corresponds to the homogeneous case ($\rho_k = \overline{\rho}$) Here the ratio of stagnation probabilities is one.

In all other instances, however, the ratio of stagnation probabilities exceeds one. This implies $\Pr[Q=0]_1 < \Pr[Q=0]_2$. That is, in the presence of nonhomogeneous demands, use of the daily average neighborhood utilization factor underestimates the actual probability of stagnation as noted in the daily quad plots of Appendix H.

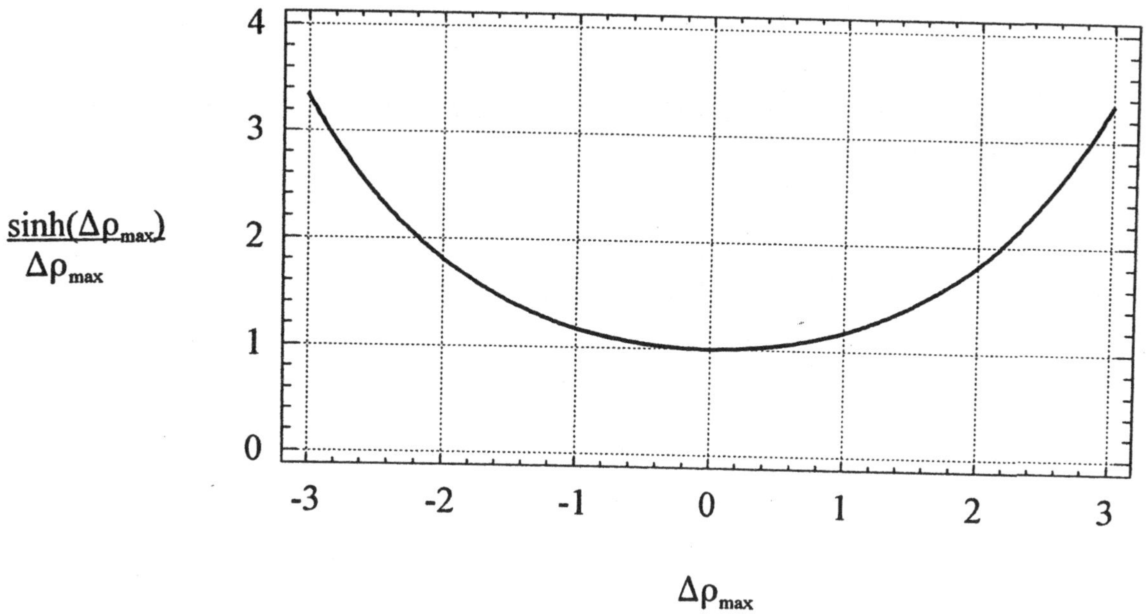

Figure 6.22 Ratio of stagnation probabilities defined in Equation 6.5.4.

Case 2. Total-Time Mean Flow

The total-time mean flow depends only on the total volume of water used, not on the temporal sequence of demands. Hence, for a fixed volume of water consumption during some specified period, the total-time mean flow rate is the same whether the customer arrivals are homogeneous or nonhomogeneous.

To illustrate, consider any arbitrary time sequence of flow pulses resulting from water demands at N homes measured at hierarchy level H-3 (for instance, see the bottom hydrograph at "Point of No-Return" in Fig 2.3). Let t_k be the duration of flow intensity q_k measured at level H-3 (including stagnant periods with q_k=0). The duration-weighted (see Table 4.5) total-time mean flow is

$$E[Q_N^*] = \frac{\sum_{k=1}^{M} t_k q_k}{\sum_{k=1}^{M} t_k} \tag{6.5.6}$$

where M is the number of times that the flow rate changes during the period of interest. The total volume consumed V and the total time elapsed T are given by

$$V = \sum_{k=1}^{M} t_k q_k \qquad \text{and} \qquad T = \sum_{k=1}^{M} t_k$$

Introducing busy-fixture time, T_B (see Table 6.4), the total-time mean flow rate for the period of duration T can be written

$$E[Q_N^*] = \left(\frac{T_B}{T_B}\right)\left(\frac{V}{T}\right) = \rho_N^* \overline{\alpha} \tag{6.5.7}$$

where ρ_N*=T_B/T is the utilization factor for the neighborhood of N homes and $\overline{\alpha}$ =V/T_B is the mean intensity of a busy fixture in the neighborhood. For instance, in Table 6.10 with T=1 month, ρ_N*=0.9527 and $\overline{\alpha}$ =2.277, so E[Q_N*] = 2.169 gpm as listed in Table 6.11. Two key points emerge:

(i) The result in Equation 6.5.7 is identical to Equation 6.13a in Table 6.9.

(ii) The total-time mean flow is not affected by the nature of the customer arrival process. For identical volumes of water consumption, homogeneous and nonhomogeneous arrivals will generate identical total-time mean demands.

Case 3. Total-Time Variance of Flow

The variance of the total-time flow depends on the sequence of demands. A heuristic argument is offered here to estimate the effect of nonhomogeneous arrivals on the variance of the total-time flow. In what follows, the mean α and variance β^2 of the flow at a busy server are assumed to be constant over time.

For a "short" time interval, during which the neighborhood utilization factor ρ_N^* is essentially constant, the mean and mean square of the total-time flow into a neighborhood of N homes are

$$E[Q_N^*] = \rho_N^* \alpha \tag{6.5.8}$$

$$E\left[\left(Q_N^*\right)^2\right] = \rho_N^*\left(\alpha^2 + \beta^2\right) + \left(\rho_N^* \alpha\right)^2 \tag{6.5.9}$$

Over a series of M time periods of duration Δt, the total-time mean flow is given by

$$E[Q_N^*] = (\alpha)\frac{\sum_{k=1}^{M} \Delta t_k \left(\rho_N^*\right)_k}{\sum_{k=1}^{M} \Delta t_k} = \alpha\, E\left[\rho_N^*\right] \tag{6.5.10}$$

Here $E[\rho_N^*]$ is the time-averaged neighborhood utilization factor. This result is consistent with Equation 6.5.7. In a similar manner, the mean square becomes

$$E\left[\left(Q_N^*\right)^2\right] = \left(\alpha^2 + \beta^2\right) E\left[\rho_N^*\right] + \left(\alpha^2\right) E\left[\rho_N^{*2}\right] \tag{6.5.11}$$

Noting $\mathrm{Var}[Q] = E[Q^2] - E[Q]^2$ and using Equations 6.5.10 and 6.5.11, the total-time variance is,

$$\begin{aligned} \mathrm{Var}\left[\left(Q_N^*\right)^2\right] &= \left(\alpha^2 + \beta^2\right) E\left[\rho_N^*\right] + \left(\alpha^2\right)\left(E\left[\rho_N^{*2}\right] - E\left[\rho_N^*\right]^2\right) \\ &= \left(\alpha^2 + \beta^2\right) E\left[\rho_N^*\right] + \left(\alpha^2\right) \mathrm{Var}\left[\rho_N^*\right] \end{aligned} \tag{6.5.12}$$

Hence, in the presence of nonhomogeneous arrivals, the variance of the total-time flow increases by an amount $\alpha^2\,\mathrm{Var}[\rho_N^*]$ over the homogeneous case. Dividing by $E[Q_N^*]^2 = \alpha^2 E[\rho_N^*]^2$ gives

$$\Theta_T^2\left[Q_N^*\right] = \frac{1 + \Theta_B^2[Q]}{E\left[\rho_N^*\right]} + \Theta^2\left[\rho_N^*\right] \tag{6.5.13}$$

where $\Theta[.]$ is the coefficient of variation.

Appendix 6.6: Stagnation Probability and Flow Variability

There is an interesting connection between stagnation probability and the relative variability of busy-time flow and total-time flow. Let Q_B denote the busy-time flow rate and Q_T denote the total-time flow rate. The relation between statistical moments of the two flows is

$$E[Q_B^n] = \frac{E[Q_T^n]}{1 - P_0} \tag{6.6.1}$$

where P_0 is the stagnation probability. If the probability of stagnation is zero, then busy-time and total-time flows are identical. Rearranging Equation 6.6.1 gives,

$$\text{For } n = 1: \quad 1 - P_0 = \frac{E[Q_T]}{E[Q_B]} \tag{6.6.2}$$

$$\text{For } n = 2: \quad 1 - P_0 = \frac{E[Q_T^2]}{E[Q_B^2]} = \frac{Var[Q_T] + E[Q_T]^2}{Var[Q_B] + E[Q_B]^2} \tag{6.6.3}$$

Squaring Equation 6.6.2 and dividing by 6.6.3 gives

$$1 - P_0 = \left(\frac{E[Q_T]}{E[Q_B]}\right)^2 \left(\frac{Var[Q_B] + E[Q_B]^2}{Var[Q_T] + E[Q_T]^2}\right) = \frac{\Theta_B^2[Q] + 1}{\Theta_T^2[Q] + 1} \tag{6.6.4}$$

where $\Theta[Q]$ is the coefficient of variation of the flow. Rearranging (6.6.4) yields,

$$P_0 = 1 - \left(\frac{\Theta_B^2[Q] + 1}{\Theta_T^2[Q] + 1}\right) \tag{6.6.5}$$

This is identical to Equation 6.15 in the text. Note that when $\Theta_B[Q] = \Theta_T[Q]$, there is no distinction between busy-time flow and total-time flow and, hence, stagnation disappears.

Appendix 6.7: Estimating PRP Parameters

In order to use the PRP model to estimate peak flows in a branching pipe or to simulate residential water demands, it is necessary to know the mean and the variance of the intensity, duration and frequency of water use. Ranges of typical values for these PRP statistics, based on nearly 365,000 indoor residential water use measurements from the dead-end study site, are summarized in Table 6.19. These values will provide realistic starting points for characterizing indoor water use in most homes across North America.

Individual households may exhibit significant deviations from the values in Table 6.19. However, the law of averages will tend to smooth out extremes, even in groups with as few as N=10 homes. Recent AwwaRF studies on Residential Water Use Patterns (Bowen *et al*, 1993) and on Residential End Uses of Water (Mayer *et al*, 1999) provide additional information to help characterize the intensity, duration and frequency of residential water demands.

Table 6.19 Typical values of pulse characteristics for indoor residential water use measured at 21 homes along the dead-end loop study site.

PRP Property	Mean		Standard Deviation		Coefficient of Variation
	symbol	typical range	symbol	typical range	
Intensity, q	α	1.0 to 4.0 gpm	β	0.5 to 2.0 gpm	0.3 to 0.7
Duration, d	τ	0.5 to 2.0 min	ω	0.5 to 3.0 min	1.0 to 2.5
Volume, p=qd	ϕ	1.0 to 4.0 gal	ψ	2.0 to 8.0 gal	1.5 to 3.0
Frequency	λ	1.3 per cap hr	–	0 to 4 per cap hr	0.0 to 3.0

Pulse volume is a derived property, defined as the product of intensity and duration. Assuming independence between intensity and duration, the mean, variance and coefficient of variation of the pulse volume are given by,

$$E[p] = E[qd] = \alpha\tau \qquad (6.7.1a)$$

$$Var[p] = Var[qd] = \alpha^2\omega^2 + \tau^2\beta^2 + \omega^2\beta^2 \qquad (6.7.1b)$$

$$\Theta^2[p] = Var[p]/E^2[p] = \Theta^2[q] + \Theta^2[d] + \Theta^2[q]\,\Theta^2[d] \qquad (6.7.1c)$$

Few utilities have the resources to continuously record water consumption at individual homes in order to estimate PRP water use parameters. Many utilities, however, are able to monitor the total rate of water flow into a service zone. This information can also be used to estimate some PRP parameters for indoor residential water use.

To illustrate, let q(i) be the flow rate measured in the main line at time t(i). First, find the sample mean and mean square from M sequential flow measurements:

$$\overline{Q} = \frac{\sum_{i=1}^{M} [q(i) - Q_L]\, \Delta t(i)}{\sum_{i=1}^{M} \Delta t(i)} \quad \textit{and} \quad \overline{Q^2} = \frac{\sum_{i=1}^{M} [q(i) - Q_L]^2\, \Delta t(i)}{\sum_{i=1}^{M} \Delta t(i)}$$

where $\Delta t(i) = t(i) - t(i\text{-}1)$ and Q_L is the (assumed constant) rate of leakage in the service zone. Define an indicator variable,

$W(i) = 0$ if $q(i) - Q_L = 0$
$W(i) = 1$ if $q(i) - Q_L > 0$

then the sample stagnation frequency is given by:

$$P_0 = \frac{\sum_{i=1}^{M} W(i)\, \Delta t(i)}{\sum_{i=1}^{M} \Delta t(i)}$$

From Equations 6.2 and 6.13, corresponding average PRP parameters for the service zone are:

$$\rho_N^* = -\ln[P_0] \tag{6.7.2}$$

$$\overline{\alpha} = \frac{\overline{Q}}{\rho_N^*} \tag{6.7.3}$$

$$\overline{\beta^2} = \frac{1}{\rho_N^*}\left[\left(\overline{Q^2}\right) - \left(\overline{Q}\right)^2\left(1 + \frac{1}{\rho_N^*}\right)\right] \tag{6.7.4}$$

Notes on the PRP parameter estimation procedure:

(1) Flow readings should be taken at least *every* second ($\Delta t \leq 1$ sec) for at least 1 hour ($M \geq 3600$) during a period of low indoor water use (early morning winter season). This frequency gives the most accurate parameter estimates and maximizes the size of the service zone which can be sampled. Service zones with up to 500 single family homes can be monitored.

(2) If $\Delta t > 1$ sec, the procedure will over-estimate ρ^*_N and under-estimate β^2.

(3) Background leakage must be identified and removed. Buchberger and Nadimpalli (2003) provide a procedure to screen and censor leakage from continuous measurements of flow.

(4) If the stagnation frequency is zero (even with all leaks removed), then the size of the service zone that is monitored must be reduced.

(5) The service zone must be completely isolated with only one feeder line in and no lines out. All measured inflow must go to residential consumers in the service zone.

Appendix 6.8: Summary of PRP Model Applied to a Branching Line

Branching mainlines are an important subset of many municipal pipe networks. Because the direction of flow in a branching pipe is known, these supply lines are amenable to detailed analysis with the PRP model. The idea is to replace conservation of *energy* with conservation of *probability*. Then in conjunction with the continuity principle, the PRP model can provide an estimate of the mean and the variability of the discharge and other features of the flow along the branching line.

To illustrate the richness and versatility of the PRP model for describing residential water use, consider the block of N homes along a branching pipe as shown in Figure 2.2. Table 6.20 summarizes several useful analytical results that follow when the PRP model is applied to the branching pipe shown at the top of Figure 2.2. These results assume a time homogeneous customer arrival rate (i.e., no diurnal variation in demand pattern). Explanatory notes for Table 6.20 are provided below.

Notes for Table 6.20:

- α and β^2 are the mean and variance of the PRP intensity at a busy server in the neighborhood.
- ϕ and ψ^2 are the mean and variance of the PRP volume at a busy server in the neighborhood.
- λ^*_N is the cumulative arrival rate for the neighborhood of N homes (see Table 6.2).
- ρ^*_N is the cumulative utilization factor for the neighborhood of N homes (see Table 6.2).
- $\Lambda^*_N = \lambda^*_N / k$ is a dimensionless rate coefficient, where k is the overall first-order reaction rate for the disinfectant.

- L_N is the volume of link N along the dead-end branch.
- The probability of stagnation follows from Equation 6.2.
- The average duration of busy and stagnant periods is derived in Appendix 6.3.
- The average number of busy and stagnant periods is derived in Appendix 6.2.
- The mean and variance of the number of busy homes follows from Appendix 6.1.
- The mean and variance of the flow rate are derived in Appendix 6.4.
- The mean and variance of the travel time through pipe link N is given by Wu (1996).
- The mean and variance of the residual disinfectant concentration at the end of pipe link N is given by Wu (1996).

Table 6.20 Some steady state results for Poisson Rectangular Pulse water demand model with homogeneous arrival process.

1. Probability of Stagnation in Pipe Link N: $P_0 = \exp[-\rho^*_N]$

2. Busy and Stagnant Periods in Pipe Link N:

Average Duration: *Busy Period:* $\frac{1}{\lambda^*_N}\left[\frac{1-P_0}{P_0}\right]$ *Stagnant Period:* $\frac{1}{\lambda^*_N}$

Avg number of busy and stagnant periods during an interval of duration T: $\lambda^*_N T P_0$

3. Mean and Variance of the Number of Busy Homes in the Neighborhood:

$$E[H] = N(1-P_0) \approx \rho^*_N \qquad Var[H] = NP_0(1-P_0) \approx \rho^*_N$$

4. Mean and Variance of Flow Rate through Pipe Link N:

$$E[Q_N] = \rho^*_N \alpha = \lambda^*_N \phi \qquad Var[Q_N] = \rho^*_N\left[\alpha^2 + \beta^2\right]$$

5. Mean and Variance of Travel Time through Pipe Link N:

$$E[T_N] = \frac{L_N}{\lambda^*_N \phi} \qquad Var[T_N] = \frac{L_N}{(\lambda^*_N)^2 \phi}\left[1 + \frac{\psi^2}{\phi^2}\right]$$

6. Mean and Variance of Residual Concentration at End of Pipe Link N:

$$E[C_N] = \exp\left[\frac{-L_N/\phi}{1+\Lambda^*_N}\right] \qquad Var[C_N] = \exp\left[\frac{-2L_N/\phi}{2+\Lambda^*_N}\right] - \exp\left[\frac{-L_N/\phi}{2+\Lambda^*_N}\right]$$

CHAPTER 7
WATER QUALITY ANALYSIS

OVERVIEW

Water quality at the entrance and the apex of the dead-end loop was sampled three ways. Continuous monitoring units (CMUs, see Figure 3.4) recorded eight physico-chemical parameters from March 8 to November 18, 1997 (256 days, corresponding roughly to the period of archived residential water demands). Weekly grab samples taken from the dead-end loop and other locations in the distribution system were analyzed for HPC, TC, AOC and TOC from January 3 to November 18, 1997 (320 days). Finally, a modified Robbins device (MRD, see Figure 3.5) provided coupon samples to quantify biofilm growth from July 13 to November 18, 1997 (130 days).

This chapter examines variations in water quality parameters observed at the entrance and apex of the dead-end loop. The utility of real-time water quality monitoring in municipal distribution systems is discussed. The presence of significant correlations between system water quality and travel times is explored. The prospect of adopting surrogate water quality parameters as reliable indicators of distribution system status is investigated. Much of material presented here is adapted from the MS thesis by Jason Carter (1998). Preliminary results on associations between HPC and water quality have been discussed by Carter, Rice, Buchberger and Lee (2000).

CONTINUOUS MONITORING UNITS

Data Recovery

Probe readings of physico-chemical parameters were taken every 10 minutes at both CMUs. During the 256-day observation period, this generated 36,864 potential readings for each water quality parameter[(1)]. The extent of water quality data recovered from the CMUs is summarized in Table 7.1.

[(1)] To minimize data redundancy and to extend the life of the monitoring units, several schemes with alternating sampling windows (e.g., 6 hours on then 6 hours off) were investigated. However, the "on-off" approach introduced complications to the water quality data set and, hence, none was selected to replace the original plan of continuous sampling at 10 minute intervals. See Appendix 7.1 for details.

Table 7.1 Data Recovery Matrix for Continuous Monitoring Units

Location and Parameter	Mar 34	Apr 1234	May 1234	June 1234	July 12345	Aug 1234	Sept 1234	Oct 12345	Nov 123
BPS: Booster Pump Station									
flowrate	AA	AAAA	AAAA	AAAB	BBBBB	BBBBB	BBBBB	BBBBB	BBB
pH	UU	UUUN	UUUU	UUUU	UPUUU	UUUU	UUUU	NUXUX	UUU
conductivity	UU	UUUN	UUUU	UUUU	UPUUU	UUUU	UUUU	NUXUX	UUU
temperature	UU	UUUN	UUUU	UUUU	UPUUU	UUUU	UUUU	NUXUX	UUU
free chlorine	UU	UUUN	UUUU	UUUU	UPUUU	UUUQ	UUUQ	NUXUX	UUU
NO_3	QQ	QQXN	XXXX	XXXX	XPXXX	XXXX	XXXX	NXXXX	XXX
turbidity	UQ	QQXN	XXXX	XXXX	XPXXX	XXXX	XXXX	NXXXX	XXX
ORP	UQ	QQQN	QQQQ	QQQQ	QPQQQ	QQQQ	QQQQ	NQQQQ	QQQ
DO	UU	UUUN	UUUU	XXQQ	QPQQQ	QQQQ	QQXX	NXXXX	XXX
DLA: Dead-end Loop Apex									
flowrate	AA	AAAA	AAAA	AAAB	BBBBB	BBBB	BBBB	BBBBB	BBB
pH	UU	NUUU	UUUU	UUUU	UUUUU	UUUU	NUUU	UUUUU	UUU
conductivity	UU	NUUU	UUUU	UUUU	UUUUU	UUUU	NUUU	UUUUU	UUU
temperature	UU	NUUU	UUUU	UUUU	UUUUU	UUUU	NUUU	UUUUU	UUU
free chlorine	UU	NUUU	UUUU	UUUU	UUUUU	UUUQ	NUUQ	UUUUU	UUU
NO_3	XX	NXXX	QXXQ	QXXQ	XXXXX	QQXX	NXXX	XXXXX	XXX
turbidity	UU	NUUU	UUUU	QQQQ	QQQQQ	QQQQ	NQQQ	QQQQQ	QQQ
ORP	UQ	NQQQ	QQQQ	QQQQ	QQQQQ	QQQQ	NQQQ	QQQQQ	QQQ
DO	UU	NUUU	UUUU	QQQQ	QQQQQ	QQXX	NXXX	XXXXX	XXX

Legend:

A	0.6 lpm bleed rate at CMU	X	Unusable data
B	1.0 lpm bleed rate at CMU	N	No data collected (equipment failure)
U	Usable data	P	Partial data
Q	Questionable data		

Very good data recovery was achieved for pH, conductivity, and temperature (each at about 85%) followed by free chlorine (~73%). However, it proved difficult to obtain reliable observations of the remaining four parameters: nitrate, ORP, dissolved oxygen (DO) and turbidity. Owing to poor data recovery, these parameters ultimately were dropped from the water quality analysis.

Problems with collecting water quality data online stemmed from a number of sources. In the case of nitrate, ORP and DO measurements, poor data recovery was attributed to the probe membranes. The nitrate probes fouled quickly causing the calibration to drift. Despite regular acid washes, the sensitivity of the ORP probes diminished over time. The DO probes worked well initially, but owing to their short service life (under six months), the DO probes expired during the early stages of the water quality data collection effort. Despite recent advancements in on-line sensor technology, probes for ORP and nitrate remain susceptible to severe drift and, therefore, are not commonly used in drinking water applications (Hargesheimer *et al*, 2002).

In the case of turbidity, the probes were confounded by the presence of calcium particles in the water. Calcium was added at the water treatment plant to create a lime scale inside the cast iron water mains. Some calcium particles accumulated in the Solomat flow-through cells and biased the turbidity readings. Installation of an upstream sedimentation chamber and weekly cleaning did not measurably improve the performance of the turbidity probes at either location.

Physico-Chemical Water Quality Analysis

Time series profiles of pH, conductivity, temperature, and free residual chlorine appear in Figures 7.1 through 7.4, respectively. These plots span a 256-day period from March 8 to November 18, 1997 and illustrate the dynamic character of physico-chemical water quality parameters in the distribution system. The pH and conductivity are conservative parameters; the temperature (heat) and free chlorine are non-conservative. Gaps in the plotted points indicate times when information was lost due to instrument calibration, system failure, or data screening(2).

Paired histograms of all four parameters at both locations are displayed in Figure 7.5 and cumulative percentages are compared in Figure 7.6. Full period statistics for pH, conductivity, temperature and free chlorine at the BPS and DLA locations are summarized in Table 7.2. Seasonal statistics for these parameters are listed in Table 7.3 for the BPS and in Table 7.4 for the DLA.

---

(2) Readings below probe detection limits were censured (see Table 3.2 for thresholds).

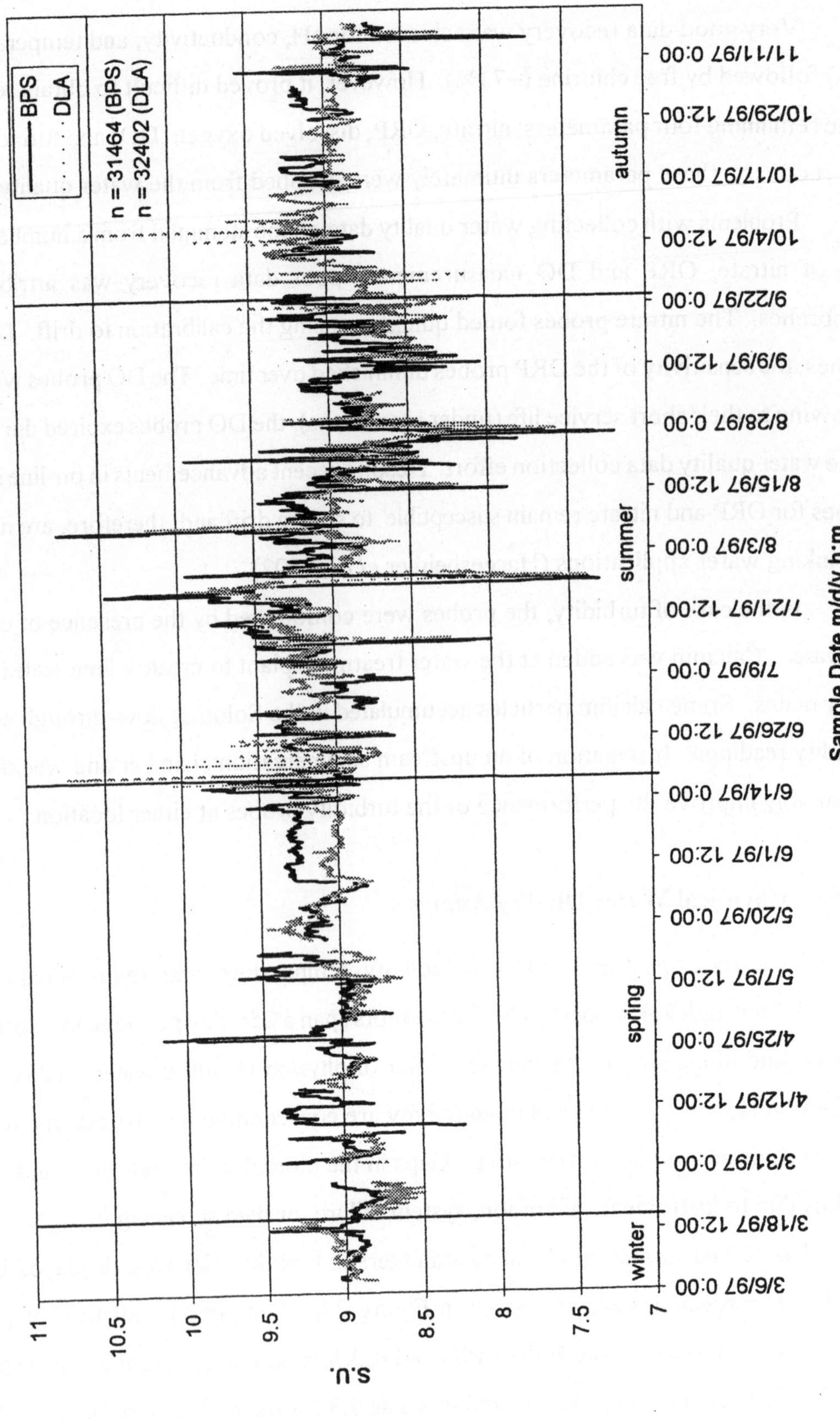

Figure 7.1 Time series of pH measured at BPS and DLA

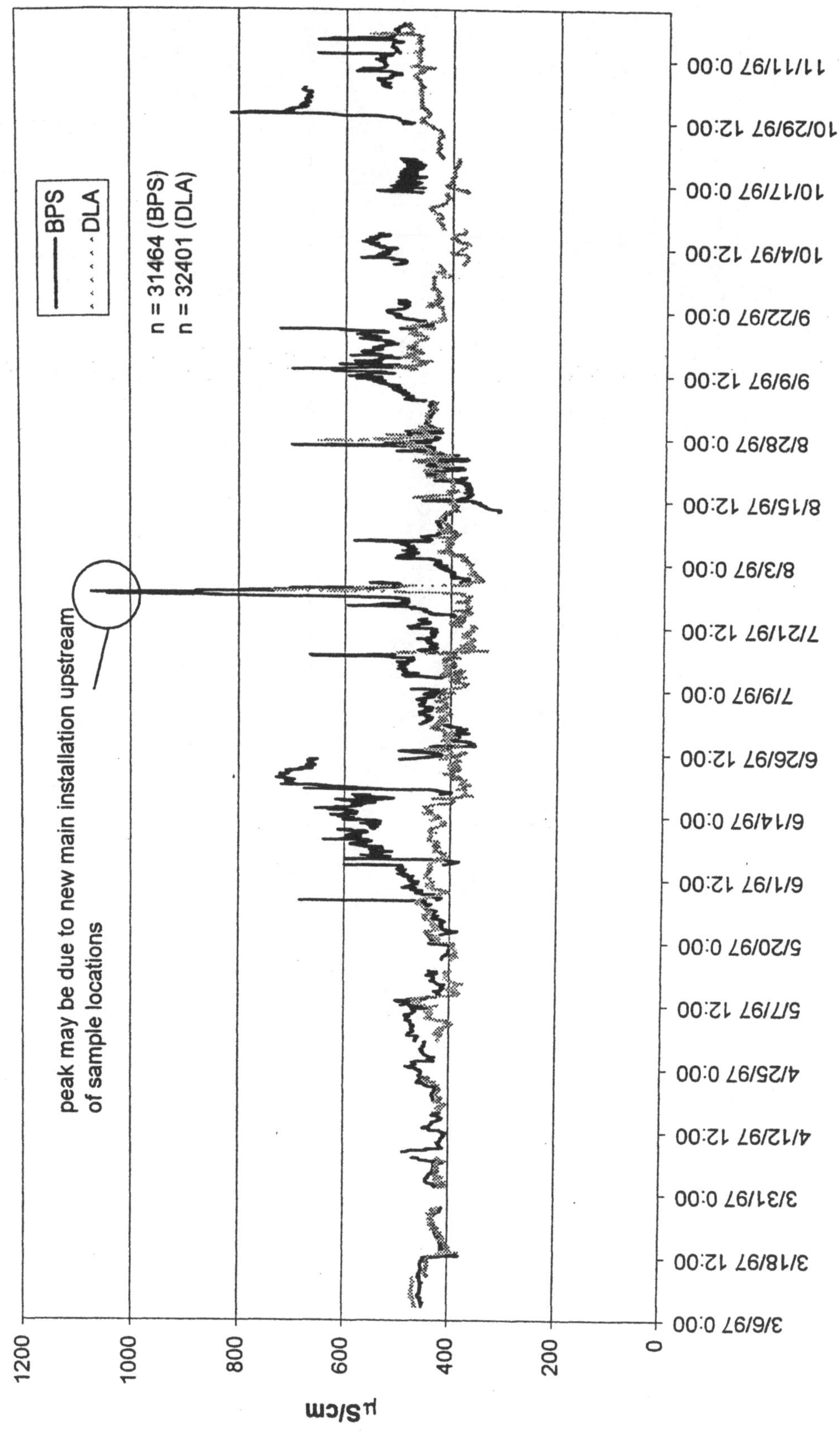

Figure 7.2 Time series of conductivity measured at BPS and DLA.

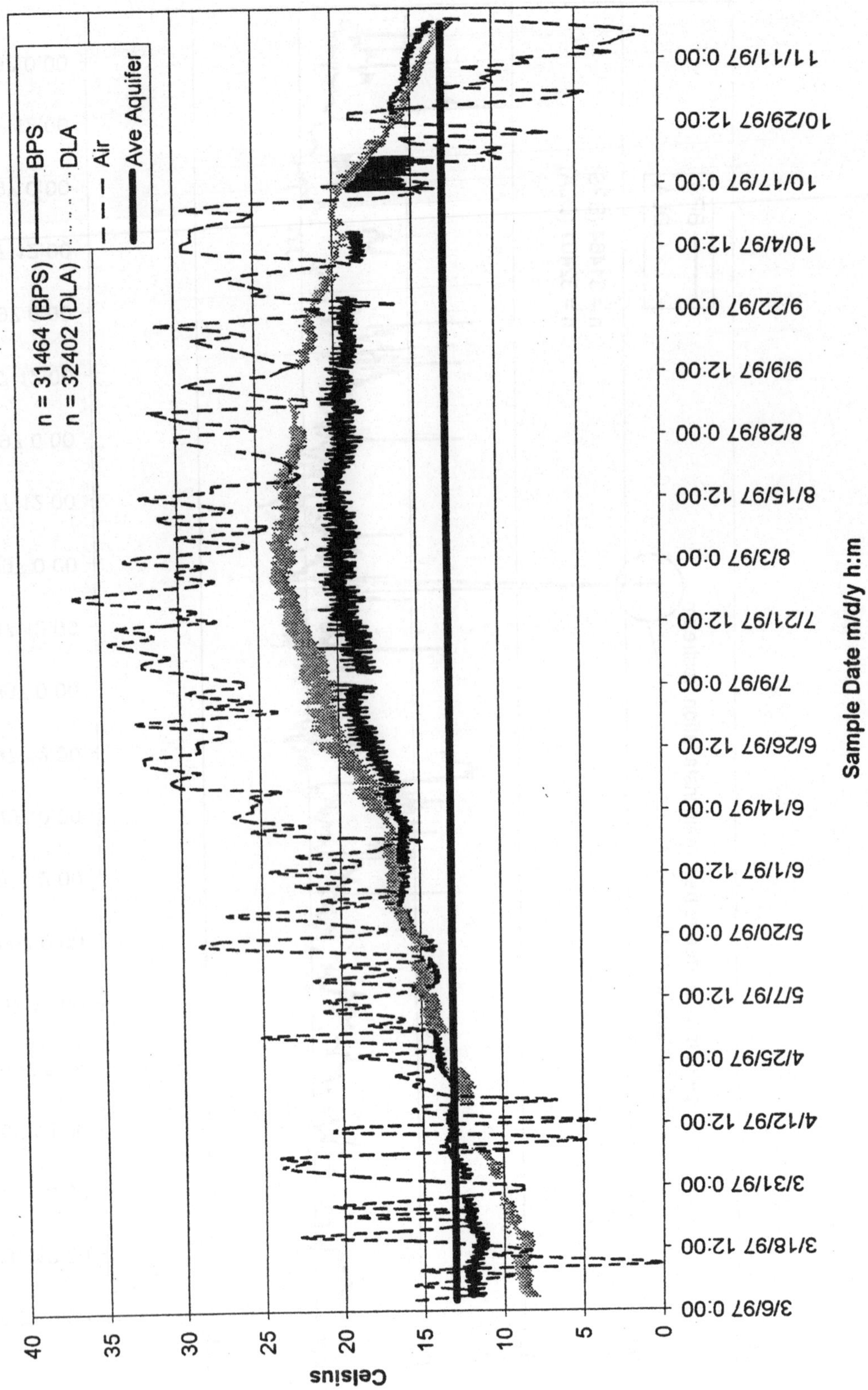

Figure 7.3 Time series of water temperature measured at BPS and DLA. Also shown are daily air temperature and annual average groundwater temperature.

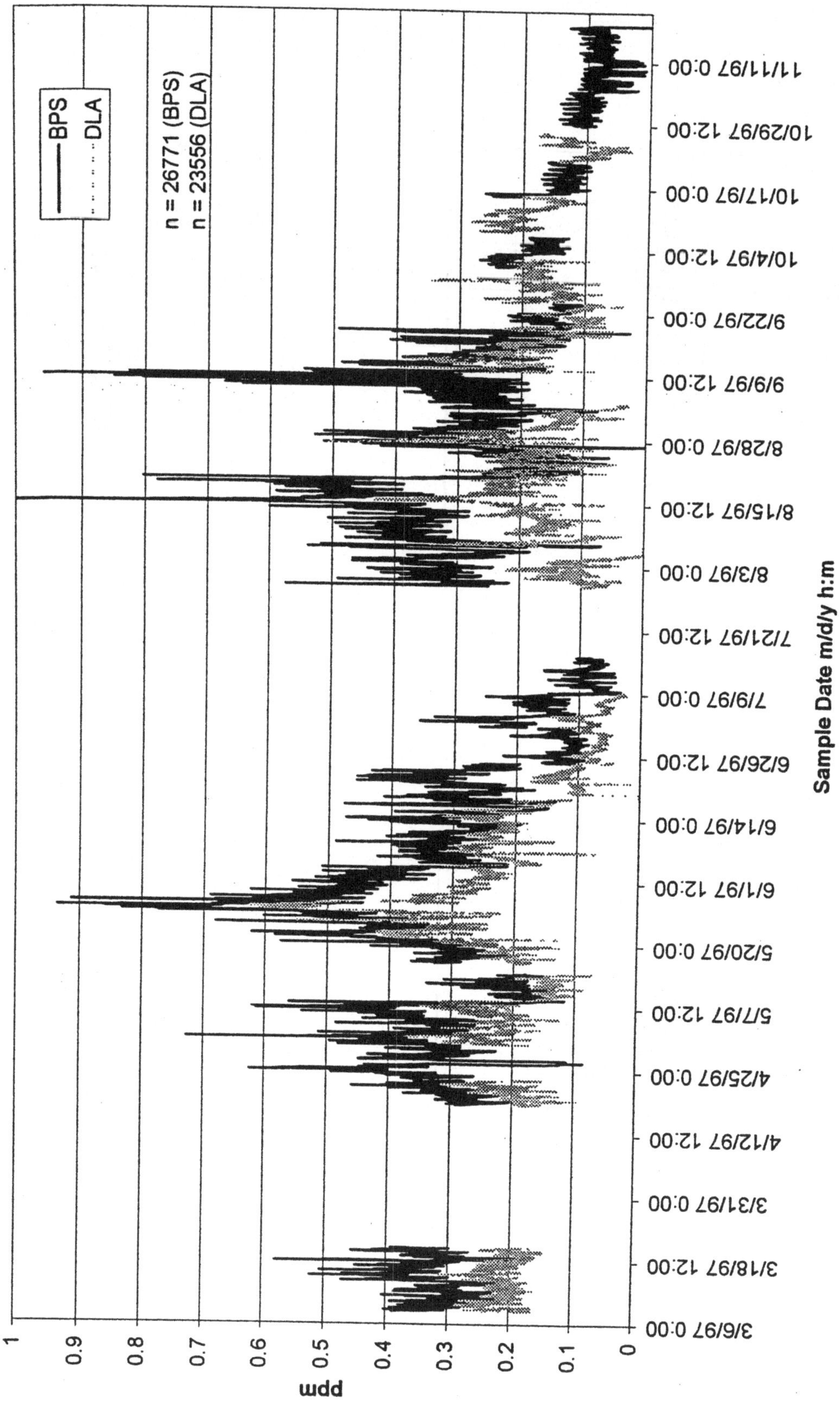

Figure 7.4 Time series of free chlorine residual at BPS and DLA.

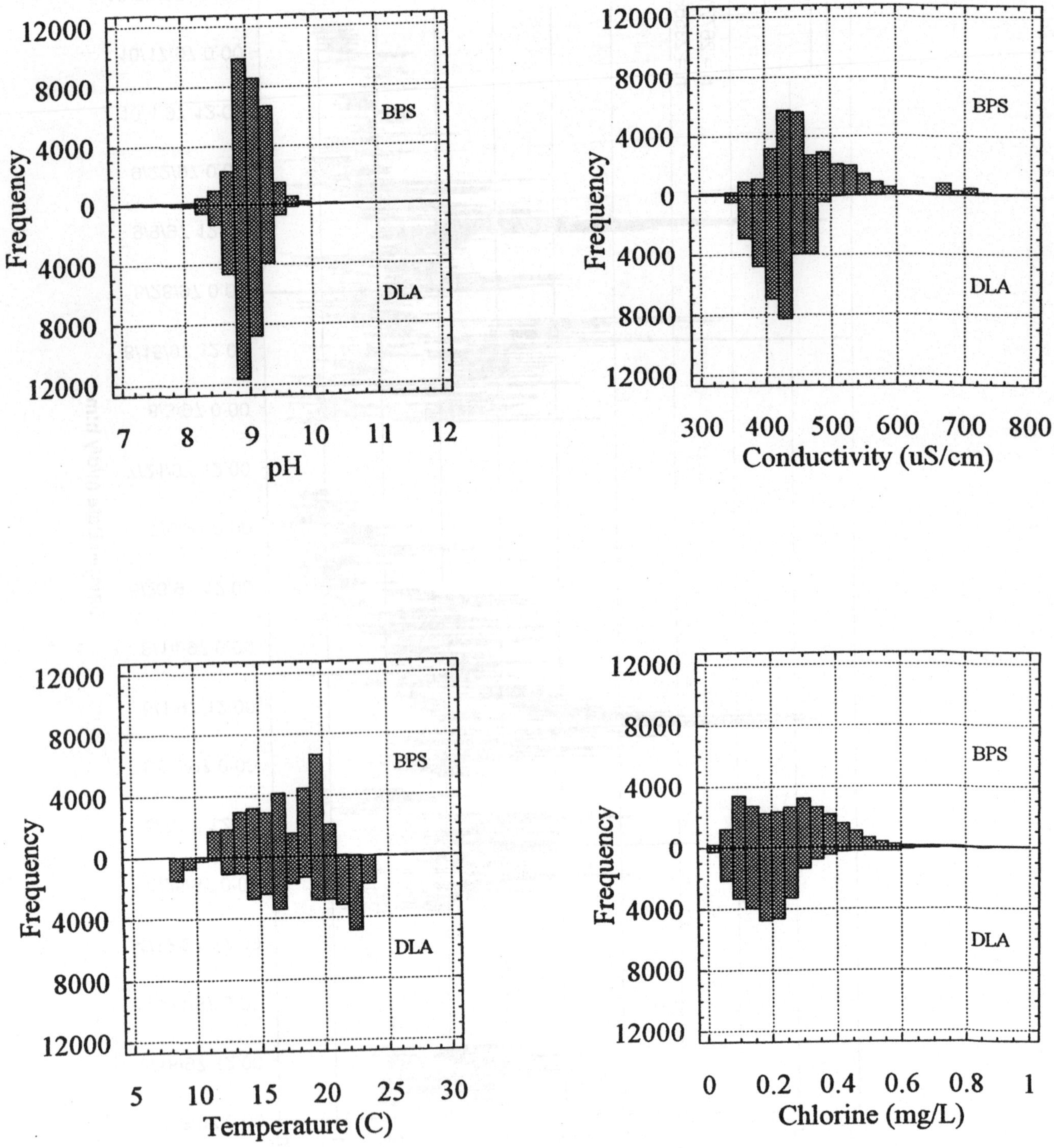

Figure 7.5 Paired frequency histograms of four physico-chemical parameters measured at the BPS and DLA during the period March 8 to November 18, 1997.

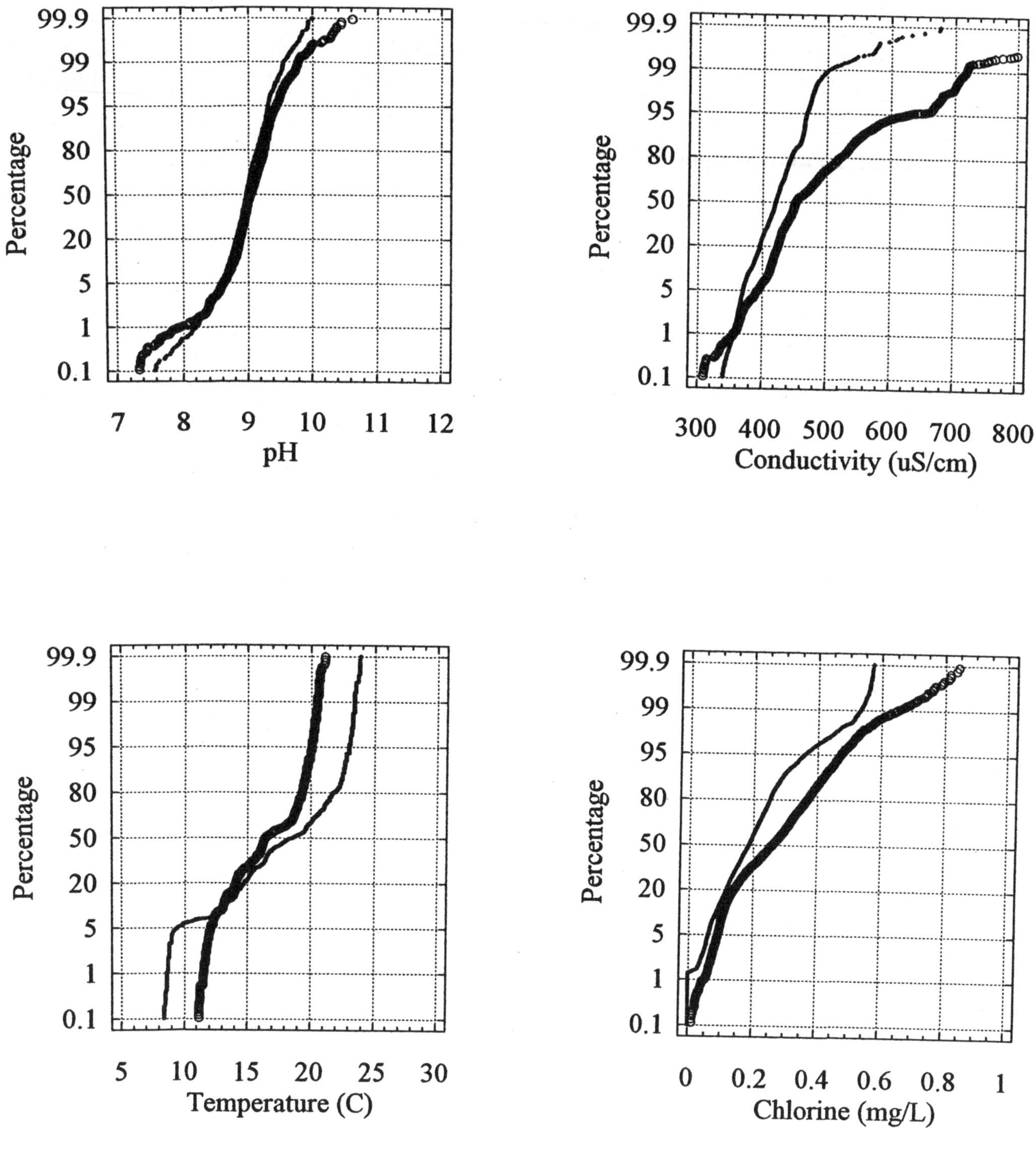

Figure 7.6 Paired cumulative probability distributions of four physico-chemical parameters measured at BPS (thick line) and DLA (thin line), March 8 to November 18, 1997.

Table 7.2 Summary statistics for physico-chemical parameters March 8 through November 18, 1997 at dead-end study site.

Location and Statistic	pH (su)	conductivity (μS/cm)	temperature (Celsius)	free chlorine (mg/L)
BPS: Booster Pump Station				
Average	9.046	478.6	16.65	0.265
Median	9.050	453.5	16.50	0.262
Mode	8.950	452.0	19.40	0.317
Standard Deviation	0.310	79.8	2.70	0.136
Minimum	7.200	307.4	11.00	0.001
Maximum	10.770	1,072.9	22.20	1.093
Skewness	-0.730	1.88	-0.29	0.660
Coefficient of Variation	0.034	0.17	0.16	0.513
Sample Size	31,464.00	31,464.00	31,464.00	26,744.00
DLA: Dead-end Loop Apex				
Average	8.965	423.0	17.98	0.181
Median	8.970	422.0	18.60	0.180
Mode	8.960	436.7	22.50	0.000
Standard Deviation	0.262	34.4	4.09	0.094
Minimum	7.390	331.4	7.80	0.000
Maximum	10.730	746.0	24.30	0.610
Skewness	-0.550	1.15	-0.60	0.790
Coefficient of Variation	0.029	0.08	0.23	0.519
Sample Size	32,402.00	32,401.00	32,402.00	27,244.00

Table 7.3
Seasonal Statistics for Physico-Chemical Parameters at BPS (Loop Entrance)

Parameter and Statistic	Winter	Spring	Summer	Autumn
pH (s.u.)				
Average	8.98	9.12	9.01	8.97
Median	8.97	9.11	9.05	8.99
Mode	9.00	8.95	9.15	9.12
Standard Deviation	0.09	0.21	0.40	0.23
Minimum	8.85	8.23	7.20	7.72
Maximum	9.49	10.16	10.77	9.56
Skewness	2.72	0.63	-0.62	-1.54
Coefficient of Variation	0.01	0.02	0.04	0.03
Sample Size	1,777.00	11,900.00	12,999.00	4,788.00
Conductivity (μS/cm)				
Average	442.61	466.01	471.75	541.53
Median	450.60	441.01	449.50	517.85
Mode	452.00	431.30	449.00	530.80
Standard Deviation	18.27	64.74	89.49	68.26
Minimum	378.90	383.80	307.40	386.40
Maximum	457.50	730.70	1072.90	813.50
Skewness	-1.76	1.73	2.26	1.46
Coefficient of Variation	0.04	0.14	0.19	0.13
Sample Size	1,777.00	11,900.00	12,999.00	4,788.00
Temperature (C)				
Average	11.86	14.62	19.32	16.25
Median	11.90	14.30	19.40	16.00
Mode	12.00	16.30	19.40	16.10
Standard Deviation	0.31	1.53	0.73	1.48
Minimum	11.00	11.00	16.70	13.40
Maximum	12.60	17.80	22.20	20.60
Skewness	-0.21	-0.03	-0.54	0.42
Coefficient of Variation	0.03	0.10	0.04	0.09
Sample Size	1,777.00	11,900.00	12,998.00	4,788.00
Free Chlorine (mg/L)				
Average	0.32	0.34	0.25	0.12
Median	0.32	0.33	0.24	0.11
Mode	0.32	0.33	0.11	0.11
Standard Deviation	0.06	0.11	0.13	0.05
Minimum	0.17	0.09	0.00	0.00
Maximum	0.58	0.94	1.09	0.27
Skewness	0.34	0.88	0.92	0.74
Coefficient of Variation	0.19	0.32	0.52	0.42
Sample Size	1,777.00	8,877.00	11,329.00	4,788.00

Table 7.4
Seasonal Statistics for Physico-Chemical Parameters at DLA (Loop Apex)

Parameter and Statistic	Winter	Spring	Summer	Autumn
pH (s.u.)				
Average	9.05	8.99	8.93	8.97
Median	9.05	8.99	8.96	8.96
Mode	9.07	8.90	8.96	8.96
Standard Deviation	0.11	0.18	0.36	0.16
Minimum	8.81	8.49	7.39	8.13
Maximum	9.47	10.73	10.19	10.54
Skewness	0.88	0.89	-0.45	0.48
Coefficient of Variation	0.01	0.02	0.04	0.02
Sample Size	1,772.00	10,111.00	12,593.00	7,926.00
Conductivity (μS/cm)				
Average	446.08	421.35	415.84	431.30
Median	442.40	423.40	409.50	428.70
Mode	466.00	433.90	419.10	436.70
Standard Deviation	22.59	20.39	42.73	31.96
Minimum	381.60	356.90	331.40	367.50
Maximum	469.10	474.00	746.00	560.10
Skewness	-0.99	-0.54	1.85	-0.03
Coefficient of Variation	0.05	0.05	0.10	0.07
Sample Size	1,772.00	10,110.00	12,593.00	7,926.00
Temperature (C)				
Average	8.81	15.21	21.89	17.36
Median	8.80	15.60	22.20	17.90
Mode	8.80	16.50	22.50	19.70
Standard Deviation	0.24	2.28	1.14	2.52
Minimum	7.80	8.30	18.10	12.30
Maximum	9.50	19.50	24.30	21.10
Skewness	-0.25	-0.94	-0.76	-0.30
Coefficient of Variation	0.03	0.15	0.05	0.15
Sample Size	1,772.00	10,111.00	12,593.00	7,926.00
Free Chlorine (mg/L)				
Average	0.21	0.23	0.15	0.18
Median	0.21	0.23	0.14	0.18
Mode	0.19	0.00	0.00	0.12
Standard Deviation	0.03	0.10	0.08	0.06
Minimum	0.15	0.00	0.00	0.03
Maximum	0.31	0.61	0.51	0.35
Skewness	0.38	0.26	0.72	0.02
Coefficient of Variation	0.14	0.43	0.53	0.33
Sample Size	1,772.00	7,978.00	9,580.00	4,266.00

pH

The time series of pH values at the BPS and the DLA (see Figure 7.1) show reasonably steady readings over the 256-day monitoring period. The mean pH at the BPS was 9.05 with a standard deviation of 0.31. The mean pH at the DLA was 8.97 with a standard deviation of 0.26. Both are near the system target value of 9.0 set at the water treatment plant. These relatively high pH values were maintained to encourage scaling and inhibit corrosion of the pipe network.

A few excursions, primarily during the summer period, sent the pH below 7.5 or above 10.5. The first incident with low pH (near the end of July 1997) coincides with a surge in conductivity (see Figure 7.2). This particular system disturbance is believed to have resulted from installation of a new water main upstream from the dead-end loop study site.

The frequency histograms and cumulative percentages (see Figures 7.5 and 7.6) show similar distributions of pH at the BPS and the DLA. Both the mean and the variance of the pH decrease slightly when moving from the BPS to the DLA. However, the change in the value of these statistics is not significant (see Table 7.6) and this implies pH is conserved around the loop.

Conductivity

The time series of conductivity values at the BPS and DLA (see Figure 7.2) are generally confined between 400 and 600 μS/cm. The mean conductivity at the dead-end entrance was 479 μS/cm; the mean at the dead-end apex was 423 μS/cm, a drop of nearly 12 percent. Conductivity values at the BPS briefly exceeded 1050 μS/cm during a peak excursion in late July. This anomaly was traced to installation of a new main upstream from the study site.

The frequency histograms and cumulative percentages (see Figures 7.5 and 7.6) show rather different conductivity distributions at the BPS and the DLA. The mean and the variance of conductivity is significantly less at the loop apex than at the loop entrance (see Table 7.6). Evidently, travel through the dead-end loop effectively smoothed (i.e., reduced) some of the variability in the conductivity of the incoming water. This seems reasonable because flow reversals would encourage mixing of water and, hence, would dampen conductivity spikes that entered the dead-end loop. It is, however, not known why the mean value of the conductivity dropped between the BPS and the DLA. The reduction in the mean conductivity implies there was a net loss of dissolved ions between the entrance and the apex of the dead-end loop during the period of observation.

Temperature

Four temperature profiles are shown in Figure 7.3. Included with CMU measurements of water temperature at the BPS and the DLA, are maximum daily air temperature and the average annual groundwater temperature. Compared to water temperatures, the air temperature fluctuates dramatically from day to day and over the course of the year. The time series of water temperatures at the BPS and DLA show prominent seasonal trends. Each rises from low values in early Spring, reaches a peak in mid Summer, and then declines through the Autumn. This pattern mimics the air temperature profile, but with a more subdued amplitude. Although much of the winter period is missing, it is clear that water temperatures would attain their annual minimum levels around February.

The travel time through the dead-end loop provides opportunity for heat exchange between the water and the pipe (or ground). During the warm summer months, the pipe network gained heat from the ground and, consequently, water temperatures increased while moving from the entrance to the apex of the loop. During the cold winter months, the network lost heat to the ground and, consequently, water temperatures decreased while moving from the entrance to the apex. This is clear in Figure 7.3 which shows DLA water temperatures above BPS water temperature in the summer but below BPS temperatures in late Autumn. The phenomenon of water temperature amplification occurs because the City draws its water supply from an alluvial aquifer having a relatively stable groundwater temperature(3) (especially compared to the daily air temperature).

At first glance, the frequency histograms and cumulative percentages (see Figures 7.5 and 7.6) indicate that the mean and the variance of water temperature increase between the entrance and the apex. Owing to the strong seasonal component in these data, extra care is needed to interpret the temperature plots. If data had been gathered for a full year, the additional measurements from the winter months (November through February) would yield DLA temperatures below those at the BPS. The net effect is that the annual mean water temperatures at the entrance and the apex of the loop would be very similar. The seasonal heating and cooling cycle would persist and, hence, the variance of the water temperature would be greater at the apex than at the entrance of the dead-end loop.

---

(3) The average annual groundwater temperature is about 13 C in the unconfined sand and gravel aquifer that supplies water to Milford. The mean annual air temperature at the study site is approximately 13 C (NOAA, 1974).

Free Chlorine

The time series of free chlorine concentrations at the BPS and the DLA are shown in Figure 7.4. Compared to the three previous water quality plots, free chlorine displays the greatest relative variability (coefficient of variation exceeds 0.50). The chlorine data also contain a few large gaps due to instrument malfunctions (e.g., loss of calibration, automated download errors) and other field problems (e.g., loss of power, flooding at the BPS).

Free chlorine is non-conservative[(4)]. Free chlorine reacts with oxidizable material in the pipe network, including biomass and pipe walls. The mean value of the chlorine concentration dropped from 0.27 mg/L (standard deviation = 0.14 mg/L) at the loop entrance to 0.18 mg/L (standard deviation = 0.09 mg/L) at the apex. The mean value of the chlorine residual at the DLA is actually less than the water quality standard of 0.20 mg/L.

Occasionally (about 1% of the time), no chlorine residual was detected at the DLA (see Figures 7.5 and 7.6). Barring instrument error, this observation corroborates the concern that long travel times in dead-end zones can lead to a complete loss of chlorine residual in the distribution system. Incidents of zero detectable chlorine residual occurred mainly during the summer when water temperatures were highest and chlorine decay was most rapid.

Figure 7.7 shows chlorine residual and water temperature at the DLA during a 50-day period from May 18 to July 7, 1997. Time series of both data sets appear in the top graph. The bottom graph shows residual concentration regressed against water temperature. The negative slope of the regression is consistent with well-known kinetic relations: chlorine loss rate increases with water temperature. To maintain a steady chlorine residual across the dead-end loop (and at other locations in the network), the disinfectant dose at the treatment plant must be increased as the water temperature rises. The increase in water temperature coupled with the decrease in chlorine residual may have affected the bacterial water quality at the study site. Warm water temperatures and low chlorine residual can enhance microbial regrowth in a water distribution system.

[(4)] Bottle tests of water from the BPS yielded an average first-order bulk reaction rate, k=0.01/hr with coefficient of variation of 25%. This is within the range 0.003 to 0.032/hr for groundwater sources reported by Vasconcelos *et al* (1997).

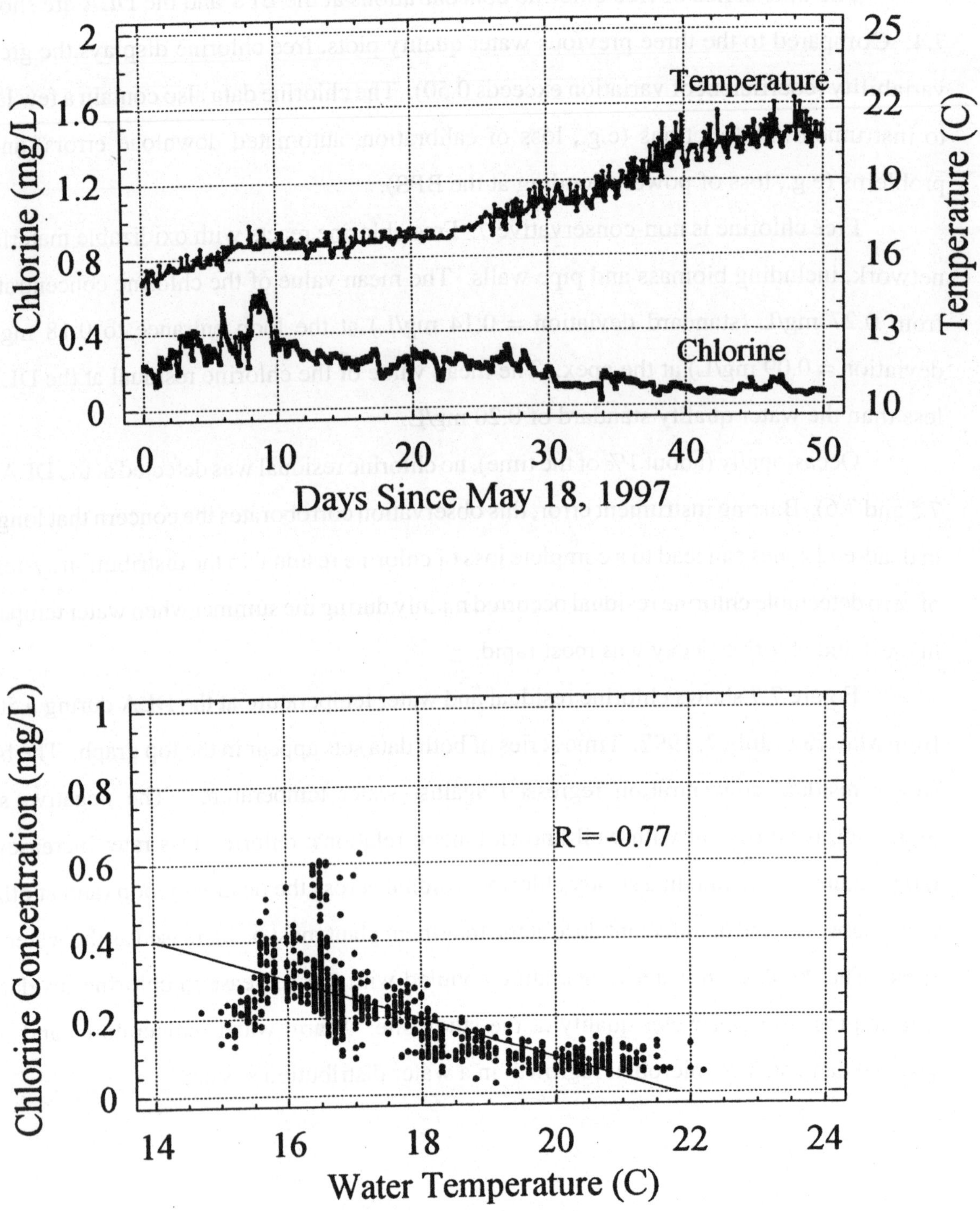

Figure 7.7 Time series of free chlorine and water temperature at the DLA (top) and regression of chlorine concentration on water temperature (bottom).

Effect of Booster Pump

The booster pumps near the entrance of the dead-end loop were used three or four times per day, typically for a period lasting between one or two hours. The booster pumps operated at a discharge of about 250 gpm and sent water out to the Edgecomb Avenue storage tank (see Figure 3.1). Activation of the pumps at the BPS caused sudden jumps (up and down) in parameter measurements at the loop entrance. This is evident in Figure 7.8 which shows three BPS pumping episodes superimposed on concurrent water quality measurements taken at the loop entrance during a 24-hour period in March 1997.

The beginning and the end of the pumping cycle are marked by relatively large shifts in the water quality readings. In this particular example, when the pumps are on, the pH drops but conductivity, temperature and chlorine levels rise. The effect of the pumps is to alter network travel times and to increase the fraction of "young" water arriving at the BPS from the treatment plant.

During the interval between pumping periods in Figure 7.8, the chlorine concentration and the water temperature (non-conservative constituents) gradually drop. The decrease in free chlorine indicates the arrival (via the bleed at 0.6 L/m) of progressively "older" water which has lost some disinfectant residual due to reactions with oxidizable material while in transit through the pipe network. The decrease in water temperature implies that the pipe network is losing heat to the ground (recall Figure 7.8 is based on data measured in early March).

During the summer months, the heat flow and, hence, the temperature trajectory are reversed. Hence, when the pumps are on in the summer, the water temperature at the BPS will decrease as cooler water arrives from the treatment plant. Conversely, when the pumps are off, the water temperature will gradually rise as the pipe network gains heat from the ground. These seasonal patterns of winter heat loss and summer heat gain during the intervals between BPS pumping periods are clearly visible in Figure 7.9. In March (top graph), the water temperature at the BPS drops when the pumps are idle; in July (bottom graph), the water temperature at the BPS rises when the pumps are idle.

Although the action of the pumps at the BPS influenced the water quality readings taken at the entrance of the dead-end loop, this effect is viewed as an artifact of normal network operation. No attempt was made to censor or modify any of the water quality data due to pump activity at the BPS.

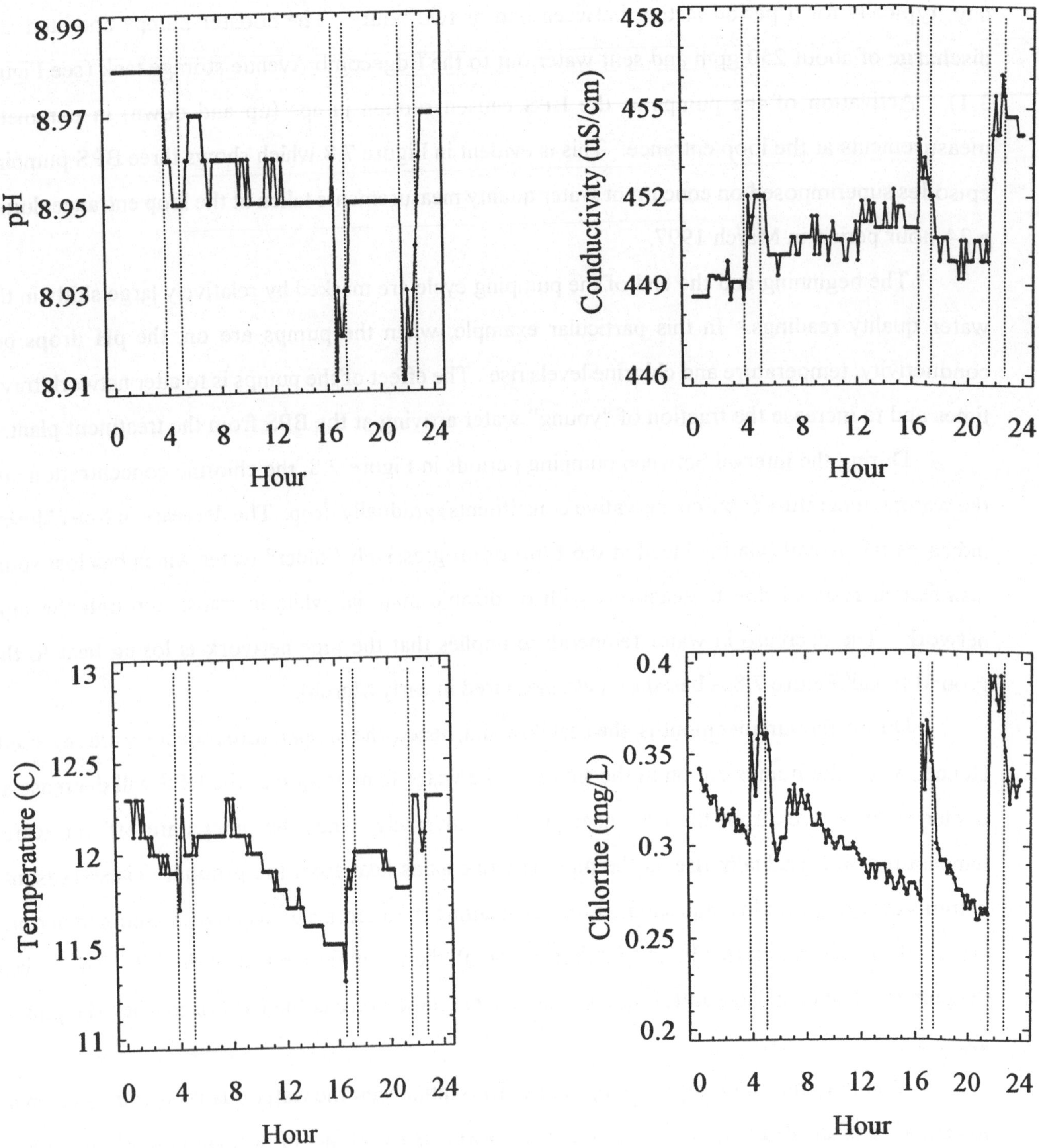

Figure 7.8 Effect of booster pump operation on readings of four physico-chemical parameters at the BPS sampling location.

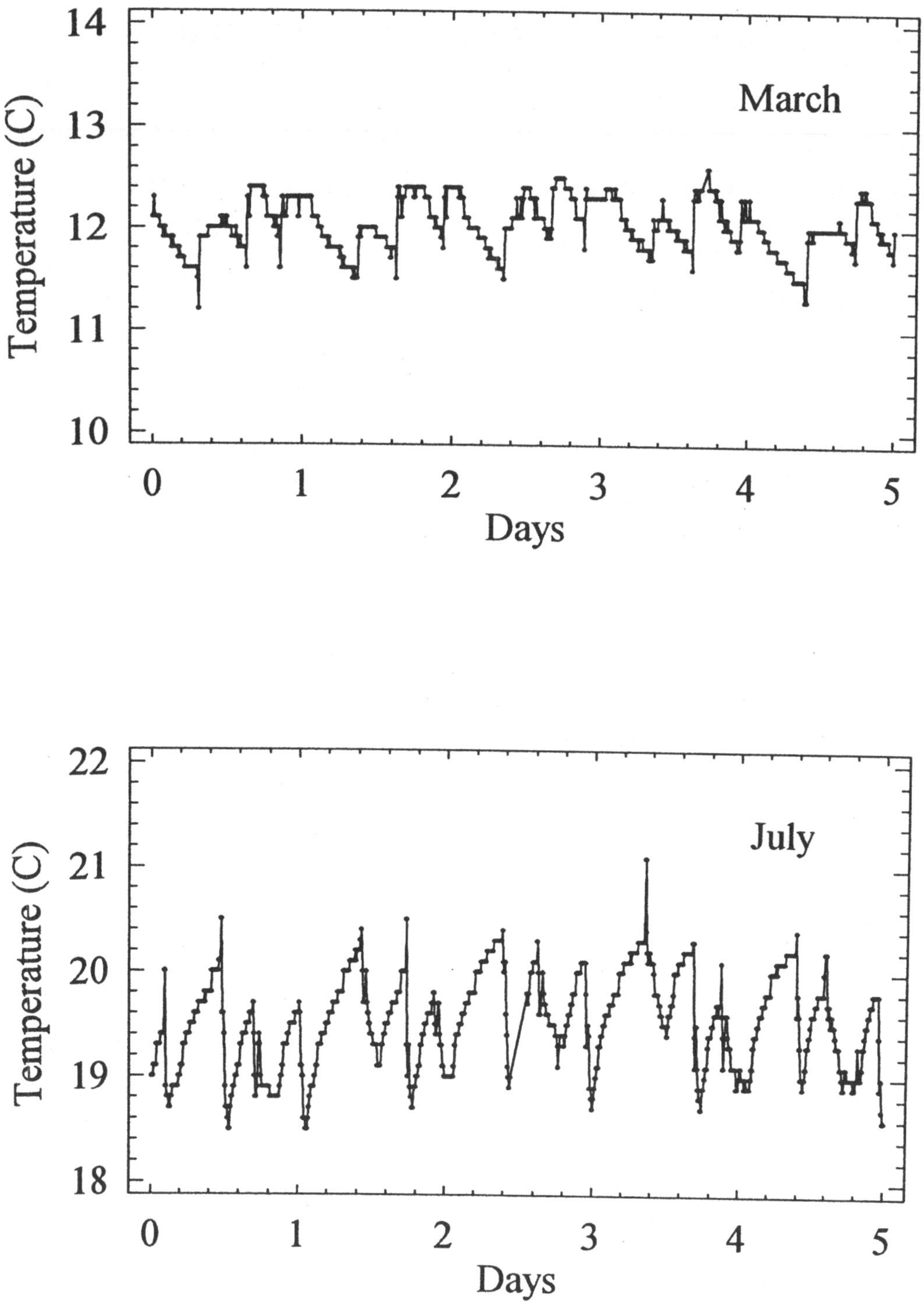

Figure 7.9 Heat flows from network to ground in cool months (top) and from ground to network in warm months (bottom); abrupt changes in temperature profile correspond to booster pump activity.

Effect of Continuous Bleed

Both CMUs required a steady supply of water. To satisfy flow requirements for the chlorine analyzer, the initial bleed rate was set to 0.60 L/m (0.16 gpm). The bleed was increased to 1.0 L/m (0.26 gpm) in June 1997 when the MRDs came on-line. The bleed through the CMU at the BPS had no perceptible effect on flow conditions. However, the bleed through the CMU at the DLA strongly influenced the low flow behavior at the study site. The presence of the bleed at the apex eliminated intermittent stagnant conditions and reduced travel times through the loop. The bleed also maintained a continuous (though unsteady) laminar flow regime throughout the pipe loop[(5)].

The bleed at the DLA was split along both branches of the loop. The clockwise (left branch) travel distance to the DLA is about 775 feet; the counter-clockwise (right branch) travel distance is about 1000 feet. From Equation 5.5 (or using Figure 5.4 with laminar flow), roughly 56% of bleed supply would take the short left route while the balance (44%) would take the long right route.

The steady bleed of 0.26 gpm translates into a water demand of 380 gallons per day. This corresponds to the daily consumption expected from two additional homes on the loop. While the DLA bleed reduced travel time, the magnitude of the reduction was not uniform across the loop.

Using the equivalent dead-end stem concept (see Chapter 5), it is possible to estimate the reduction in travel time induced by the steady bleed at the loop apex. Suppose the number of connections on a dead-end stem increases from N-1 to N with the addition of a new home at the downstream terminal point (representing the steady bleed). From Eq 5.15, it follows that the percent reduction in cumulative travel time through the first m links of the dead-end main is:

$$\text{Percent reduction} = 1 - \frac{\text{H.M.}[N-m, N-1]}{\text{H.M.}[N-m+1, N]} = 1 - \frac{\sum_{k=N-m+1}^{N} \left(\frac{1}{k}\right)}{\sum_{k=N-m}^{N-1} \left(\frac{1}{k}\right)} \qquad (7.1)$$

---

(5) As indicated in Table 5.1, even the maximum bleed rate of 0.26 gpm was well within the limits of laminar flow for the 6-inch supply main. During the busiest periods, the flow would have migrated into the transition regime and on rare occasions may have approached turbulent conditions in links near the loop entrance (see also Figure 5.2).

Table 7.5 Effect of Continuous Bleed on Dimensionless Mean Travel Time Through Links of Equivalent Dead-End Stem in a Homogeneous Neighborhood.

(1) Link Number (m)	(2) Dimensionless Cumulative Travel Time (N=10 homes)	(3) Dimensionless Cumulative Travel Time (N=11 Homes)	(4) Percent Reduction in Cumulative Travel Time Due to Continuous Bleed
1	0.100	0.091	9.0
2	0.211	0.191	9.5
3	0.336	0.302	10.1
4	0.479	0.427	10.9
5	0.646	0.570	11.8
6	0.846	0.737	12.9
7	1.096	0.937	14.5
8	1.429	1.187	16.9
9	1.929	1.520	21.2
10	2.929	2.020	31.0
11	na	3.020	na

Notes: Col (1): link m=1 is closest to the dead-end entrance (see also col 1 in Table 5.7).
Col (2) is from Eq 5.15 with N=10 (without DLA bleed).
Col (3) is from Eq 5.15 with N=11 (with active DLA bleed).
Col (4) is 1 – [col(3)/col(2)]. Results can be obtained from Equation 7.1 using N=11 homes.

where in Equation 7.1, H.M.[a,b] is the harmonic mean of consecutive integers between a and b,

$$H.M.[a,b] = \frac{b-a+1}{\sum_{k=a}^{b}\left(\frac{1}{k}\right)}$$

Results in Table 7.5 indicate that, at homes close to the entrance (say pipe links 1 and 2), the steady bleed at the apex reduced travel times by about 10 percent. However, at the apex (near link 10), the effect of the steady bleed is to reduce loop travel times by at least 30 percent.

Effect of Autocorrelation

The CMUs recorded water quality measurements every 10 minutes. The advantage of adopting a high frequency sampling scheme is that it can detect sudden changes in water quality more readily than a low frequency scheme. One disadvantage, however, is that consecutive readings for any particular constituent are not independent. A measurement at one instant will likely be close in value to the previous measurement. This persistence in the data set is clear in Figures 7.1 through 7.4 (see especially conductivity and temperature).

The autocorrelation functions (ACF) shown in Figure 7.10 for *hourly* values of pH, conductivity, temperature and chlorine illustrate the strength of the memory in each data set. If these water quality data were independent over time, the ACF would be close to zero for all positive lags. Assuming that the data follow a first-order autoregressive process, the equivalent number of independent measurements, N_E, in a series of N observations is given by (Hirtzel *et al*, 1982)

$$N_E = \frac{N^2(1-\rho_1)^2}{N(1-\rho_1^2)-2\rho_1(1-\rho_1^N)} \tag{7.2}$$

where ρ_1 is the lag-1 autocorrelation coefficient[6]. For large N, this expression can be written

$$N_E \approx N\left(\frac{1-\rho_1}{1+\rho_1}\right) \tag{7.3}$$

Typical values of the lag-1 autocorrelation coefficient for water quality data recorded at the dead-end loop ranged from ρ_1=0.975 to 0.999, depending on the constituent and the season. Adopting a representative value of ρ_1=0.99, Eq 7.3 gives $N_E \approx 0.005N$. With N=32,000, this gives an equivalent independent sample size of 160, or less than one independent reading per day.

---

[6] The sample estimate of the lag-1 autocorrelation coefficient for a series of observations x_i is,

$$\rho_1 = \frac{\sum_{i=1}^{N-1} x_i x_{i+1} - \frac{1}{N-1}\left(\sum_{i=1}^{N-1} x_i\right)\left(\sum_{i=1}^{N-1} x_{i+1}\right)}{\sqrt{\sum_{i=1}^{N-1} x_i^2 - \frac{1}{(N-1)^2}\left(\sum_{i=1}^{N-1} x_i\right)^2}\ \sqrt{\sum_{i=1}^{N-1} x_{i+1}^2 - \frac{1}{(N-1)^2}\left(\sum_{i=1}^{N-1} x_{i+1}\right)^2}}$$

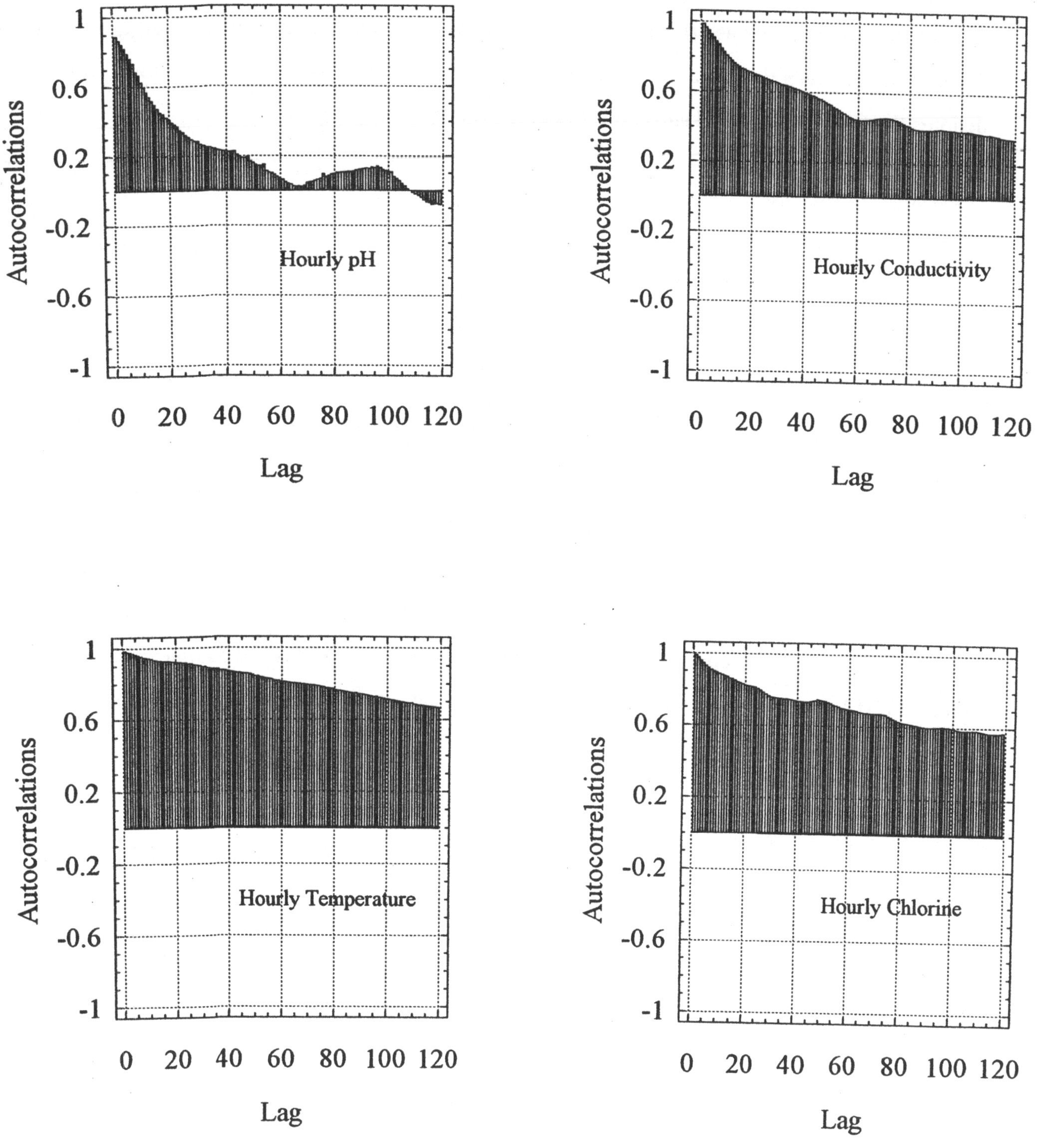

Figure 7.10 Hourly readings of four physico-chemical water quality parameters; each exhibits strong autocorrelation in the time series of observations.

Spatial Variability of CMU Data

Using equivalent independent sample sizes, several classical statistical tests were performed to determine whether or not the mean and the variance of each water quality constituent changed significantly between the entrance and the apex of the dead-end loop. In what follows, subscript 1 denotes the BPS location and subscript 2 denotes the DLA location.

Differences in Means

The null hypothesis, H_0: $\mu_1 = \mu_2$, versus the alternative, H_A: $\mu_1 \neq \mu_2$ is tested with the statistic

$$T^* = \frac{\Delta\overline{X}}{S_{\Delta\overline{X}}} \tag{7.4}$$

where $\Delta\overline{X} = \overline{X}_1 - \overline{X}_2$ is the difference in sample means between the BPS and the DLA; the denominator is a pooled standard deviation of differences between means, given by

$$S_{\Delta\overline{X}} = \sqrt{\frac{N_1 + N_2}{N_1 N_2 (N_1 + N_2 - 2)}\left[(N_1 - 1)S_1^2 + (N_2 - 1)S_2^2\right]}$$

where N_1 and N_2 are the equivalent independent sample sizes and S^2 is the sample variance. H_0 is rejected if $|T^*| > t_{1-p/2,\,\nu}$ where t is a variate from Student's-t distribution with $\nu = N_1 + N_2 - 2$ degrees of freedom and significance level, p.

Differences in Variances

The null hypothesis, $H_0: \sigma_1^2 = \sigma_2^2$, versus the alternative, $H_A: \sigma_1^2 \neq \sigma_2^2$ is tested with the ratio

$$F^* = \frac{S_1^2}{S_2^2} \tag{7.5}$$

where S_1^2 is the larger sample variance. H_0 is rejected if $F^* > F_{1-p,\,N1-1,\,N2-1}$ where F is a variate from an F distribution with N_1-1 and N_2-1 degrees of freedom. Results for both tests, summarized in Table 7.6, show that only the pH remained unchanged between the entrance and the apex of the loop.

Table 7.6 Tests for Differences in CMU Water Quality Parameters Measured at the Booster Pump Station and at the Dead-end Loop Apex.

	pH (su)	Conductivity (µS/cm)	Temperature (C)	Chlorine (mg/L)
lag-1 autocorrelation, ρ_1	0.99	0.99	0.99	0.99
approximate sample size, N	32,000	32,000	32,000	27,000
equivalent sample size, N_E	160	160	160	135
Student's t-Test for Differences in Means (p=1%) (H_0: $\mu_{DLA} = \mu_{BPS}$ versus H_A: $\mu_{DLA} \neq \mu_{BPS}$)				
Degrees of Freedom, ν	318	318	318	268
Mean at BPS	9.046	478.6	16.65	0.265
Mean at DLA	8.965	423.0	17.98	0.181
Difference in Means, $\Delta \bar{X}$	0.081	55.6	-1.33	0.084
Stan Dev of Diff, $S_{\Delta X}$	0.032	6.882	0.386	0.014
Test Statistic (T*, Eq 7.4)	2.52	8.08	-3.45	5.90
Critical T Statistic	2.58	2.58	2.58	2.59
Decision	**accept H_0**	reject H_0	(*)	reject H_0
F-Test for Differences in Variances (p=1%) (H_0: $\sigma^2_{DLA} = \sigma^2_{BPS}$ versus H_A: $\sigma^2_{DLA} \neq \sigma^2_{BPS}$)				
Degrees of Freedom, ν	159	159	159	134
Variance at BPS	0.096	6,399	7.254	0.019
Variance at DLA	0.069	1,178	16.56	0.009
Test Statistic (F*, Eq 7.5)	1.39	5.43	2.28	2.09
Critical F Statistic	1.45	1.45	1.45	1.50
Decision	**accept H_0**	reject H_0	reject H_0	reject H_0

(*) Some care is needed to interpret the t-test of the temperature data. The absolute value of the test statistic ($|T^*|=3.45>2.58$) indicates that there is indeed a significant difference in the mean water temperature between the entrance and the apex of the dead-end loop. However, owing to the seasonal cycle in the temperature data, this conclusion is valid only for the 256-day period of observation. It is expected that over the course of a full year, there would not be a statistically significant difference in mean water temperatures between the entrance and apex of the loop. Note, however, that the difference in the variance of the water temperatures would persist even with data from the full year.

Correlations with Travel Time

Water age often is viewed as a surrogate measure of water quality. With this in mind, the intent here was to check for the presence of significant correlations between travel times through the loop and corresponding changes in water quality at the entrance and the apex. Travel times of individual water parcels that had reached the DLA were found using a "back-tracking" method similar in concept to the "forward-tracking" scheme described in Chapter 5. This step provided the times at which a particular parcel entered and exited the dead-end loop. Then, knowing these clock times, the corresponding change in water quality could be estimated by comparing the parameter measurements made at the BPS and the DLA.

In principle, this was a straightforward exercise; however, in practice, this proved to be very challenging. Three main problems were encountered:

(1) **Booster pump station** – The BPS presented two complications. First, it was located 120 feet downstream from the loop entrance (see Figure 3.2). Second, the pumps were idle about 75% of the time. When the pumps were active, water reaching the BPS and water drawn into the dead-end loop were indistinguishable. When the pumps were idle, water still flowed into the dead-end loop, but water in the 8 inch mainline at the BPS was essentially stagnant (only flow was CMU bleed of 1 L/m). Water quality measurements made under nearly stagnant conditions at the BPS may not be representative of the flow drawn into the dead-end loop. To compensate, attempts were made to analyze only the water quality data corresponding to active pumping periods at the BPS.

(2) **Missing water demands** – Water demand recovery ranged from 90 to 95% (see Chapter 4). Nonetheless, 5 to 10% missing demand data translates into 1 or 2 hours per day when the flow pattern and, hence, the flow path at the loop is unknown. Uncertainty in the flow path (left branch or right branch) can introduce large errors in the train of water parcels that arrive at the apex.

(3) **Continuous bleed at apex** – The CMU at the apex had a continuous bleed ranging from 0.6 to 1.0 L/m. The bleed effectively replaced intermittent stagnant flow with unsteady continuous laminar flow throughout the loop. Well developed laminar flow in a pipe implies Taylor-type dispersion which would encourage mixing between adjacent parcels. This effect was not considered in the travel time analysis.

Chlorine was the prime candidate for water quality tracking because it changed dramatically between the entrance and the apex (see Figure 7.4). Several attempts were made to track water parcels through the loop and identify changes in chlorine between the entrance and the apex. While each had some degree of success, no definitive reproducible correlation between travel time and chlorine residual could be established for *individual parcels* moving through the loop. Owing to the uncertainty in water quality inputs at the loop entrance, to questions about residential water demands, and to changing flow regimes across the loop, the precise correspondence between parcels at the entrance and parcels at the apex remained elusive.

It was, however, possible to relate the change in the overall mean chlorine concentration at the entrance (0.27 mg/L) and the apex (0.18 mg/L) to the mean travel time. Using the forward-tracking method, the mean travel time from the entrance to the apex was about 30 hours during Summer 1997. This is considerably less than the estimate of about 77 hours during Winter 1996 (see Tables 5.5 and 5.6). The decrease in mean travel time resulted from increasing flows attributed to three factors: (i) the number of occupied homes on the loop grew from 18 in 1996 to 21 in 1997, (ii) the continuous bleed at the apex added a level of demand equivalent to two additional homes on the loop and (iii) water demands now included seasonal outdoor use.

Using a first-order model with reaction rate constant, k= 0.01/hr and a mean travel time of 30 hours leads to a predicted mean chlorine concentration at the apex of 0.20 mg/L[7]. This is reasonably close to the observed mean value of 0.18 mg/L. Over-prediction could imply the mean travel time to apex exceeds 30 hours or the decay constant exceeds 0.01/hr or both. Owing to wall demand (not considered here) the overall decay rate could increase, though this component is not large under laminar flow conditions.

---

(7) The value of the decay constant is from footnote 4 in this Chapter.

$$\frac{C_{DLA}(t)}{C_{BPS}(0)} = \exp[-kt]$$

so

$$C_{DLA}(30\ \text{hr}) = (0.27\ \text{mg / L})\ \exp[-(0.01\ /\ \text{hr})\ (30\ \text{hr})] = 0.20\ \text{mg / L}$$

Summary of Continuous Monitoring

Two monitoring units (see Figure 3.4) provided nearly continuous readings of eight drinking water quality parameters: pH, conductivity, temperature, chlorine, turbidity, nitrate, DO and ORP. Both units operated from March 8, 1997 to November 18, 1997 near the entrance and the apex of a residential dead-end loop in the water distribution system of the City of Milford, Ohio. With readings taken every 10 minutes, about 245,000 water quality readings were obtained during this 9 month observation period.

Three of the probes were found to be unreliable and therefore were excluded from the data analysis. Turbidity readings were confounded by the presence of excess lime in the water main. The nitrate probes could not maintain a consistent calibration under the continuous flow operation. The ORP probes gradually lost their sensitivity. In addition, the dissolved oxygen probes expired during the early phase of the continuous flow monitoring and, therefore, DO was excluded from the data analysis.

It was hoped that ORP readings would serve as a convenient surrogate for the concentration of free chlorine. Initially, ORP readings did appear to correlate with *large* fluctuations in free chlorine; however, the relatively low sensitivity of the ORP probe coupled with its progressive loss of resolution effectively eliminated ORP as a viable alternate measure of disinfectant residual in this distribution system.

The remaining four probes (pH, conductivity, water temperature and free chlorine) consistently provided reliable readings during the continuous flow operation. Data recovery rates ranged from 73 to 85 percent. On average, each probe generated over 30,000 observations at both monitoring locations (the entrance and the apex of the loop). The high frequency sampling scheme detected several water quality excursions, one of which was attributed to replacement of a water main upstream from the dead-end loop. Most water utilities use a grab sample approach. However, owing to their sporadic nature, grab samples would be unlikely to detect a real-time excursion and, hence, may not accurately characterize water quality variability in the distribution system.

The continuous bleed maintained by the monitoring unit at the dead-end apex had a significant influence on the flow conditions at the study site. The bleed effectively eliminated intermittent stagnation, the distinguishing feature of dead-end flow. Instead a continuous (but unsteady) laminar flow prevailed across the loop. The daily demand induced by the bleed was equivalent to the water consumption expected from two additional homes on the loop. This additional flow reduced travel times from the entrance to the apex of the loop by at least 30 percent.

Owing to extremely high autocorrelation in the time series of water quality measurements (ρ_1>0.99), the equivalent number of *independent* observations is estimated to be only about 0.5 percent of the gross sample size (see Eq 7.3). Based on the equivalent number of independent observations, several statistical tests were performed to check for significant differences in the mean and the variance of water quality parameters across the loop.

The distribution of pH did not change between the entrance and the apex of the loop. However, there were statistically significant decreases in the mean and the variance of conductivity and chlorine. The drop in the mean conductivity between the entrance and the apex is unexplained, but it implies a loss of dissolved ions along the loop. The decrease in the variability of conductivity is likely due to mixing that accompanies bi-directional water movement and frequent flow reversals in the dead-end loop.

Chlorine is reactive. Thus, reductions in free chlorine concentration are expected, particularly considering that the travel time from the entrance to the apex ranges from hours to several days or more. Moving from the entrance to the apex, the mean chlorine concentration dropped 33 percent, from 0.27 mg/L to 0.18 mg/L. Using a mean travel time of 30 hours and a first-order reaction rate constant of k=0.01/hr (measured from bottle tests), the predicted concentration at the apex is 0.20 mg/L. Discrepancies between observed and predicted mean chlorine concentrations can be attributed to uncertainties in the travel time, in the reaction rate, and the reaction kinetics. The rate of chlorine loss increased with water temperature.

Water temperature variations were strongly affected by seasonal weather trends. Vigorous heat exchange occurred in the loop. During the winter, the pipe network lost heat to the ground and hence water temperatures dropped. Conversely, during the summer, the network gained heat and temperatures rose. Because most of the water quality data were collected during the warmer part of the year, the mean water temperature tended to increase between the loop entrance and the apex. Over a full annual cycle, however, it is expected that the mean water temperature would not change significantly across the loop.

GRAB SAMPLES

Data Recovery

Weekly grab samples were collected at four locations in the distribution system (WTP-water treatment plant, WST-Wallace Street tank, BPS-booster pump station and DLA-dead-end loop apex; see Figure 3.1). These samples were analyzed for HPC, TC, AOC and TOC according to the procedures outlined in Chapter 3. The tests for total coliforms failed to register a single positive response throughout the study. Consequently, TC is not considered further.

As summarized in Table 7.7, sample recovery rates for the period January 3, 1997 to November 18, 1997 were quite good. TOC achieved 100 percent data recovery, while AOC and HPC approached 85 and 88 percent recovery, respectively. The two primary reasons for gaps in the grab sample data are: (i) the sampling location was temporarily inaccessible or (ii) the sample was mishandled in transit. All results from the lab analysis for HPC (162 values by R2A media), AOC (78 values) and TOC (119 values) are listed in Appendix 7.2. Full period statistics for HPC, AOC and TOC are summarized in Tables 7.8 through 7.10, respectively.

Heterotrophic Plate Counts

HPCs are an indirect measure of the number of viable microbes in a bulk fluid sample. When applied to a water distribution system, the HPC reflects those organisms that have survived treatment, or that have emerged via regrowth or that have exploited other network vulnerabilities (e.g., careless repairs, cross connections, transitory contamination). The numbers of bacteria persisting along a pipe network depend on nutrient levels and disinfectant residual. Generally, bacterial numbers tend to increase with distance (i.e., travel time) from the water treatment plant.

The median HPC value obtained from 162 samples at four locations was 180 CFU/mL. This value is low compared to readings that range from 1 to over 10,000 CFU/mL in other water distribution systems (Reasoner and Rice, 1993).

The HPC count varied dramatically with location in the Milford distribution system (Figure 7.11). Finished water at the treatment plant had the lowest HPC levels with a median of 50 CFU/mL (Table 7.11). Because Milford's supply source is groundwater with low levels of natural organic matter that would serve as nutrients for HPC bacteria, it is reasonable to conjecture that the HPC counts would remain low throughout the network. Surprisingly, however, median HPC levels rose

Table 7.7 Data Recovery Matrix for Grab Samples

Location and Parameter	Jan 1234	Feb 1234	Mar 1234	Apr 1234	May 1234	June 1234	July 12345	Aug 1234	Sept 1234	Oct 12345	Nov 123
WTP											
HPC	0000	XX00	0000	000X	0000	0000	0X000	0000	0000	0000	000
AOC	0 0	0 X	0 0	0 0	0 00	X 0	X 0	0 X	X 0	0 00	000
TOC	X 0	0 00	0000	000	0 00	0000	0 00	0 X	X 0	0 00	X 0
WST											
HPC	X00X	XX0X	0000	000X	0000	0000	XX000	0000	0000	0000	000
AOC	X 0	0X	00	00	000	X0	00	0 X	X 0	0 00	000
TOC	X 0	00	0000	000	000	0000	00	0 X	X 0	0000	X 0
BPS											
Biofilm							00	0000	0000	0000	000
HPC	00X0	XX00	0000	000X	0000	0000	0X000	0000	0000	0000	000
AOC	0 0	X 0	0 0	0 0	0 00	X 0	0 0	0 X	X 0	0 00	000
TOC	X 0	X 0	0 0	000	0 00	0000	X 00	0 X	0 0	0 00	X 0
DLA											
Biofilm							00	0000	0000	0000	000
HPC	0000	XX00	0000	000X	0000	0000	0X000	0000	000X	0000	000
AOC	0 0	X 0	0 0	0 0	0 00	X0	0 0	0 0	X 0	0 00	000
TOC	X 0	X 0	0 0	0 0	0000	0000	0	0 X	0 0	0 00	X 0

Legend
O Usable data
X Missing data (see Appendix 7.2)

Table 7.8 Summary Statistics for HPC (on R2A) January to November 1997

Variable	WTP	WST	BPS	DLA
Sample size	42	38	41	41
Mean (mg/L)	221.7	498.2	900.7	269.5
Median (mg/L)	50	330	410	100
Mode (mg/L)	20	200	150	70
Standard deviation (mg/L)	593.5	627.0	1537.1	937.7
Minimum (mg/L)	1	1	10	10
Maximum (mg/L)	3800	3100	8600	6100
Skewness	5.62	3.08	3.64	6.31
Coefficient of Variation	2.68	1.26	1.71	3.48

Table 7.9 Summary Statistics for AOC January to November 1997

Variable	WTP	WST	BPS	DLA
Sample size	19	19	20	20
Mean (mg/L)	0.088	0.191	0.144	0.147
Median (mg/L)	0.08	0.134	0.105	0.09
Mode (mg/L)	0.08	0.07	0.09	0.10
Standard deviation (mg/L)	0.075	0.196	0.137	0.201
Minimum (mg/L)	0.002	0	0.01	0.01
Maximum (mg/L)	0.30	0.88	0.64	0.96
Skewness	1.44	2.74	2.68	3.83
Coefficient of Variation	0.85	1.03	0.95	1.37

Table 7.10 Summary Statistics for TOC January to November 1997

Variable	WTP	WST	BPS	DLA
Sample size	30	28	31	30
Mean (mg/L)	0.801	0.850	0.837	0.798
Median (mg/L)	0.80	0.80	0.83	0.76
Mode (mg/L)	0.80	1.12	0.80	0.86
Standard deviation (mg/L)	0.143	0.171	0.175	0.186
Minimum (mg/L)	0.56	0.56	0.54	0.48
Maximum (mg/L)	1.12	1.13	1.39	1.19
Skewness	0.52	0.41	1.01	1.67
Coefficient of Variation	0.18	0.20	0.21	0.23

to 330 CFU/mL at the Wallace Street tank, less than a half mile from the water treatment plant (Table 7.11), most likely because the tank allows water to be nearly stagnant for a log period.

The WST is an elevated reservoir fed by a single fill and drain pipe. Hence, the last water added when the tank is filling will be the first water removed when the tank is draining. The WST holds about 650,000 gallons and, for fire protection needs, the water surface is confined to the top five feet of the tank. The long residence time coupled with weak mixing allows the disinfectant residual to diminish to levels that may not inhibit microbial growth. To some degree, this may explain the sharp increase in bacterial numbers observed between the treatment plant and the storage tank.

The HPC median reached 410 CFU/mL at the BPS, the maximum level observed among the four locations (Tables 7.8 and 7.11). This high value might result from the piping arrangement and pumping schedule at the BPS. There was often no flow through the 8-inch mainline at the BPS[(8)]. When the BPS pumps were activated (about four times per day), the flow jumped to 250 gpm and velocity surged from 0 to 1.6 ft/s. These periodic sudden changes in hydraulic conditions may have scoured biofilm and other material from the inside wall of the pipe. The BPS had many elbows, valves and other appurtenances which could create stagnant pockets offering a protective niche to shield microbes from disinfectant residual. In addition, small amounts of substrate material may have leached from the rubber fittings around the pumps. All of these factors may have combined to elevate the HPC count at the BPS, even though the chlorine residual usually exceeded 0.2 mg/L.

Surprisingly, the median HPC dropped to 100 CFU/mL at the apex of the dead-end loop. Because the loop entrance was 120 feet *upstream* from the booster pump station, water diverted into the loop may not have experienced the local conditions that encouraged high HPC values at the BPS. In fact, aside from one summer excursion (DLA HPC levels reached 6100 CFU/mL on June 11, 1997) the HPC distribution at the DLA more closely resembles the HPC distribution of the finished water at the treatment plant than any other sampling site (see Figure 7.11).

Seasonal variations of HPC at a given site tended to be less pronounced than spatial variations in HPC among sites (Table 7.14). Interestingly, summer coincided with lowest median HPC level at the WTP, but with the highest HPC medians at the BPS and the DLA. Time series graphs and box-whisker plots illustrating these trends appear in Figures 7.12 and 7.13.

[(8)] The booster station pumps were idle up to 16 hours per day. During idle periods, the only flow through the 8-inch line at the BPS was the 0.26 gpm bleed at the water quality CMU.

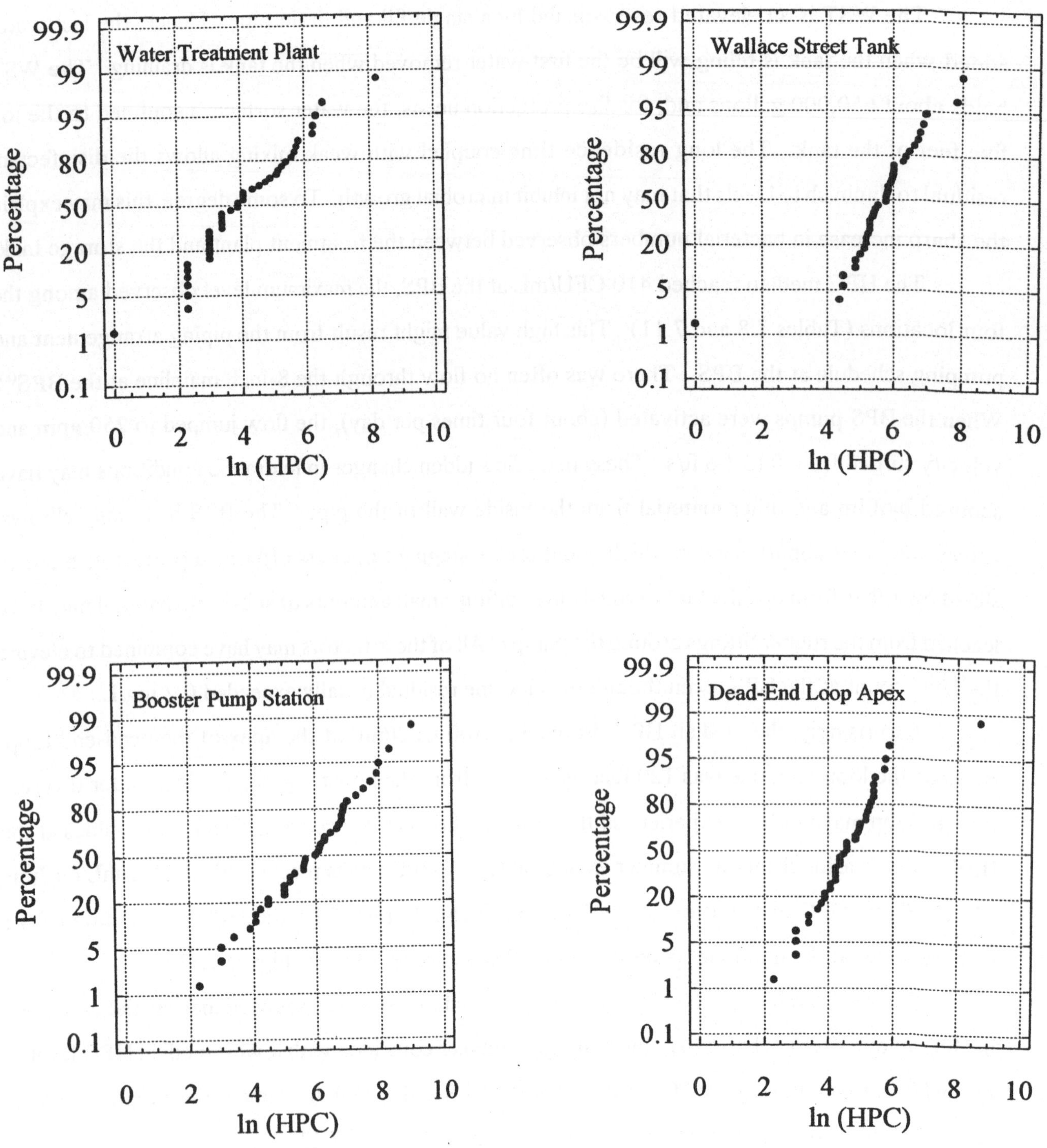

Figure 7.11 Cumulative distribution of HPC at four sampling locations.

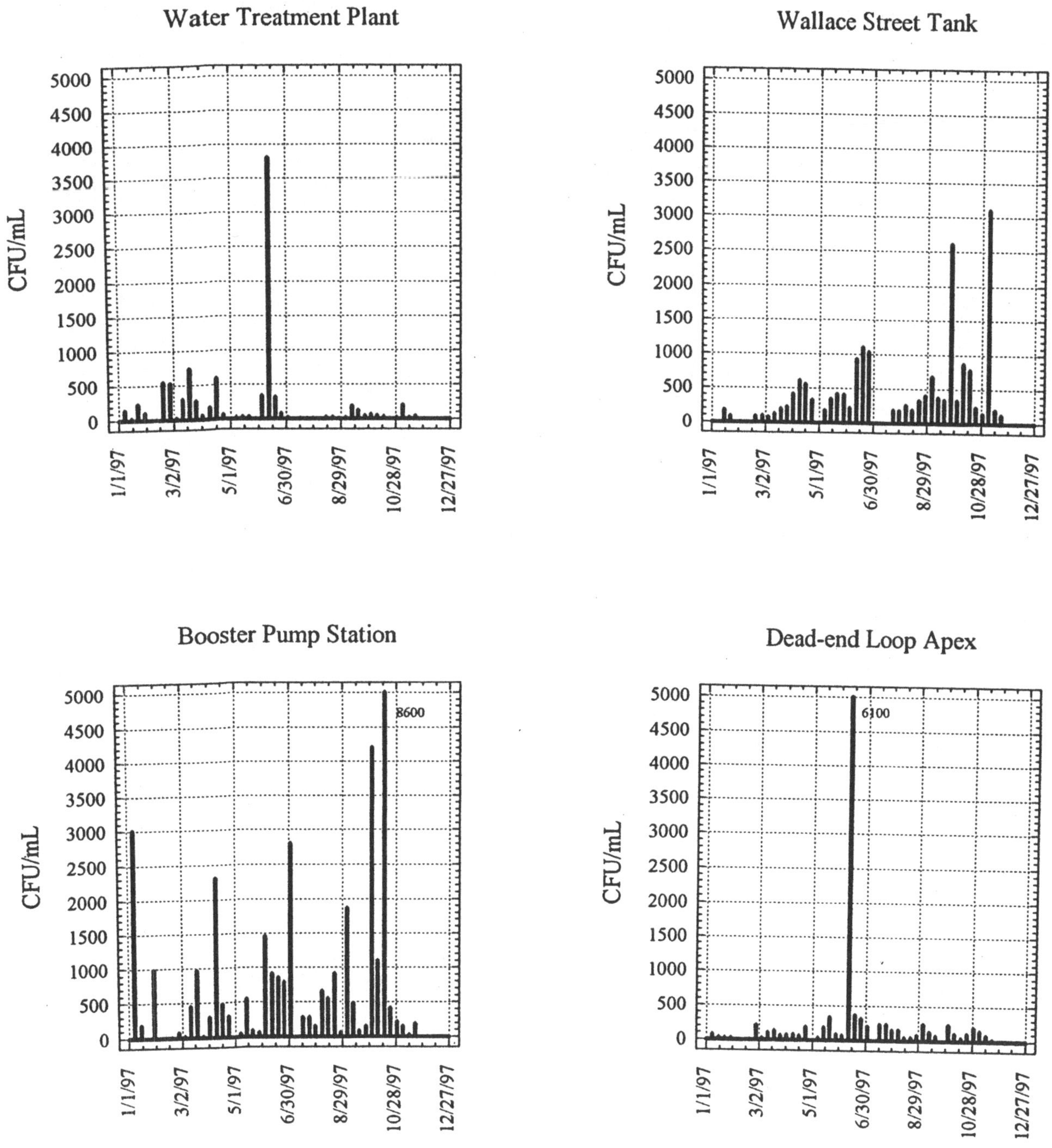

Figure 7.12 Time series of HPC at four sampling locations.

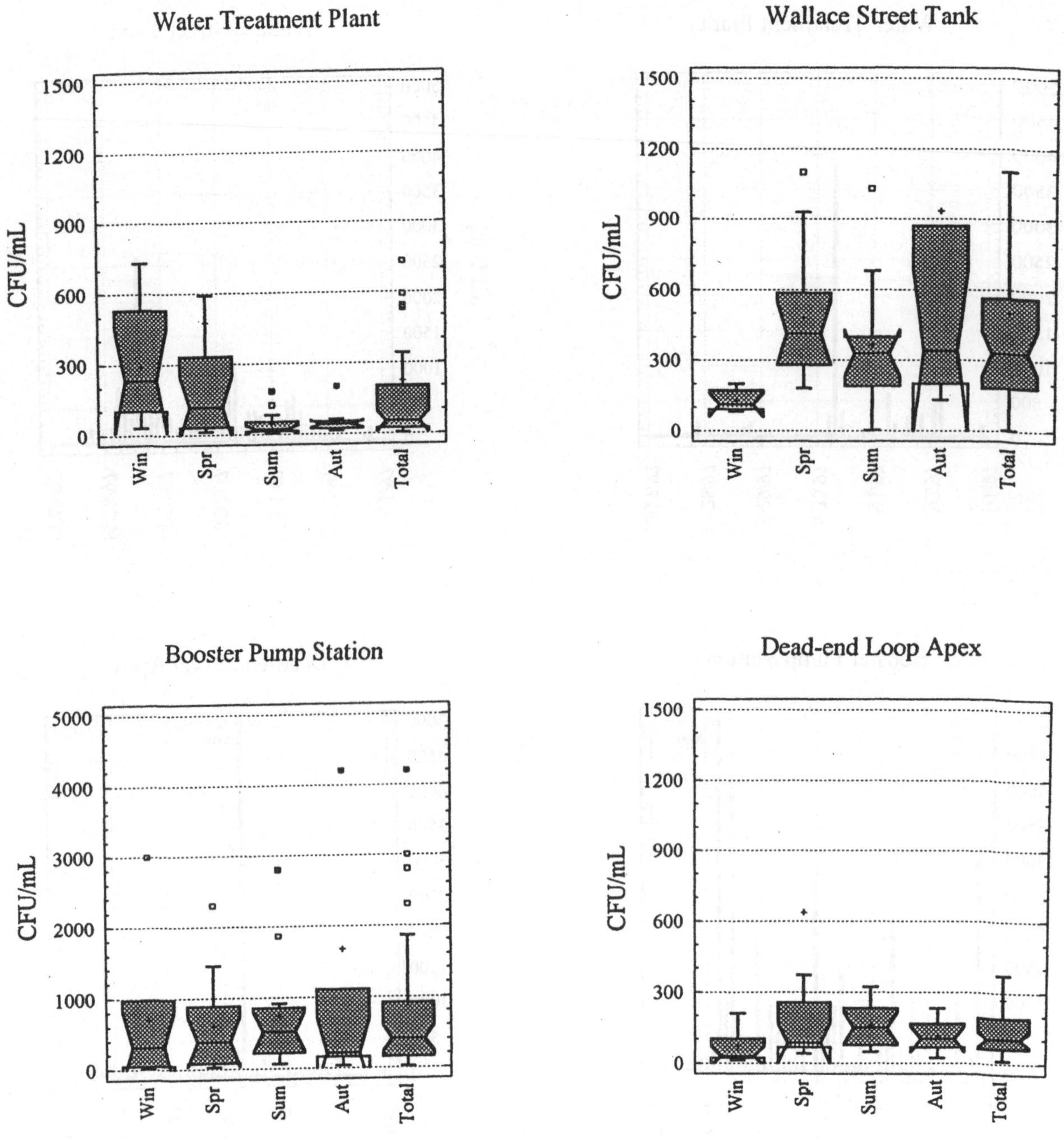

Figure 7.13 Seasonal box-whisker plots of HPC at four sampling locations.

Table 7.11
Seasonal statistics for heterotrophic plate count (on R2A).

Location / Statistic	Winter	Spring	Summer	Autumn	Annual
WTP: Water Treatment Plant					
Sample Size	9.00	12.00	12.00	9.00	42
Mean (CFU/mL)	292.22	475.00	41.75	53.33	221.7
Median (CFU/mL)	230.00	115.00	20.00	40.00	50.0
Stand Dev (CFU/mL)	252.77	1062.15	55.83	57.45	593.5
Minimum (CFU/mL)	30.00	10.00	1.00	10.00	1
Maximum (CFU/mL)	730.00	3800.00	180.00	200.00	3800
Skewness	0.65	3.30	1.84	2.53	5.62
Coef of Variation	0.87	2.24	1.34	1.08	2.68
WST: Wallace Street Tank					
Sample Size	6.00	12.00	11.00	9.00	38
Mean (CFU/mL)	126.67	480.00	363.64	934.44	498.2
Median (CFU/mL)	110.00	415.00	330.00	340.00	330
Stand Dev (CFU/mL)	49.26	283.13	279.40	1125.23	627.0
Minimum (CFU/mL)	80.00	180.00	1	130.00	1
Maximum (CFU/mL)	200.00	1100.00	1030.00	3100.00	3100
Skewness	0.69	1.28	1.47	1.45	3.08
Coef of Variation	0.39	0.59	0.77	1.20	1.26
BPS: Booster Pump Station					
Sample Size	8.00	12.00	12.00	9.00	41
Mean (CFU/mL)	708.75	610.83	741.67	1670.00	900.7
Median (CFU/mL)	315.00	385.00	515.00	210.00	410
Stand Dev (CFU/mL)	1007.38	685.36	814.49	2915.48	1537.1
Minimum (CFU/mL)	10.00	20.00	60.00	30.00	10
Maximum (CFU/mL)	3000.00	2300.00	2800.00	8600.00	8600
Skewness	2.04	1.56	1.84	2.13	3.64
Coef of Variation	1.42	1.12	1.10	1.75	1.71
DLA: Dead-end Loop Apex					
Sample Size	9.00	12.00	12.00	8.00	41
Mean (CFU/mL)	68.89	635.83	157.50	113.75	269.5
Median (CFU/mL)	30.00	85.00	147.50	100.00	100
Stand Dev (CFU/mL)	66.98	1724.22	87.58	70.09	937.7
Minimum (CFU/mL)	10.00	40.00	45.00	20.00	10
Maximum (CFU/mL)	210.00	6100.00	320.00	230.00	6100
Skewness	1.34	3.44	0.30	0.50	6.31
Coef of Variation	0.97	2.71	0.56	0.61	3.48

Assimilable Organic Carbon

AOC is the fraction of TOC that is readily available as substrate to bacteria. The amount of AOC in the distribution system generally decreases with increasing distance (i.e., travel time) from the treatment plant. This reduction is due largely to consumption by bacteria. AOC removal during water treatment restricts the regrowth of bacteria in the pipe network by limiting nutrients. Potential sources of AOC in a distribution system include reservoirs that may be open to birds or rodents. In addition, some oxidants (e.g., free chlorine) may convert a portion of the TOC to AOC. This reaction continues throughout the distribution system and, hence, AOC readings at the point of water treatment may not represent the AOC levels that prevail across the pipe network.

A total of 78 AOC measurements were made at the four sampling locations within the distribution system. The overall AOC median concentration was 0.10 mg-C/L. Depending on the source water and the treatment process, AOC concentrations in municipal water distribution systems have been observed to vary by a factor of 100, ranging from 0.01 to 1.0 mg-C/L (Gibbs et al., 1993; van der Kooij et al., 1995).

Cumulative distributions, time series and box-whisker plots of AOC values are shown in Figures 7.14, 7.15 and 7. 16, respectively. All AOC values were less than 1.0 mg-C/L. In contrast to the disparate HPC readings, the distribution of AOC is quite similar among the four locations (Figure 7.14). Some seasonal variations in AOC values are evident (Figure 7.16), but these fluctuations are not statistically significant (Table 7.14).

Finished water at the WTP had a median AOC of 0.08 mg-C/L, the lowest value among the four locations. The median AOC increased to 0.13 mg-C/L at the WST, highest among the sample sites. Although this increase is not statistically significant, the presence of substrate may explain the high HPC readings observed at the WST. The exact origin(s) of this AOC were not found, but potential sources might include wildlife access to a vent in the storage tank and oxidation of TOC in the finished water from the treatment plant. The median AOC dropped to 0.11 mg-C/L at the BPS. This decline in substrate follows patterns observed in other systems. Paradoxically, though, the HPC count rose between the WST and the BPS (see Table 7.8). This may support the speculation that BPS HPC values were inflated by biofilm sheared from the pipe walls during the pumping cycles at the BPS. At the DLA, the median AOC value continued to drop, reaching 0.09 mg-C/L. This reduction suggests there was some (albeit small) substrate consumption along the dead-end loop.

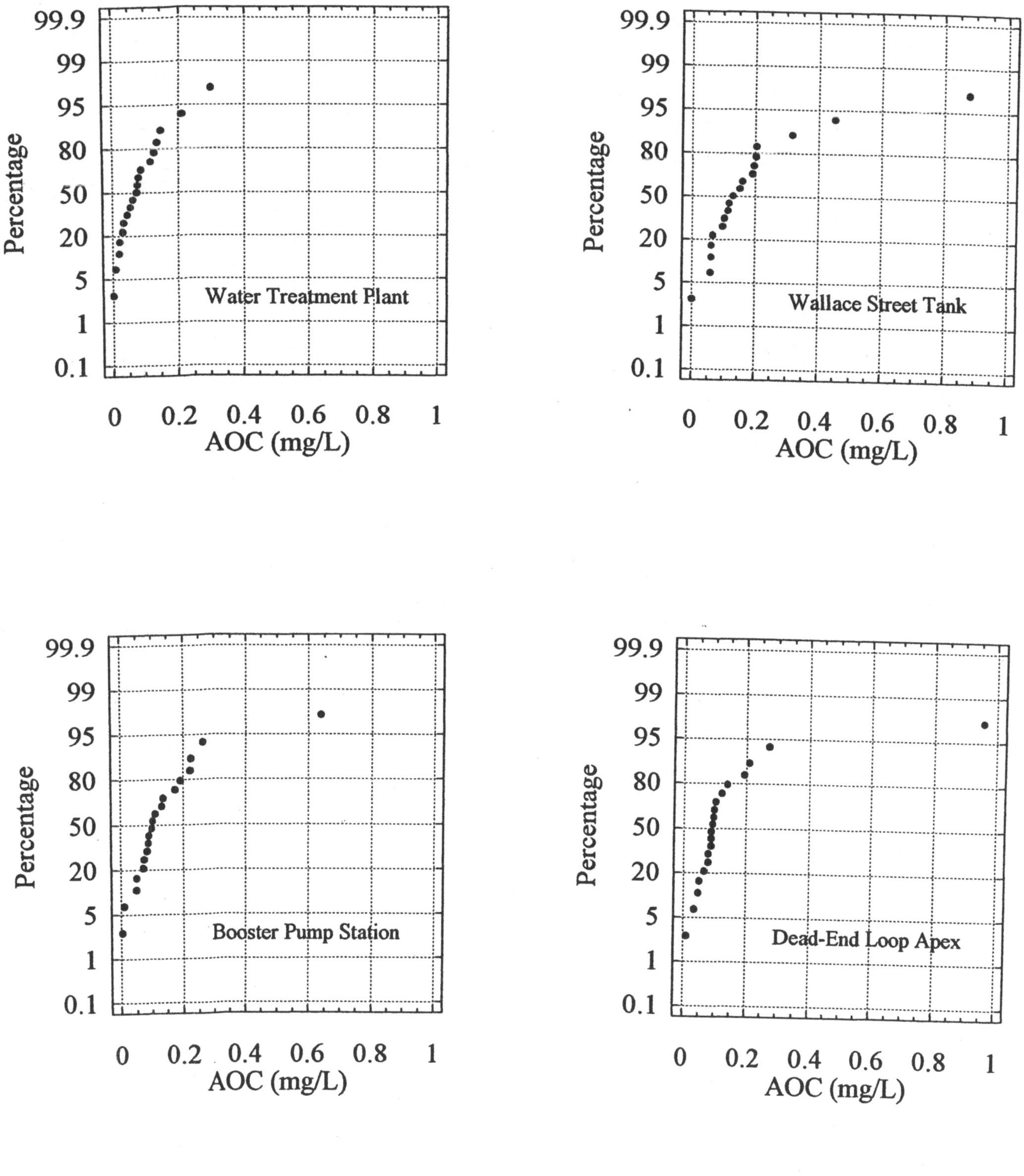

Figure 7.14 Cumulative distribution of AOC at four sampling locations.

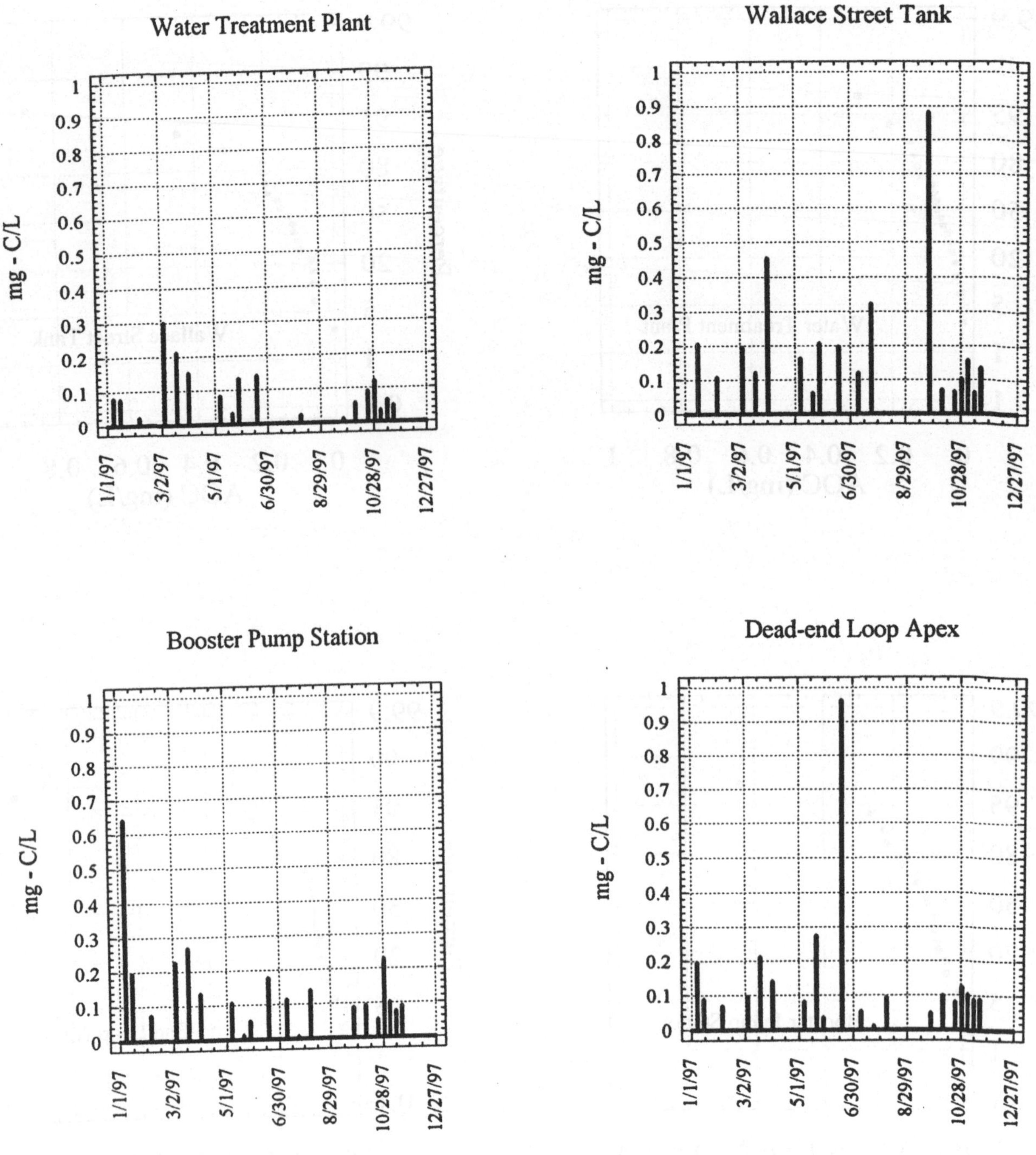

Figure 7.15 Time series of AOC at four sampling locations.

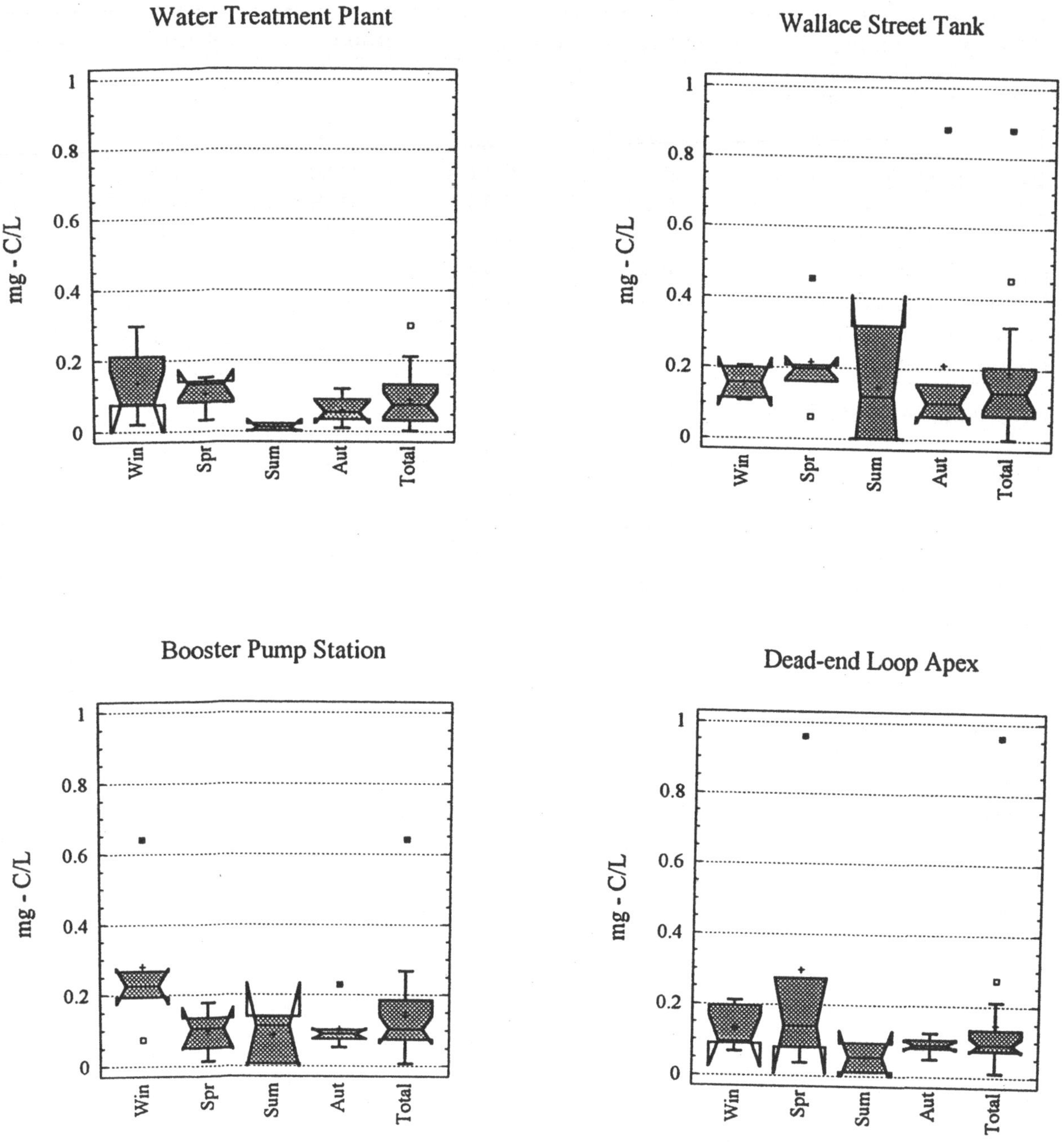

Figure 7.16 Seasonal box-whisker plots of AOC at four sampling locations.

Table 7.12
Seasonal statistics for assimilable organic carbon.

Location / Statistic	Winter	Spring	Summer	Autumn	Annual
WTP: Water Treatment Plant					
Sample size	5.00	5.00	2.00	7.00	19.00
Mean (mg-C/L)	0.14	0.11	0.01	0.06	0.09
Median (mg-C/L)	0.08	0.13	0.01	0.06	0.08
Stand Dev (mg-C/L)	0.11	0.05	0.01	0.04	0.08
Minimum (mg-C/L)	0.02	0.03	0.00	0.01	0.00
Maximum (mg-C/L)	0.30	0.15	0.02	0.12	0.30
Skewness	0.73	-1.04	na	0.41	1.44
Coef of Variation	0.79	0.46	1.00	0.67	0.89
WST: Wallace Street Tank					
Sample size	4.00	5.00	3.00	7.00	19.00
Mean (mg-C/L)	0.16	0.22	0.15	0.21	0.19
Median (mg-C/L)	0.16	0.20	0.12	0.10	0.13
Stand Dev (mg-C/L)	0.05	0.14	0.16	0.30	0.20
Minimum (mg-C/L)	0.11	0.06	0.00	0.06	0.00
Maximum (mg-C/L)	0.21	0.45	0.32	0.88	0.88
Skewness	0.10	1.25	0.72	2.57	2.74
Coef of Variation	0.31	0.64	1.07	1.43	1.05
BPS: Booster Pump Station					
Sample size	5.00	5.00	3.00	7.00	20.00
Mean (mg-C/L)	0.28	0.10	0.09	0.10	0.14
Median (mg-C/L)	0.23	0.11	0.11	0.09	0.11
Stand Dev (mg-C/L)	0.21	0.07	0.07	0.06	0.14
Minimum (mg-C/L)	0.07	0.01	0.01	0.05	0.01
Maximum (mg-C/L)	0.64	0.18	0.14	0.23	0.64
Skewness	1.54	-0.26	-1.36	2.19	2.68
Coef of Variation	0.75	0.70	0.78	0.60	1.00
DLA: Dead-end Loop Apex					
Sample size	5.00	5.00	3.00	7.00	20.00
Mean (mg-C/L)	0.13	0.30	0.05	0.09	0.15
Median (mg-C/L)	0.10	0.14	0.05	0.09	0.09
Stand Dev (mg-C/L)	0.07	0.38	0.04	0.02	0.20
Minimum (mg-C/L)	0.07	0.04	0.01	0.05	0.01
Maximum (mg-C/L)	0.21	0.96	0.09	0.12	0.96
Skewness	0.50	1.96	0.00	-0.83	3.83
Coef of Variation	0.54	1.27	0.80	0.22	1.33

Total Organic Carbon

TOC is the amount of organically bound carbon. Examples of TOC in a distribution system include organic matter, such as humic or fulvic acids, that have passed through the treatment process. Generally, groundwater sources have a low TOC content. Fortunately, laboratory analytical methods for detecting TOC are fairly accurate. Because AOC is a fraction of TOC, loss of AOC should parallel the changes TOC.

A total of 119 TOC measurements were made at four locations in the distribution system. The overall TOC median concentration was 0.80 mg/L. This value falls well in the range of typical TOC concentrations (0.1 to 2 mg/L) for distribution systems supplied by groundwater sources (Montgomery Consulting Engineers, 1985).

Cumulative distributions, time series and box-whisker plots of TOC values are shown in Figures 7.17, 7.18 and 7. 19, respectively. All TOC values fell in the narrow range 0.40 to 1.5 mg/L. Similar to the AOC readings, the distribution of TOC is quite similar among the four locations (Figure 7.17). Seasonal variations in TOC values at each location are very slight (Figure 7.19) and not statistically significant (Table 7.14). TOC levels were not static during the course of the year; however the minor variations in the measured values can be attributed to random fluctuations.

Samples from both the WTP and the WST had a median TOC value of 0.80 mg/L. This is in stark contrast to AOC which experienced a 70 percent *increase* moving from the treatment plant to the storage tank. This suggest that the rise in the AOC levels between the WTP and the WST is probably due to the oxidation of the background TOC. If there were an external source of AOC (such as access through a vent in the standpipe), then there would have been a concomitant increase in the gross TOC levels at the WST.

At the BPS the median TOC value was 0.83 mg/L; at the DLA it was 0.76 mg/L. These values are not significantly different from the median TOC values at the point of treatment. The slight decrease in TOC between the BPS and the DLA is consistent with the trend in AOC.

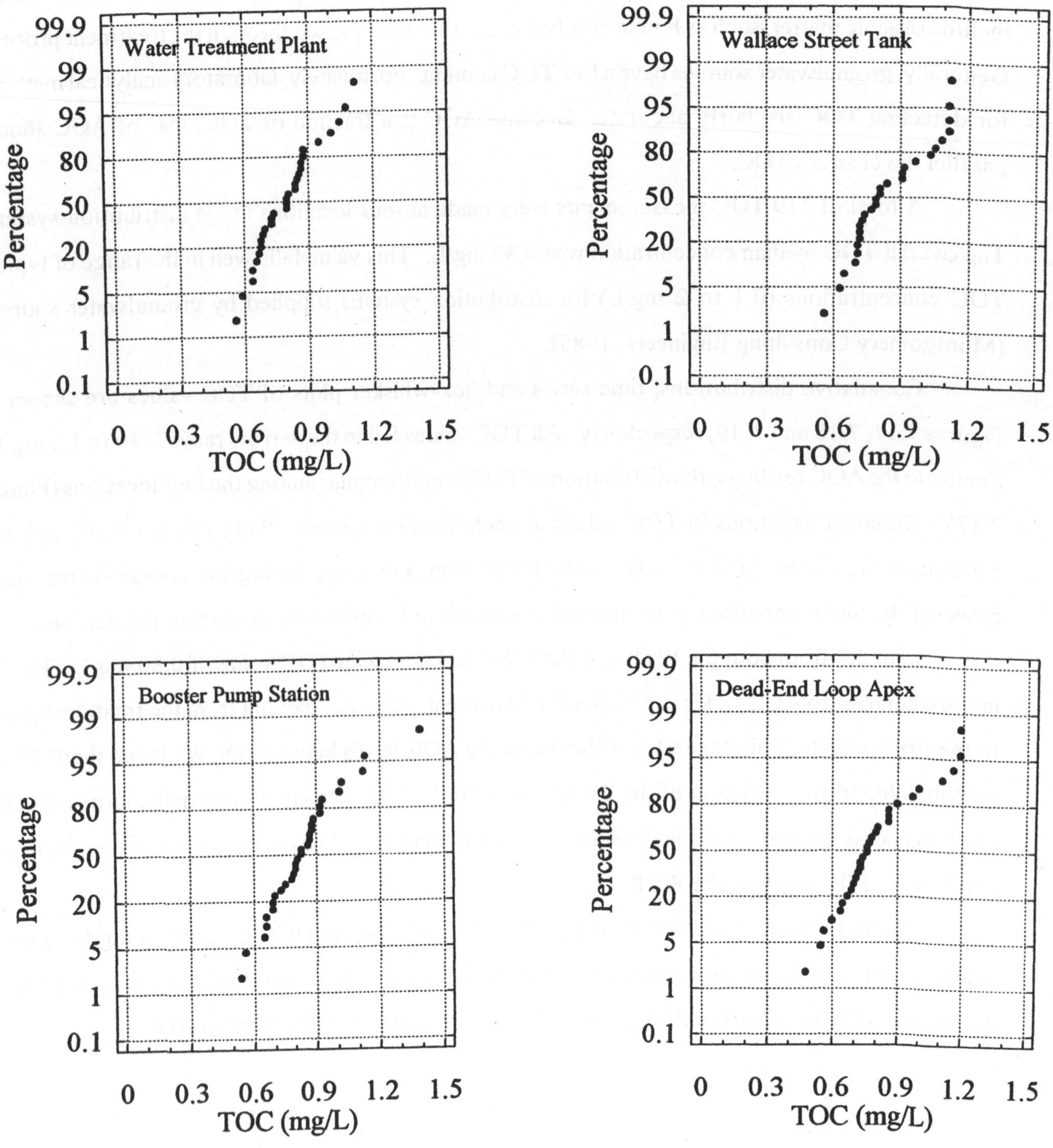

Figure 7.17 Cumulative distribution of TOC at four sampling locations.

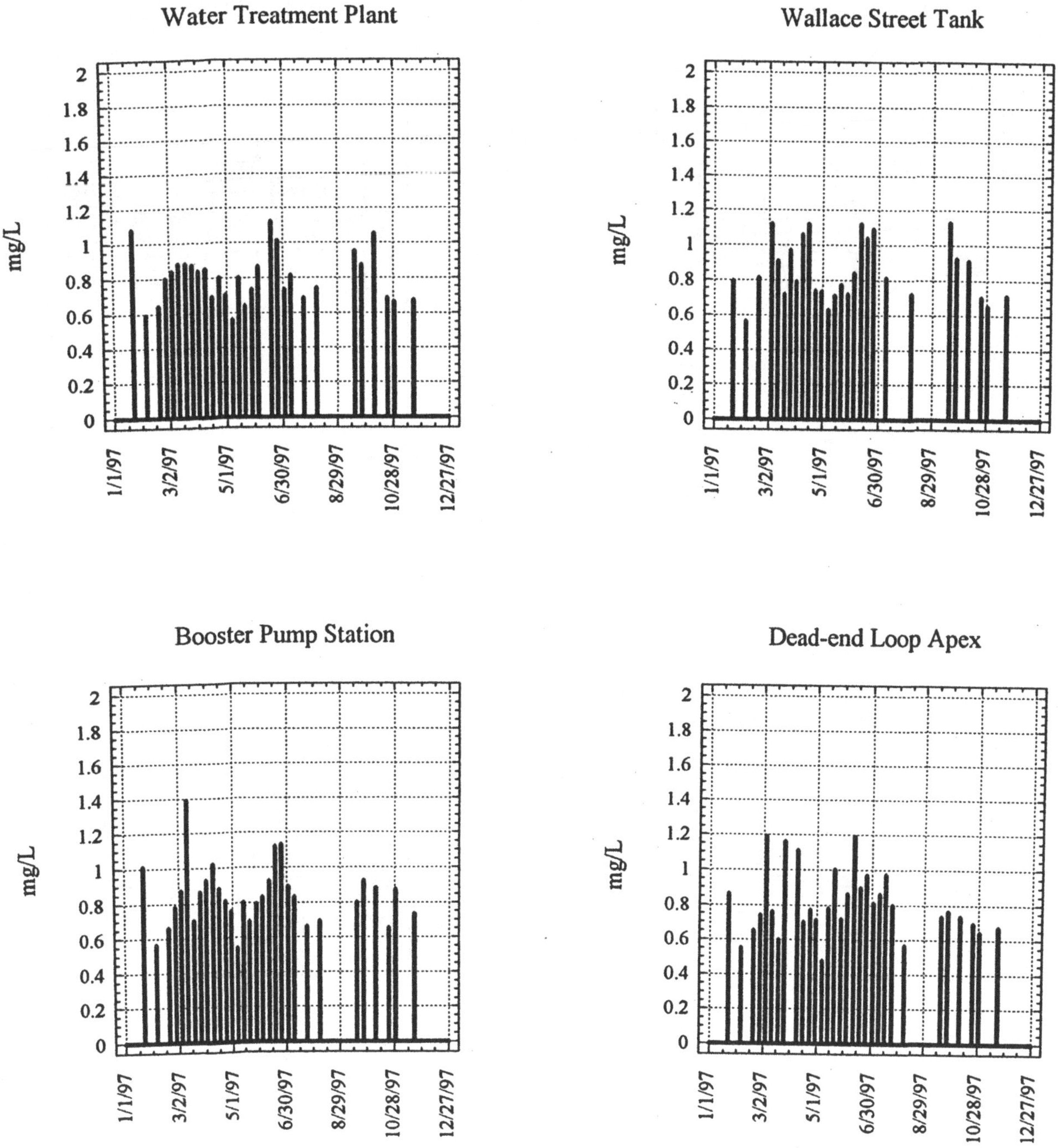

Figure 7.18 Time series of TOC at four sampling locations.

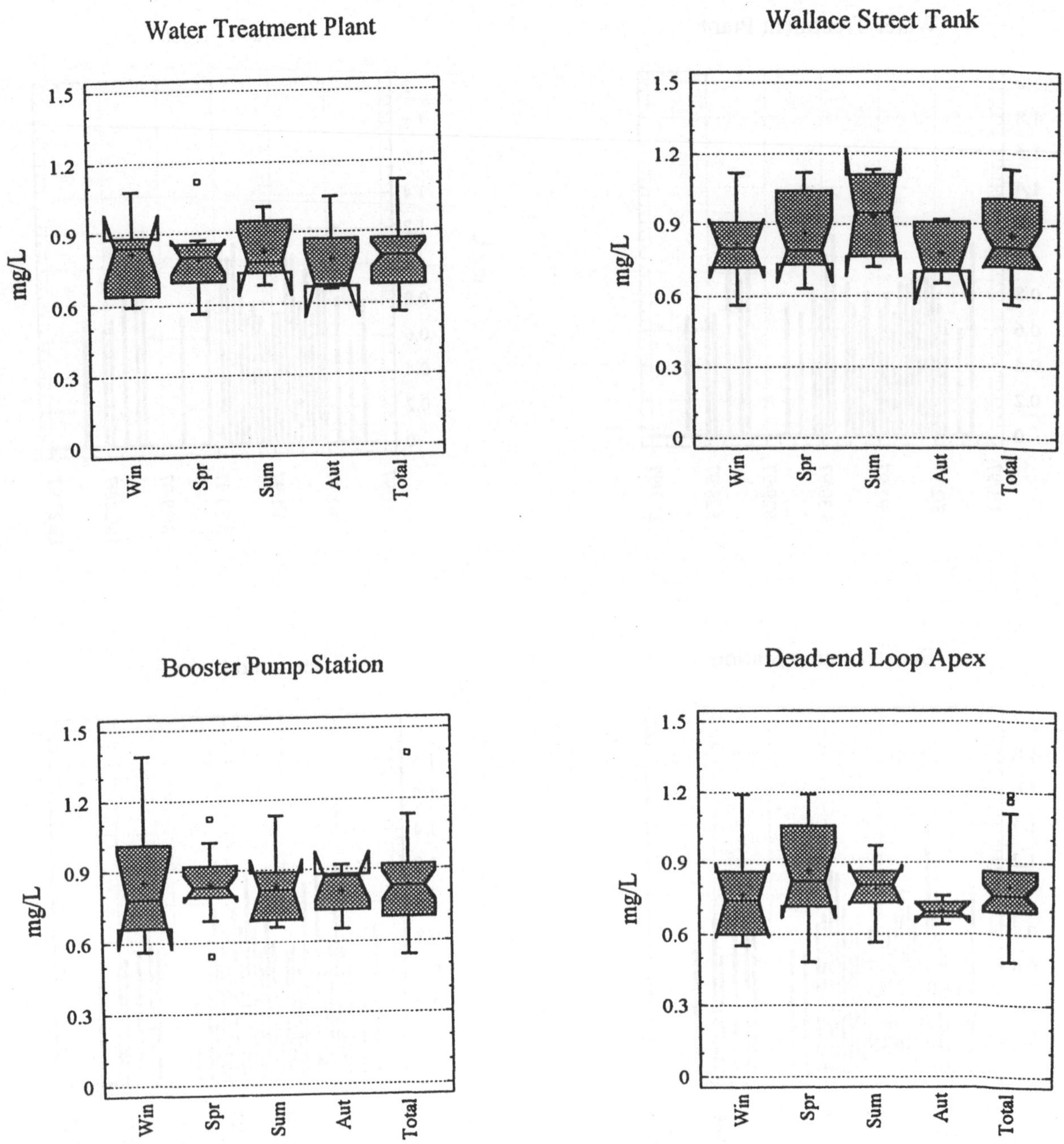

Figure 7.19 Seasonal box-whisker plots of TOC at four sampling locations.

Table 7.13
Seasonal statistics for total organic carbon.

Location / Statistic	Winter	Spring	Summer	Autumn	Annual
WTP: Water Treatment Plant					
Sample size	7.00	12.00	6.00	5.00	30
Mean (mg-/L)	0.82	0.79	0.82	0.79	0.80
Median (mg-/L)	0.84	0.80	0.78	0.68	0.80
Stand Dev (mg-/L)	0.16	0.14	0.13	0.17	0.14
Minimum (mg-/L)	0.59	0.56	0.68	0.66	0.56
Maximum (mg-/L)	1.08	1.12	1.01	1.05	1.12
Skewness	0.11	0.78	0.66	1.18	0.52
Coef of Variation	0.20	0.18	0.16	0.22	0.18
WST: Wallace Street Tank					
Sample size	6.00	13.00	4.00	5.00	28
Mean (mg-/L)	0.82	0.86	0.94	0.78	0.85
Median (mg-/L)	0.80	0.79	0.95	0.71	0.80
Stand Dev (mg-/L)	0.19	0.17	0.20	0.13	1.12
Minimum (mg-/L)	0.56	0.63	0.72	0.65	0.56
Maximum (mg-/L)	1.12	1.12	1.13	0.92	1.13
Skewness	0.45	0.42	-0.13	0.46	0.41
Coef of Variation	0.23	0.20	0.21	0.17	0.20
BPS: Booster Pump Station					
Sample size	7.00	13.00	6.00	5.00	31
Mean (mg-/L)	0.85	0.84	0.83	0.81	0.84
Median (mg-/L)	0.78	0.83	0.82	0.87	0.83
Stand Dev (mg-/L)	0.28	0.15	0.17	0.11	0.18
Minimum (mg-/L)	0.56	0.54	1.11	0.65	0.54
Maximum (mg-/L)	1.39	1.12	0.66	0.92	1.39
Skewness	1.33	-0.11	1.13	-0.75	1.01
Coef of Variation	0.33	0.18	0.20	0.14	0.21
DLA: Dead-end Loop Apex					
Sample size	7.00	12.00	6.00	5.00	30
Mean (mg-/L)	0.76	0.87	0.79	0.70	0.80
Median (mg-/L)	0.74	0.82	0.81	0.69	0.76
Stand Dev (mg-/L)	0.21	0.22	0.14	0.05	0.19
Minimum (mg-/L)	0.55	0.48	0.56	0.64	0.48
Maximum (mg-/L)	1.19	1.19	0.97	0.76	1.19
Skewness	1.46	0.08	-0.66	0.21	1.67
Coef of Variation	0.28	0.25	0.18	0.07	0.23

Spatial and Seasonal Variability of Grab Samples

A nonparametric Kruskal-Wallis analysis of variance (ANOVA) by ranks test was performed to determine whether or not the median values of the grab sample parameters changed significantly with respect to location and to season. In what follows, the four sample locations (WTP, WST, BPS and DLA) are denoted by subscripts 1, 2, 3 and 4, respectively and θ is the sample median.

The null hypothesis, H_0: $\theta_1 = \theta_2 = \theta_3 = \theta_4$, versus the alternative, H_A: Not H_0 is tested with the statistic (Sheskin, 1997)

$$H^* = \frac{12}{N(N+1)} \sum_{j=1}^{k} \left[\frac{\left(\sum R_j\right)^2}{n_j} \right] - 3(N+1) \tag{7.6}$$

where $N = \Sigma n_j$ is the total number of samples, n_j is the number of samples from location j, R_j is the rank order of the sample from location j, and k=4 (either four locations or four seasons). H_0 is rejected if $H^* > \chi^2_{p,v}$ where χ^2 is a variate from chi-square distribution with v=k-1 degrees of freedom and significance level, p.

Results of this ANOVA test, summarized in Table 7.14, indicate that there are no statistically significant differences in the median values for AOC and TOC either among the four sampling locations or among the four seasons (p=0.01). This was not the case, however, with HPC.

Seasonal differences in the HPC median were detected at both the water treatment plant (p=0.002) and the Wallace Street tank (p=0.007), but not at the booster pump station or the dead-end loop apex. In addition, there is a significant difference in the overall HPC median among at least two of the locations. The finding of a significance difference indicates that the variability in the sample medians is more than can be expected under random chance. Instead, some of the variability in the HPC readings probably reflect actual changes in the biological characteristics of the water quality at each sampling location.

Table 7.14
Kruskal-Wallis ANOVA Results for Grab Sample Parameters.

Parameter / Location	Test Statistic and Significance	Decision
HPC: Heterotrophic Plate Count on R2A–Changes with Season		
Water Treatment Plant	H*=14.5 (p=0.002)	reject H_0
Wallace Street Tank	H*=12.0 (p=0.007)	reject H_0
Booster Pump Station	H*=0.55 (p=0.908)	accept H_0
Dead-end Loop Apex	H*=6.67 (p=0.083)	accept H_0
HPC: Heterotrophic Plate Count on R2A–Changes with Location		
WTP→WST→BPS→DLA	H*=40.9 (p=0.007)	reject H_0
AOC: Assimilable Organic Carbon–Changes with Season		
Water Treatment Plant	H*=6.60 (p=0.086)	accept H_0
Wallace Street Tank	H*=1.75 (p=0.626)	accept H_0
Booster Pump Station	H*=4.80 (p=0.187)	accept H_0
Dead-end Loop Apex	H*=4.30 (p=0.232)	accept H_0
AOC: Assimilable Organic Carbon–Changes with Location		
WTP→WST→BPS→DLA	H*=6.96 (p=0.073)	accept H_0
TOC: Total Organic Carbon–Changes with Season		
Water Treatment Plant	H*=0.60 (p=0.901)	accept H_0
Wallace Street Tank	H*=2.60 (p=0.451)	accept H_0
Booster Pump Station	H*=0.30 (p=0.954)	accept H_0
Dead-end Loop Apex	H*=4.30 (p=0.233)	accept H_0
TOC: Total Organic Carbon–Changes with Location		
WTP→WST→BPS→DLA	H*=2.19 (p=0.534)	accept H_0

Null Hypothesis:	H_0: $\theta_{WTP} = \theta_{WST} = \theta_{BPS} = \theta_{DLA}$	(all sample medians are same)
Alternate Hypothesis:	H_A: Not H_0	(at least 2 medians differ)

At p=0.01 with ν=3, $\chi^2_{.01,3} = 11.3$. Hence reject H_0 if H*>11.3 where H* is defined in Eq 7.6.

Water Quality Correlations

Up to this point, all reported grab sample HPC values are based on the *low* nutrient R2A medium. Grab samples were also incubated using two *high* nutrient media: PCA and TSA-SB. HPC values from all three plate agars were examined for correlations with other system water quality parameters.

A nonparametric Spearman rank-order correlation analysis was used to identify monotonic relationships among HPC, AOC and TOC. In addition, the Spearman correlation analysis explored associations between HPC levels and several physico-chemical parameters measured at the CMUs. Water quality parameters recorded at the CMUs were averaged over each week of the study period. The weekly averaged values were then used in the correlation analysis.

Correlations having a significance level $p \leq 0.10$ were judged to be statistically significant. The null hypothesis, H_0: $\rho_s = 0$, versus the alternative, H_A: $\rho_s \neq 0$, is tested with the statistic (Sheskin, 1997)

$$R^* = 1 - \frac{6\sum_{j=1}^{n}\left(d_j\right)^2}{n\left(n^2-1\right)} \tag{7.7}$$

where n is the sample size and d is the difference in ordered ranks for each observation in a group. H_0 is rejected if $R^* \leq R_{s,p}$ the critical value from a tabulation of the Spearman's Rho at significance level, p.

pH and Conductivity

A significant positive correlation was detected between pH and R2A plate counts at both the BPS and DLA (Table 7.15). Negative correlations were detected between conductivity and R2A plate counts, but the association was statistically significant only at the DLA. Although bacteria in the pipe network scavenge ions from the drinking water, this effect probably can not be measured with the detection method employed in this study. Neither TSA-SB nor PCA plate counts were correlated with pH or conductivity.

Table 7.15 Spearman Rank Correlations for Bulk Fluid HPCs from 3 Nutrient Plate Agars.

Location/ Plate Agar	pH	Conductivity	Water Temperature	Free Chlorine	AOC	TOC	PCA	TSA-SB
WTP--Water treatment Plant								
R2A	–	–	–	–	0.26 (0.28) 18	***0.47 (0.02) 27***	***0.57 (0.04) 13***	0.27 (0.35) 13
TSA-SB	–	–	–	–	0.00 (1.00) 7	-0.08 (0.83) 8	0.70 (0.15) 13	
PCA	–	–	–	–	0.61 (0.13) 7	-0.11 (0.77) 8		
WST--Wallace Street Tank								
R2A	–	–	–	–	0.09 (0.73) 17	***0.40 (0.05) 25***	***0.52 (0.01) 13***	0.41 (0.15) 13
TSA-SB	–	–	–	–	0.18 (0.66) 7	***0.67 (0.10) 7***	***0.79 (0.01) 13***	
PCA	–	–	–	–	-0.43 (0.29) 7	0.51 (0.21) 7		
BPS--Booster Pump Station								
R2A	***0.29 (0.09) 36***	-0.12 (0.94) 36	0.20 (0.22) 36	0.01 (0.96) 32	0.33 (0.17) 18	***0.45 (0.02) 27***	***0.54 (0.06) 13***	0.47 (0.11) 13
TSA-SB	-0.02 (0.95) 17	0.14 (0.57) 17	-0.07 (0.78) 17	-0.12 (0.62) 17	0.55 (0.18) 7	0.59 (0.12) 8	***0.92 (0.00) 13***	
PCA	-0.13 (0.61) 17	0.07 (0.76) 17	-0.15 (0.54) 17	-0.14 (0.58) 17	0.53 (0.20) 7	***0.66 (0.08) 8***		
DLA--Dead-end Loop Apex								
R2A	***0.53 (0.00) 35***	***-0.37 (0.03) 35***	***0.29 (0.09) 36***	***-0.43 (0.02) 30***	0.32 (0.21) 17	0.27 (0.18) 26	0.18 (0.55) 12	0.17 (0.58) 12
TSA-SB	0.02 (0.92) 17	0.10 (0.68) 17	-0.30 (0.20) 17	0.33 (0.19) 12	-0.45 (0.27) 7	0.13 (0.73) 8	***0.88 (0.00) 12***	
PCA	0.20 (0.43) 17	-0.23 (0.35) 17	-0.03 (0.89) 17	-0.16 (0.51) 17	-0.09 (0.82) 7	0.06 (0.87) 8		

Notes: (1) Entries in each cell are Spearman correlation R*, significance level (p) and sample size, n

(2) Significant correlations are in ***bold italics*** and correspond to $p \le 0.10$.

(3) – means no CMU sample available.

Water Temperature

Bacterial growth rates generally increase with a rise in temperature. High bacterial numbers often confound maintenance of disinfectant residual within water mains during summer months. A significant positive correlation between the water temperature and R2A plate counts was found at the DLA site (Table 7.15). Due to a long retention time in the dead-end loop (30 hours on average), the effect of water temperature on bacterial numbers may have been magnified. Bacterial counts on TSA-SB and PCA media were not correlated with water temperature at the BPS or DLA.

Free Chlorine

Disinfectant is usually added before distribution of treated water to inhibit regrowth of bacteria in the pipe network. A free chlorine residual at the DLA was found to significantly suppress the bulk fluid bacterial numbers from R2A media (Table 7.15). The lack of a similar correlation at the BPS site reflects the confounding effects due to the diurnal mixing of fresh water from the WTP and aged water released from the WST. Neither the TSA-SB nor the PCA bacterial counts were found to be significantly correlated with the free chlorine residual (Table 7.15).

AOC

The level of nutrients in a distribution system can have a significant impact on bacterial numbers. Water containing a high concentration of AOC may experience significant bacterial regrowth when compared to water with low levels of AOC. In this study, however, there were no significant correlations between AOC measurements and bulk fluid bacterial numbers (Table 7.15). This lack of association may have been due to the relatively low concentrations of AOC throughout the pipe network (see Table 7.12).

TOC

A significant positive correlation was detected between the amount of bulk fluid TOC and R2A HPCs at the WTP, WST, and BPS locations (Table 7.15). However, there was not a significant correlation between TOC and R2A at the DLA, where TOC levels were lowest. The absence of an association between TOC and HPCs at the apex may indicate the effect of a "threshold" TOC value. TOC levels were significantly correlated with TSA-SB plate counts only at the WST and with PCA plate counts only at the BPS.

R2A, PCA and TSA-SB

The low nutrient R2A yielded HPC populations (CFU/mL) which were strongly correlated with CMU physico-chemical parameters (pH, conductivity, temperature, chlorine) at the DLA. Conversely, the high nutrient agars (PCA and TSA-SB) displayed no correlation with any of the CMU readings at either the entrance or the apex of the dead-end loop. The PCA results were often significantly correlated with HPCs from one or both of the other nutrient plate agars. For instance, there were three cases of significant correlation between PCA and R2A (at WTP, WST, BPS); similarly there were three cases of significant correlation between PCA and TSA-SB (at WST, BPS, DLA). Interestingly, however, the TSA-SB results showed no statistically significant association with R2A results at any location.

Summary of Grab Samples

Bulk fluid grab samples were taken at four locations in the City of Milford's water distribution system and analyzed for HPC, TC, AOC and TOC. The sampling period began January 8, 1997 and continued through November 18, 1997. The overall data recovery rate exceeded 90 percent. A total of 359 readings were obtained to characterize the biological water quality of the bulk fluid.

All tests for total coliform were negative. HPC values fluctuated dramatically with location and with season. HPC results ranged from a summer low of 1 CFU/mL at the WTP to an autumn high of 8600 CFU/mL at the BPS. By contrast, AOC and TOC observations were quite stable. All AOC readings were under 1 mg-C/L; all TOC readings were in the range 0.4 to 1.5 mg/L. These AOC and TOC values are typical for distribution systems fed by a groundwater source.

A Kruskal-Wallis ANOVA test found no significant differences in the median values for both AOC and TOC with respect to location or season. Median HPC levels, however, were found to have statistically significant differences with respect to location. In addition, median HPC levels at the WTP and the WST were found to vary significantly with the season.

Spearman rank correlations showed consistent significant associations between standard physico-chemical water quality parameters and bulk fluid HPC levels at the DLA but not at the BPS. This finding underscores the realization that multiple confounding factors affect the viability of bacteria within a municipal water distribution system.

MODIFIED ROBBINS DEVICE

Biofilm samples were collected from July 13 to November 18, 1997. The samples were sonicated from plugs extracted from Modified Robbins Devices (MRDs) installed at the BPS and DLA. Samples were plated in triplicate on three media: R2A, PCA and TSA-SB. The PCA and TSA-SB plates represent a high nutrient environment, perhaps similar to the human body. Following incubation, the plates were analyzed for mean number of colony forming units per square centimeter (CFU/cm^2) according to the standard methods mentioned in Chapter 3. Results were used to quantify the biofilm growth potential at the entrance and the apex of the dead-end loop.

The MRDs were designed and operated (to the extent possible) to mimic conditions in the distribution system. However, there were two key differences between the experimental MRDs and the actual pipe, namely, wall material and flow regime. The real pipe was a 15 cm cast-iron conduit subject to unsteady bi-directional laminar flow. In contrast, the MRDs were constructed with 10 cm PVC pipe and subject to steady laminar flow moving in one direction only. These distinctions make it difficult to infer directly the extent of biofilm growth inside an undisturbed cast-iron main[(9)].

Nevertheless, the MRDs did experience the same water as the actual pipe loop. In addition, a tracer test with Rhodamine WT dye indicated that there was no preferential flow paths along the MRD. Hence, the entire inside perimeter of the MRD reactor could be expected to receive equal exposure to biological and chemical agents in the water. A Kruskal-Wallis ANOVA test indicated that the location of the sampling port around the circumference of the MRD did not appreciably influence the resulting CFU reading (Carter, 1998).

Biofilm Growth Potential

Bacteria in the water distribution system required approximately five weeks to colonize the inside wall of the MRD and attain a "psuedo" steady population. This stage signifies that bacteria have reached a near equilibrium among the processes of attachment, detachment, growth and death. Following this initial colonization period, bacterial densities in the MRD biofilm samples plated on

---

[(9)] Three ductile iron plugs were placed in the MRDs at the BPS and DLA for a 5-week period. The biofilm growth on the ductile iron (measured as CFU count) was at least 1000 times lower than the growth on the CPVC plugs.

R2A and PCA approached 10^6 CFU/cm^2 (Figure 7.20). Because the reproduction rate of bacterial colonies in a water distribution system is very slow, the majority of colonies counted on the MRD samples probably represent bacteria migrating from the bulk fluid to the pipe wall.

Summary statistics for biofilm growth potential for each nutrient agar at both locations are listed in Table 7.16. The low nutrient R2A results show similar plate counts at the BPS and the DLA. However, the high nutrient PCA and TSA-SB results show a decrease in mean and median plate counts between the entrance and the apex of the loop.

Spatial Variability of MRD Data

A nonparametric Mann–Whitney U test(10) was performed to determine whether or not the median values of the biofilm growth potential changed significantly with respect to location for each of the nutrient agars. As before, subscript 1 denotes the BPS, subscript 2 denotes the DLA and θ is the sample median. The null hypothesis, H_0: $\theta_1 = \theta_2$, versus the alternative, H_A: $\theta_1 \neq \theta_2$, is tested with the statistics (Sheskin, 1997)

$$U_1 = n_1 n_2 + \frac{n_1(n_1+1)}{2} - \sum_{j=1}^{n_1} (R_1)_j$$

$$U_2 = n_1 n_2 + \frac{n_2(n_2+1)}{2} - \sum_{j=1}^{n_2} (R_2)_j \tag{7.8}$$

where n is the sample size and R is the rank for each observation in a group. Let $U^* = \min\{U_1, U_2\}$. Then H_0 is rejected if $U^* \leq U_p(n_1,n_2)$, the critical value from a tabulation of the Mann-Whitney U Statistic at significance level, p.

---

(10) The Kruskal-Wallis ANOVA test with 2 samples reduces to the Mann-Whitney U test.

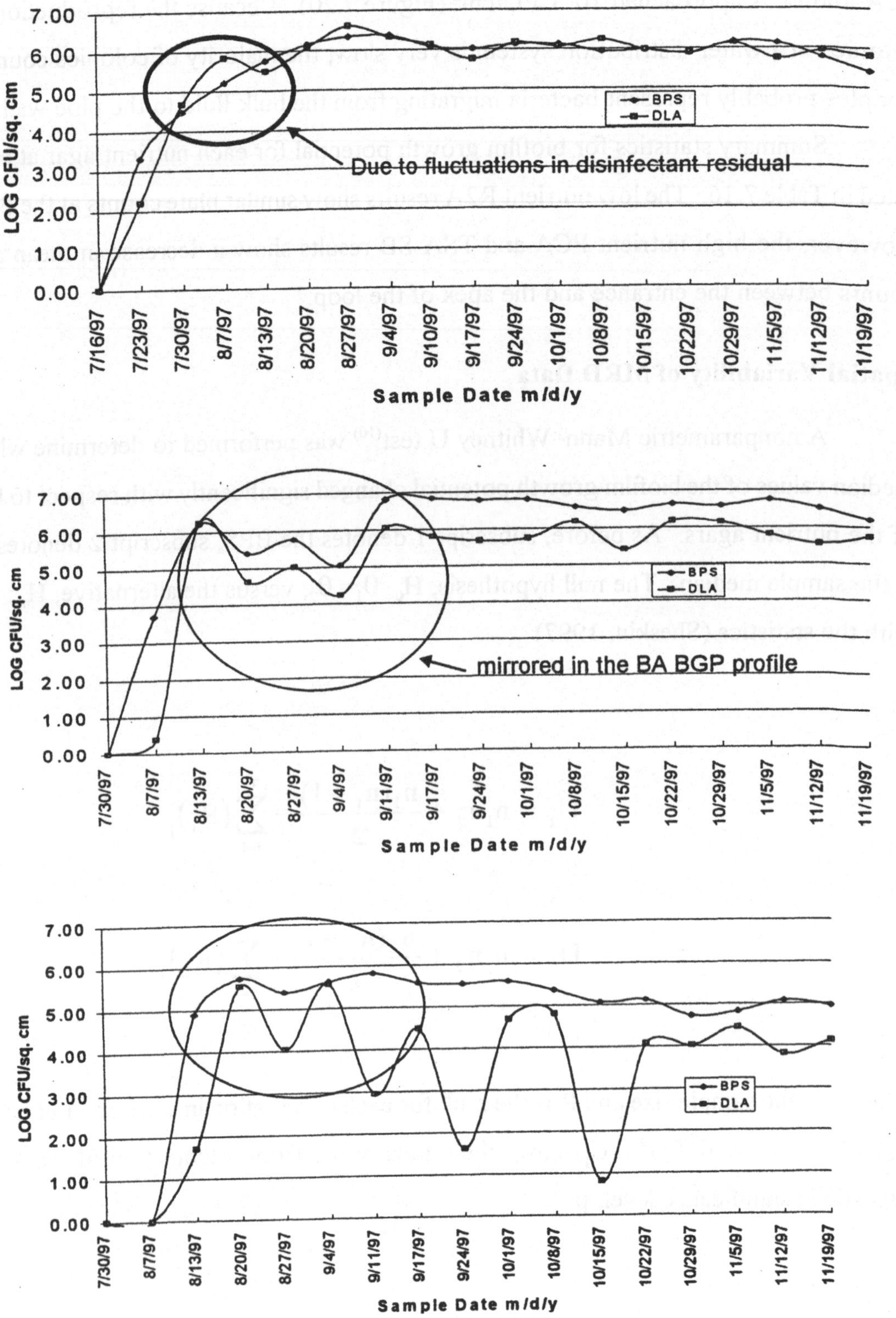

Figure 7.20 Biofilm growth potential on three nutrient agars at the entrance and the of the dead-end loop (top: R2A; middle: PCA; bottom: TSA-SB).

Table 7.16 Summary Statistics for Biofilm Growth Potential at the Modified Robbins Devices.

	low nutrient R2A		high nutrient PCA		high nutrient TSA-SB	
	BPS	DLA	BPS	DLA	BPS	DLA
Sample Size	15	15	15	15	15	15
Mean	1.18	1.28	3.58	0.72	0.26	0.06
Median	1.17	1.15	3.83	0.55	0.23	0.01
Standard Deviation	0.58	0.93	2.36	0.58	0.19	0.12
Minimum	0.26	0.30	0.11	0.02	0.05	0.00
Maximum	2.38	4.10	6.90	1.78	0.65	0.37
Skewness	0.67	2.19	-0.09	0.57	0.74	2.31

Note: Units are 10^6 CFU/cm

Table 7.17 Mann-Whitney U Test for Biofilm Growth Potential.

Agar and Location	Average Rank	Test Statistic, U*	Decision
R2A at BPS R2A at DLA	15.6 15.4	min{111, 114}	accept H_0
PCA at BPS PCA at DLA	20.8 10.2	min{33, 192}	reject H_0
TSA-SB at BPS TSA-SB at DLA	21.6 9.4	min{21, 204}	reject H_0

Null Hypothesis: H_0: $\theta_{BPS} = \theta_{DLA}$ (sample medians are the same)
Alternate Hypothesis: H_A: $\theta_{BPS} \neq \theta_{DLA}$ (sample medians are not the same)

At significance level, p=0.01 with $n_1 = n_2 = 15$, the critical Mann Whitney U value is: $U_{0.01} = 51$. Hence, reject H_0 if $U^* \leq 51$, where $U^* = \min\{U_1, U_2\}$ with U_1 and U_2 defined in Equation 7.8.

Results of this test, summarized in Table 7.17, indicate that there is not a statistically significant difference (p=0.01) between the median plate counts at the BPS and the DLA for the low nutrient agar (R2A). However, in the case of the high nutrient agars (PCA and TSA-SB), the differences between the median plate counts at the BPS and the DLA are statistically significant (p=0.01). Large differences in the biofilm plate counts from the high nutrient agar between the BPS and the DLA may indicate that the high water quality in the dead-end loop tends to suppress opportunistic pathogens.

Summary of Modified Robbins Device

Two MRDs, installed near the entrance and the apex of a dead-end loop, provided weekly biofilm samples during the period July 13 to November 18, 1997. Samples were collected in triplicate and plated on R2A, PCA and TSA-SB nutrient media. Within five weeks, the biofilm reached a pseudo steady condition inside the MRD with counts approaching 10^6 CFU/cm^2 on CPVC plugs plated in R2A. In contrast, the counts obtained from ductile iron plugs plated in R2A did not exceed 10^3 CFU/cm^2. The biofilm growth potential observed for the low nutrient R2A media was similar at the entrance and the apex of the dead-end loop. However, the biofilm growth potential measured for the high nutrient agars (PCA and TSA-SB) showed a significant drop between the entrance and the apex of the loop.

Appendix 7.1 CMU Sampling Schemes

To extend the life of the monitoring units and to minimize data redundancy, several water quality monitoring schemes were investigated. Sampling windows of three, four and six-hour durations were tested. Figure 7.21 shows readings for pH, temperature and conductivity obtained every 10 minutes during a six-hour sampling window in March 1997 at the CMU near the apex of the dead-end loop. Between sampling periods, the flow (0.6 L/m) for the continuous bleed through the CMU was stopped. This "on-off" approach introduced complications to the water quality data set.

For example, during the pause (about 9 hours in Figure 7.21), the temperature of the stagnant water in the CMU rose from 9 to 17 °C. The conductivity also increased from 409 to 417 µS/cm, about a 2% rise. The pH remained stable at just under 9 whether the bleed was on or off. When the bleed was started again, readings from the temperature and conductivity probes immediately dropped, but required nearly 30 minutes to stabilize.

While the on-off sampling approach did reduce the data set, the large gaps and spurious fluctuations in the time series introduced complications in data analysis. Because the probes were still deployed in the CMUs, this arrangement probably did not appreciably extend the life of the probes. In the end, the trade-off was judged to be not worthwhile. Hence, the on-off sampling schemes were abandoned in favor of the original plan of continuous sampling at 10-minute intervals.

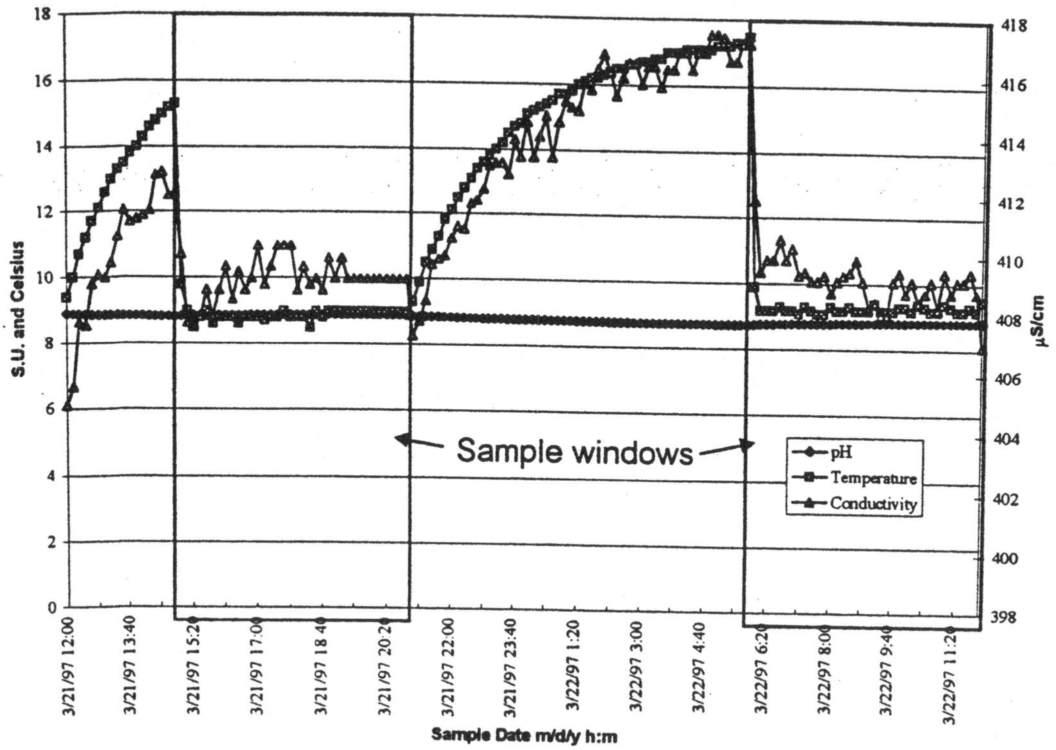

Figure 7.21 Water quality readings during a six-hour window in March 1997 at the DLA.

Appendix 7.2: Grab Sample Data

Summarized in Table 7.18 are the raw data from 359 grab samples taken from the Milford water distribution system during the period January 8, 1997 to November 19, 1997. All lab analyses were performed by the U.S. Environmental Protection Agency in Cincinnati, Ohio.

Table 7.18 Raw data from water quality grab samples.

Date in 1997	HPC (CFU/mL)				AOC (mg/L)				TOC (mg/L)			
	WTP	WST	BPS	DLA	WTP	WST	BPS	DLA	WTP	WST	BPS	DLA
01/08	140	(a)	3000	70	0.08		0.64	0.20				
01/16	30	170	180	30	0.08	0.21	0.19	0.09				
01/22	230	90	(c)	20					1.08	0.79	1.01	0.86
01/29	100	(a)	980	20								
02/13	(a)	(a)	(b)	(b)	0.02	0.11	0.07	0.07	0.59	0.56	0.56	0.55
02/18	(a)	(a)	(a)	(b)								
02/20	540	90	10	10					0.64	0.81	0.66	0.65
02/27	530	(a)	70	210					0.80		0.78	0.74
03/06	30	80	20	30	0.30	0.19	0.23	0.10	0.84	1.12	0.87	1.19
03/14	300	130	450	100					0.88	0.91	1.39	0.76
03/20	730	200	960	130	0.21	0.12	0.27	0.21	0.88	0.72	0.70	0.60
03/27	270	230	20	60					0.87	0.97	0.86	1.16
04/03	60	420	290	70	0.15	0.45	0.14	0.14	0.84	0.79	0.93	
04/10	170	610	2300	80					0.85	1.06	1.02	1.11
04/18	590	560	470	60					0.69	1.12	0.88	0.70
04/22	60	330	300	180					0.80	0.74	0.81	0.77
04/29	(b)	(b)	(b)	(b)					0.70	0.73	0.75	0.71
05/08	20	180	50	40	0.08	0.16	0.11	0.08	0.56	0.63	0.54	0.48
05/14	30	350	540	180					0.80	0.71	0.80	0.78
05/21	30	420	90	330	0.03	0.06	0.01	0.27	0.64	0.77	0.69	1.00
05/28	10	410	60	90	0.13	0.21	0.05	0.04	0.73	0.72	0.79	0.72
06/04	340	220	1450	70					0.86	0.84	0.83	0.86
06/11	3800	930	910	6100						1.12	0.92	1.19
06/20	320	1100	850	370	0.14	0.20	0.18	0.96	1.12	1.04	1.12	0.90
06/25	80	1030	790	320					1.01	1.09	1.13	0.97
07.02	20	(b)	2800	200					0.73		0.89	0.81
07.09	(c)	(c)	(c)	(d)		0.12	0.11	0.05	0.81	0.81	0.83	0.86
07/16	10	1	280	230								
07/23	10	190	280	230	0.01	0.32	0.01	0.01	0.68		0.66	0.80
07/30	10	180	150	150								

(continued on next page)

Table 7.18 Raw data from water quality grab samples.
(cont'd)

Date in 1997	HPC (CFU/mL)				AOC (mg/L)				TOC (mg/L)			
	WTP	WST	BPS	DLA	WTP	WST	BPS	DLA	WTP	WST	BPS	DLA
08/07	1	260	650	145	0.02	0	0.14	0.09	0.74	0.72	0.69	0.56
08/13	20	200	550	45								
08/20	20	330	910	50								
08/27	10	400	60	80								
09/04	20	680	1860	240								
09/10	180	380	480	130								
09/17	120	350	90	70					0.95	**<u>1.13</u>**	0.80	0.73
09/24	50	2600	150	(b)	0.01	**<u>0.88</u>**	0.09	0.05	0.87	0.92	0.92	0.76
10/01	60	340	4200	230								
10/08	50	870	1100	100	0.06	0.07	0.09	0.10	1.05	0.91	0.88	0.73
10/15	30	780	**<u>8600</u>**	50								
10/22	10	240	410	100	0.09	0.07	0.05	0.08	0.68	0.70	0.65	0.69
10/29	20	150	210	190	0.12	0.10	0.23	0.12	0.66	0.65	0.87	0.64
11/05	200	**<u>3100</u>**	150	140	0.04	0.16	0.10	0.10				
11/12	20	200	30	80	0.06	0.06	0.08	0.09				
11/19	40	130	180	20	0.05	0.13	0.09	0.09	0.67	0.71	0.73	0.67
Size	42	38	41	41	19	19	20	20	30	28	31	30
359	162				78				119			

Note: Maximum value in each column is **<u>bold underline</u>**

Legend for missing HPC data:

(a) = site was not accessible
(b) = sample was mishandled in transit
(c) = no lab results available
(d) = too numerous to count

CHAPTER 8
SUMMARY AND CONCLUSIONS

SUMMARY

Residential Water Demands

Over 490,000 residential water demands were recorded during two separate sampling periods at 21 homes along a dead-end loop in Milford, Ohio. The first monitoring exercise captured about 125,000 water demands during February and March 1996. The second monitoring exercise captured nearly 365,000 water demands during the seven month period April through October 1997.

Residential water demands were converted to Single Equivalent Rectangular Pulses (SERPs). The SERPs preserved the start time, duration, volume, and average intensity of the original water demand. Each SERP from the 1997 observation period was carefully screened, categorized as an indoor or outdoor use, archived in text files and then stored on compact disc (see Appendix A for order information). These field data represent one of the largest collections of high resolution residential water demands ever compiled. They were used to perform several detailed analyses, including:

- Characterize statistical properties and patterns of residential water use.
- Run a hydraulic network model to simulate flows on the dead-end loop.
- Derive bi-directional travel time distributions to all homes on the loop.
- Test the hypothesis that residential water use is a Poisson pulse process.

Water Quality Readings

Water quality at the project site was sampled three ways. Continuous monitoring units recorded eight physico-chemical parameters (pH, conductivity, temperature, chlorine, nitrate, turbidity, ORP and DO) at the entrance and the apex of the dead-end loop. Weekly grab samples taken from the dead-end loop and two other locations in the distribution system were analyzed for HPC, TC, AOC and TOC. A modified Robbins device provided samples to quantify biofilm growth at the entrance and apex of the loop. In total, over 245,000 water quality measurements were obtained at the study site during the period January to November 1997.

CONCLUSIONS FROM WATER DEMAND MONITORING

Single Equivalent Rectangular Pulses (SERPs)

Any residential water demand can be expressed as a Single Equivalent Rectangular Pulse (SERP) which preserves the starting time, the duration, the volume and the average intensity of the original water demand. At a given residence, SERP statistics for indoor water use remain remarkably consistent from month to month. This implies that indoor SERPs can be adequately characterized with residential demand monitoring programs that last only a few weeks (rather than a few months). The log normal distribution provides a reasonably good description of indoor SERP intensity, duration and volume. Across the entire dead-end loop, the mean volume of an indoor pulse was 1.71 gallons with a mean duration of 0.77 minutes and a mean intensity of 2.22 gpm. Per capita consumption at the dead-end loop study site is consistent with national norms.

Flow Regimes

Current network models assume that *quasi-steady continuous turbulent flow* exists throughout the water distribution system. However, in dead-ends zones, the predominant flow regime is likely to be *random intermittent laminar flow*. At the test site, this behavior is clearly evident in the daily quad plots (see Appendix H) which reveal highly variable bursts of (mainly) laminar flow interspersed with sporadic periods of stagnation. Accurate knowledge of the flow regime is essential to properly simulate the hydraulic conditions which govern the fate and transport of chlorine and other constituents moving through the pipe network.

Stagnant Conditions

Stagnant conditions exist along a pipe when the flow rate drops to zero. Mainlines in dead-end zones experience stagnation when all downstream servers are idle. A very good estimate of the probability of stagnation in a neighborhood of N homes is given by: $P_0(t)=\exp[-N\,\overline{\rho}]$, where $\overline{\rho}$ is the average utilization factor for a typical single family home in the service zone (see Tables 6.3 and 6.4). At the study site, the average daily utilization factor is estimated to be, $\overline{\rho}\approx 0.045$. Hence, during the course of a day, the probability of stagnation at a single family residence (N=1) is, $P_0=\exp[-0.045] = 95\%$. Across the study site with N=21 active homes in 1997, stagnant conditions in the dead-end loop occurred on average about 40 percent of the time, $P_0=\exp[-21\times 0.045] = 39\%$.

Hydraulic Modeling

Current network models that perform extended period simulations are driven by average *total-time* demands applied at the neighborhood scale (see hierarchy level H-3 in Figure 6.4 and Table 6.9). To improve water quality predictions, hydraulic models should adopt event-driven *busy-time* demands at the household scale (hierarchy H-2). This degree of resolution is the only way to mimic the abrupt random shifts among flow regimes (i.e., stagnant, laminar, transition, turbulent) that occur in peripheral zones of the water distribution system. In service zones where laminar flow is common (and hence Taylor mixing may be important), it may be prudent for water quality models to incorporate dispersive transport along with advective transport.

Travel Times

The theoretical mean travel time for an entire dead-end neighborhood is given by the ratio of total pipe volume to average water demand (T=V/Q). However, the mean travel time to individual households in dead-end zones is a strong function of location. Consumers near the entrance of the dead-end have mean travel times that are less than the neighborhood average. Conversely, consumers near the terminal end of a stem or near the apex of a loop have mean travel times that are much longer than the neighborhood average. In the case of a dead-end stem (where only uni-directional flow is permitted), the dramatic increase in mean travel time for homes near the terminal end is due to the decrease in demand from a dwindling consumer base (essentially T goes up because Q goes down). In the case of a dead-end loop (where bi-directional flow is allowed), the increase in travel time for homes near the apex is related to the large number of flow reversals that effectively trap a parcel of water as it migrates between active nodes (essentially T goes up because the *apparent* V goes up).

Travel times from the loop entrance to all homes at the study site were estimated with a computer program that tracked the age of small parcels of water moving through the 6-inch cast iron supply line to the demand nodes. Based on residential demands from Winter 1996, the overall flow-weighted mean travel time in the neighborhood was about 30 hours. On an individual household basis, the mean travel time through the supply line ranged from 8 hours for homes near the entrance to almost 96 hours for homes near the apex. The coefficient of variation for travel times stayed relatively constant at about 0.30. This implies that the mean and the standard deviation of travel times grew at nearly the same rate.

Poisson Rectangular Pulse (PRP) Hypothesis

The hypothesis that residential water use behaves as a time dependent Poisson rectangular pulse (PRP) process is confirmed. Busy fixtures have a Poisson probability distribution; busy homes have a binomial probability distribution. The difference between these two models is very slight. Practically speaking, the more convenient Poisson model can be used to describe the probability distribution of busy servers (fixtures and homes) in any pressure zone at any time and at any location in the pipe network (the PRP model for busy servers is not restricted to dead-end regions). The compelling body of evidence supporting the PRP hypothesis includes excellent agreement between observed and predicted characteristics of busy servers (e.g., probability distributions, state transitions, and autocorrelation functions) and flow rates (e.g., mean, variance and stagnation).

The PRP model is parsimonious, requiring only three parameters: (i) α–the average demand intensity at a *busy* server, (ii) β^2–the variance of the demand intensity at a *busy* server, and (iii) $\rho(t)$–the time dependent utilization factor for a typical single family residence. Values of α and β^2 are stable at a given home and can be obtained directly from measured SERPs (examples are given in Appendix C). The utilization factor displays a characteristic diurnal pattern (i.e., the residential "rho profile") that reflects the time-dependence of the PRP model (an example appears in Figure 6.3).

Applying concepts pioneered by 20th century communication utilities (who investigated demands on telephone exchanges), the PRP model provides a rigorous framework for describing residential water demands in pipe networks. The PRP approach complements existing deterministic network models for solving problems on design and operation of water distribution systems.

Branching Pipes

Branching mainlines (including dead-ends) are an important subset of many municipal water distribution systems. Because the direction of flow in a branching pipe is known, these lines are amenable to analysis with the PRP model. The basic idea is to replace conservation of energy with conservation of probability. Then, in conjunction with the continuity principle, the PRP analysis provides, not just a single flow rate, but rather an estimate of the probability distribution of flow rates through the branching line. In low flow regions, the PRP model indicates the probability of stagnation at any point along the branching pipe at any time of day. In high flow regions, PRP analysis provides an estimate of the risk that actual water demands will exceed a peak design value.

Applications of Residential Water Demand Research

The versatility of the PRP process for residential water demands represents a significant breakthrough in efforts to model water consumption in municipal distribution systems. The PRP approach offers water utilities, for the first time, a single unified method to quantify water demands over various time scales and spatial scales (from seconds to days in a single family home or across an entire service zone). Some immediate applications of the PRP model include:

❖ Determine the magnitude of the peak flow in a branching supply line feeding any number of homes for a specified level of reliability (see Equations 6.20 and 6.21).

❖ Estimate the rate of network leakage in service zones. This algorithm can be applied anywhere in the network, provided the service zone is completely isolated with only one feeder line in and no lines out (see Figures 6.16 and 6.17 for concept). Buchberger and Nadimpalli (2003) provide a detailed example.

❖ Generate stochastic water demands at pipe nodes in a network model. These demands can drive a water quality model to assess the vulnerability of the distribution system if attacked via deliberate bio-chemical intrusion (see Fig 6.18 for concept). A preliminary version of a computer program to generate instantaneous and time-averaged indoor and outdoor PRP residential water uses at demand nodes is available from the authors.

❖ Evaluate the effectiveness of proposed water conservation measures. The PRP model provides a sound mechanistic description of residential water use based on the intensity, duration and frequency of individual demands. Utilities can systematically explore how installing new fixtures (i.e., controlling intensity and duration) or forcing behavior modification (i.e., controlling frequency) may impact overall water consumption.

Two other important observations for utilities arise from the water demand portion of this study:

❖ Avoid dead-end lines. Even though branching lines proliferate throughout most existing pipe networks, one-way pipelines should not be permitted in new construction where practicable. Dead-end lines lack redundancy and allow stagnation. As a consequence, average travel times to downstream consumers along dead-end lines increase exponentially (see Figure 5.7). This inadvertently extends opportunities for the quality of finished water to deteriorate before reaching the consumer.

❖ Recognize that peripheral areas in water distribution systems often experience highly variable intermittent laminar flow. This hydraulic regime deviates drastically from the steady plug flow assumption used in current water quality models. Hence, discrepancies between model predictions and field measurements of water quality (e.g., chlorine residual) will be confounded by the uncertainty in the prevailing network hydraulics.

Future PRP Projects

The PRP model provides a reasonable, rigorous and robust framework for helping to solve contemporary problems that arise with municipal water distribution systems. While space limitations preclude further details on other potential PRP applications, the list of other possibilities covers such wide ranging topics as:

- ❖ Derive and test the mean, variance and probability distribution of travel times along branching lines in the pipe network.
- ❖ Derive the mean, variance and probability distribution for the concentration of disinfectant residual, disinfection by-products, and other constituents moving to remote points of consumption in the distribution system.
- ❖ Extend the PRP approach to describe other components of municipal water use including, commercial, industrial, institutional, government and other demands.
- ❖ Extend the PRP approach to looped areas of the network.
- ❖ Determine the most appropriate temporal and spatial resolutions needed for accurate predictions of water quality in the pipe network.

These issues are important in view of increasingly stringent regulations promulgated under the Safe Drinking Water Act and heightened concerns about the vulnerability of water distribution systems to bio-chemical terrorism. Basic principles from the Poisson rectangular pulse model can be used to investigate these problems. Clearly the PRP approach offers a new stochastic counterpart to complement existing deterministic models which are now indispensable tools for designing and operating distribution systems to meet water demands and quality standards.

CONCLUSIONS FROM WATER QUALITY SAMPLING

During the course of this study, the City of Milford consistently produced and delivered a high quality finished water to all residences along the dead-end loop. Specific conclusions about water quality related to the continuous monitoring units, the grab samples, and the modified Robbins devices follow:

Continuous Monitoring Units

Real time continuous monitoring of many physico-chemical parameters in municipal water distribution systems is feasible for SCADA programs. Over a nine month period, four probes (pH, conductivity, water temperature and free chlorine) at the entrance and at the apex of the dead-end loop provided reliable readings with data recovery rates approaching 85 percent. Even though three other probes (turbidity, nitrate, ORP) proved unreliable during data collection in 1997, continuing advances in probe design and membrane technology should circumvent these shortcomings.

Due to problems with the sampling probe, data were insufficient to test whether or not ORP could be used as a viable surrogate for the concentration of disinfectant residual in this distribution system. As expected, pH was conserved between the entrance and the apex of the loop. However, there were statistically significant decreases in the mean and the variance of conductivity and chlorine. The drop in the mean conductivity is unexplained, but it implies a loss of dissolved ions across the loop. Moving from the entrance to the apex, the mean chlorine concentration dropped 33 percent, from 0.27 mg/L to 0.18 mg/L. Using an average travel time of 30 hours and a first-order reaction rate constant of $k=0.01/hr$, the predicted chlorine concentration at the apex is 0.20 mg/L. Discrepancies between observed and predicted mean chlorine concentrations can be attributed to uncertainties in the travel time, the reaction rate, and the reaction kinetics. The rate of chlorine loss increased with water temperature. No definitive reproducible correlation between travel time and chlorine residual could be established for *individual water parcels* moving through the loop. Owing to uncertainties in water quality inputs, to questions about residential water demands, and to changing flow regimes at the loop, the precise correspondence between parcels at the entrance and parcels at the apex remained elusive. Substantial heat exchange occurred at the dead-end loop. During winter, the network lost heat to the ground and hence water temperatures dropped. Conversely, during summer, the network gained heat and water temperatures rose.

Grab Samples

A total of 359 bulk fluid grab samples were taken from January to November 1997 at four locations in Milford's water distribution system. Each was analyzed for HPC, TC, AOC and TOC. All tests for total coliform were negative. HPC values fluctuated dramatically with location and with season, ranging from a summer low of 1 CFU/mL at the WTP to an autumn high of 8600 CFU/mL at the BPS. Bulk fluid HPC levels (in low nutrient R2A media) were consistently higher in the dead-end region than at the water treatment plant. By contrast, AOC readings (<1 mg-C/L) and TOC (0.4 to 1.5 mg/L) were stable and in the range expected for distribution systems fed by a groundwater source.

Spearman rank correlations showed significant associations between standard physico-chemical water quality parameters and bulk fluid HPC levels at the apex but not at the entrance to the loop. This finding underscores the realization that multiple confounding factors affect the viability of bacteria within a municipal water distribution system. The combined impact of warm temperatures and long residence times on the biological quality of finished water should be considered when designing and operating pipe networks and storage tanks. A slight drop in AOC levels without a concomitant increase in bulk fluid HPCs between the entrance and the apex may be indirect evidence of biofilm activity through the dead-end loop. These observation suggests that bacterial regrowth can occur even in a pipe network fed by low nutrient groundwater.

Modified Robbins Devices

Two MRDs, installed near the entrance and the apex of a dead-end loop, provided weekly biofilm samples during the period July 13 to November 18, 1997. Samples were collected in triplicate and plated on R2A, PCA and TSA-SB nutrient media. Within five weeks, the biofilm reached a pseudo steady condition inside the MRD with counts approaching 10^6 CFU/cm^2 on CPVC plugs plated in R2A. In contrast, the counts obtained from ductile iron plugs plated in R2A did not exceed 10^3 CFU/cm^2 . The biofilm growth potential observed for the low nutrient R2A media was similar at the entrance and the apex of the dead-end loop. However, the biofilm growth potential measured for the high nutrient agars (PCA and TSA-SB) showed a significant drop between the entrance and the apex of the loop. TSA-SB plate counts were not correlated with R2A plate counts. Hence, TSA-SB plate counts could complement counts from conventional R2A media in this distribution system.

Many of the species of bacteria found in water are not recovered by traditional plating procedures. However, determining levels of culturable HPC bacteria can provide useful information regarding the microbiological quality during the treatment, storage and distribution of drinking water. In this study statistically significant relationships were observed between physical and chemical water quality parameters and levels of HPC bacteria at the apex of the dead-end loop, particularly counts enumerated using R2A media. These types of data may provide useful information for water utilities involved in the operation and maintenance of water distribution systems.

RECOMMENDATIONS ON CONTINUOUS MONITORING

Design and operation of a water quality monitoring program should be tailored to the specific water utility. Parameters of importance will depend on the source quality, treatment train and applicable regulations. To identify which water quality parameters are most important for a particular distribution system, it is essential to have accurate seasonal analyses of the source water and treated water.

Probes designed for continuous operation can be expensive and sometimes problematic to maintain. For example, ion specific electrodes (ISE) work well in the laboratory. When deployed in the field, however, ISEs are susceptible to membrane fouling and calibration drift. If neglected, these factors will compromise the data and shorten the life span of the probe. Other standard probes (e.g., pH, temperature, conductivity) are more durable and reliable in a continuous flow environment. Accurate chlorine monitors are available, though it is important to correct for the hydrogen ion when the pH rises above 7.5.

CMUs require periodic lab tune-ups to ensure optimal field operation. The utility may need to train staff or hire new personnel dedicated to the continuous monitoring program. These individuals would calibrate instruments, clean membranes, perform tune-ups and troubleshoot minor glitches. During the planning stages, the utility should inform the equipment manufacturer about the overall objectives of the monitoring program. Often, the manufacturer can provide insights to improve instrument usage, to extend equipment life, or to upgrade current systems.

The location of the CMU in the pipe network is a critical factor in securing representative water quality data. Important logistical considerations include: site access, power availability, mainline accessibility, safety and security issues, and wastewater drainage (if the CMU requires a bleed line). Other relevant factors may include history of consumer complaints, proximity to storage tanks, frequency of stagnant flows and zones of multi-source blending. Hydraulic modeling of the water distribution system can identify other service areas that may warrant some attention (zones with frequent flow reversals, long travel times, unusual demand patterns, etc).

Sampling frequency should reflect operational objectives. If the intent is to quickly detect a water quality excursion, then sampling frequency should be very high. This is best accomplished with an on-line meter. However, some on-line probes (e.g., ORP and nitrate) are susceptible to severe drift and, therefore, are not suitable for drinking water applications. If the intent of the sampling program is to characterize finished water at various locations in the network (say for regulatory reporting), then the sampling frequency can be much lower.

Water quality observations spaced close in time have extremely high autocorrelations. With a sampling frequency of 6/hr at the study site, the autocorrelation often exceeded 99% (see Fig 7.10). High autocorrelation implies that consecutive water quality measurements contain redundant information. Consequently, the equivalent number of statistically independent observations is relatively small. Grab sample programs used by many utilities have the advantage of yielding independent water quality measurements. However, infrequent grab samples may not accurately characterize water quality variability in the network nor identify real-time excursions occurring in the distribution system.

Continuous monitoring units will generate extensive sets of water quality data. Exploratory data analysis and hypothesis testing of these field data can be performed automatically with various commercially available statistical software packages, with advanced data storage and retrieval functions. To aid in the data analysis and interpretation, the statistical program should offer convenient graphical options and report generation capabilities. Like any new initiative, some start-up time will be needed to iron out minor wrinkles. Once operational, however, real time water quality data from a CMU can provide many benefits to the water utility, including:

- ❖ Dynamic snap-shot on the status of finished water at strategic points in the network.
- ❖ Data to estimate travel times and dispersion rates in pipe networks.
- ❖ Data to assist in calibration of hydraulic and water quality models.
- ❖ Early warning of water quality excursions and anomalies in the network.
- ❖ Detection of trends and problems which require maintenance intervention.

Please note: There are no pages missing. The Appendices are on the CD enclosed on the back cover. They begin on p. 258 and end on p. 450.

REFERENCES

This section is divided into 2 parts:

(1) Bibliography on Residential Water Use (5 pages)

and

(2) Bibliography on Water Quality (8 pages)

Both bibliographies contain citations to all materials (journal papers, conference proceedings, textbooks, theses, regulations, etc) that have been referenced in this report. In addition, both bibliographies include citations to many of the references used in the MS theses by Trent Schade (1996) and Jason Carter (1998) which were performed as part of the work in this investigation.

BIBLIOGRAPHY on RESIDENTIAL WATER USE

Aher, A., A. Chouthai, L. Chandrasekar, W. Corpening, L.Russ, and B. Vijapur. 1991. *East Bay Municipal Utility District Water Conservation Study*. Oakland California. Report No. R 219. Steven Institute of Technology, 114 pages.

Anderson, D.L., D. Mulville-Friel, and W.L. Nero. 1993. The Impact of Water Conserving Fixtures on Residential Water Use Characteristics in Tampa, Florida. *Proceedings of Conserve 93*. Las Vegas, AWWA.

Anderson, J.S. and K.S. Watson. 1967. Patterns of household usage, *Journal of the American Water Works Association*, 59(10):1228-1237.

AWWA Research Committee. 1974. Water distribution research and applied development needs, *Journal of AWWA*, 66(6):385-390.

AWWA. 1975. Sizing Water Service Lines and Meters, *American Water Works Association, Manual M22*, 100 pages.

Barkdoll, B.D. 1992. *Modeling instantaneous residential demands in municipal water distribution systems*, MS thesis, University of Cincinnati, 100 pages.

Bobee, B., P. Boucher and H. Demard. 1980. A method for estimating peak water demand of multifamily residences, *Water Resources Bulletin*, 16(4):729-737.

Boland, J.R., A.A. McPhall and E.M. Opitz. 1990. *Water Demand of Detached Single-Family Residences*, prepared for Metropolitan Water District of Southern California, Los Angeles, California, 56 pages with appendices.

Bowen, P.T., J.F. Harp, J.W. Baxter, R.D.Shull. 1993. *Residential Water Use Patterns*, AWWA Research Foundation, Denver, CO, 105 pages.

Bras, R.L. 1990. *Hydrology*, Addison Wesley, Reading, MA, 643 pages.

Brown and Caldwell Consulting Engineers. 1984. *Residential Water Conservation Projects - Summary Report*. HUD-PDR-903. Dept Housing & Urban Development, Wash, D.C.

Buchberger, S.G. 1995. Conditional frequency analysis of autocorrelated lake levels, *ASCE Journal of Water Resources Planning and Management*, 121(2):158-170.

Buchberger, S. G. and L. Wu. 1995. A model for instantaneous residential water demands, *ASCE Journal of Hydraulic Engineering*, 121(3):232-246.

Buchberger, S. G. and G.J. Wells. 1996. Intensity, duration and frequency of residential water demands, *ASCE J of Water Resources Planning and Management*, 122(1):11-19.

Buchberger, S.G. and T.G. Schade. 1997. Poisson rectangular pulse model for residential water use, *Proceedings XXVII IAHR Congress*, San Francisco, California, Aug 10-15, 1997

Buchberger, S.G., J.T. Carter and Y.H. Lee. 1998. Travel times in deadend mains. 1998. *Proceedings ASCE 25th Annual Conference on Water Resources Planning and Management*, Chicago, Illinois, June 6-10, 1998

Buchberger, S.G. and Y. Lee. 1999. Evidence supporting the Poisson pulse hypothesis for residential water demands. *Proceedings CCWI: International Conference on Computing and Control for the Water Industry*, Exeter, United Kingdom, September 13-15, 1999

Buchberger, S.G. and G.Nadimpalli. 2003 New method to estimate leakage rates in drinking water distribution systems, *Proceedings of the International Conference on Pumps, Electro-mechanical Devices and Systems Applied to Urban Water Management (PEDS) 2003*, Valencia, Spain, April 22-25, 2003.

Butler, D. 1991. A small-scale study of wastewater discharges from domestic appliances, *Journal of the Institution of Water Engineering and Management*, 5(4):178-185.

Carter, J.T. 1998. Biofilms and Planktonic Bacteria in a Drinking Water Distribution System, MS Thesis, University of Cincinnati, 89 pages.

Chan, W.Y. 1981. Estimation of water demand in apartment buildings, *Plumbing Engineer*, 9(6)26-29.

Chan, W.Y. 1985. An analytical method for estimating water demand in buildings, *ASPE Journal of Engineered Plumbing*, 1(2):131-144.

Chan, W.Y. and L.K. Wang. 1980. Reevaluating Hunter's model for residential water demand, *Journal of the American Water Works Association*, 72(8):446-449.

Clark, R.M. and J.A. Coyle. 1990. Measuring and modeling variations in distribution system water quality, *Journal of the American Water Works Association*, 82(8):46-53.

Cronauer J.A. and J.S. Gidley. 1985. Time series analysis of hourly domestic water demand, *in Proc of Computer Applications in Water Resources*, Buffalo, NY, pp 391-400..

Danielson, L.E. 1979. An analysis of residential demand for water using micro time-series data, *Water Resources Research*, 15(4):763-767.

DeOreo, W.B., J.P. Heaney and P.W.Mayer. 1996. Flow trace analysis to assess water use, *Journal AWWA*, 88(1):79-90.

Dziegielewski, B. and J.J. Boland. 1989. Forecasting urban water use: the IWR-MAIN model, *Water Resources Bulletin*, 25(1):101-109.

Erlang, A.K. 1917-18. Solution of some problems in the theory of probabilities of significance in automatic telephone exchanges, *Post Office Electrical Engrs*, J., 10, 189-197.

Feth, J.H. 1973. *Water Facts and Figures for Planners and Managers*, U.S. Geological Survey Circular 601-I, Arlington, VA, 30 pages.

Flack, J.E. 1982. *Urban Water Conservation: Increasing Efficiency-in-Use Residential Water Demand*, ASCE, New York, 99 pages.

Gilbert, J.B., W.J. Bishop, and J.A. Weber. 1990. Reducing water demand during drought years. *Journal AWWA*. 82(5):34-39.

Gross, D. and C.M. Harris. 1985. *Fundamentals of Queueing Theory*, John Wiley and Sons, New York, 587 pp.

Hanke, S.H. and L. de Maré, 1984. Municipal water demands, in *Modeling Water Demands*, edited by J. Kindler and C.S. Russell, Academic Press, pp 149-240.

Helms, B.P. and J.F. Vallery. 1980. *Treated Water Demand and the Economics of Regionalization, Volume 1 The Residential Demand for Treated Water*, EPA-600/2-80-162, Cincinnati, OH, 83 pages.

Hirtzel, C.S., J.E. Quon, and R.B. Corotis. 1982. Mean of autocorrelated air quality measurements, *ASCE Journal of Environmental Engineering*, 108(3):488-501.

Howe, C.W. and E.P. Linaweaver Jr. 1967. The impact of price on residential water demand and its relation to system design and price structure, *Water Resources Research*, 3(1):13-32.

Hunter, R.B. 1940. *Methods of Estimating Loads in Plumbing Systems*, Report BMS65, National Bureau of Standards, U.S. Department of Commerce, 21 pages.

Kennedy, M.S., S. Sarikelle and K. Suravallop. 1991. Calibrating hydraulic analyses of distribution systems using fluoride tracer studies, *Journal AWWA*, 83(7):54-59.

Kleinrock, L. 1975. *Queueing Systems, Vol 1: Theory*, Wiley Interscience, NY, 417 pg.

Kleinrock, L. 1976. *Queueing Systems, Vol 2: Computer Applications*, Wiley Interscience, New York, 549 pages.

Konen, T.P., and W.Y. Chan. 1979. An investigation of the frequency and duration of use parameter in sizing water distribution systems, American Society of Plumbing Engineers Research Foundation, Inc, ASPERFI Report No. 102-79, Stevens Institute of Technology, Hoboken, NJ.

Kramer, A.K., M.H. Griese and A. Shanker. 1994. Experience with calibrating a water quality model, in *Proceedings AWWA Engineering Design Conference*, Cincinnati, Ohio, pp 179-191.

Kubik, H.E. 1992. Annual extreme lake elevations by total probability theorem, *Journal of Great Lakes Research*, 18(1):202-206.

Lee,Y.H. and S.G. Buchberger. 1999. Modeling indoor and outdoor residential water use as the superposition of two Poisson rectangular pulse processes. *Proceedings ASCE 26th National Conference on Water Resources Planning and Management*, Tempe, Arizona, June 6-9, 1999

Lewis, E.E. 1996. *Introduction to Reliability Engineering*, Wiley, NY, 435 pages.

Linaweaver Jr., F.P., J.C. Geyer and J.B. Wolff. 1966. *Residential Water Use, Final and Summary Report*, Johns Hopkins University, 79 pages with appendices.

Lingireddy, S., Wood, D.J., and Nelson, A. (1998) Modified Network Model for Rural Water Systems, Journal of Water Resources Planning and Management, 124(5):296-299.

Litke, D.W. and L.F. Kauffman.1993. *Analysis of residential use of water in the Denver metropolitan area, Colorado 1980-87*, USGS-WRI Report 92-4030.

Maddaus, W.O. 1987. *Water Conservation*, American Water Works Association, Denver, Colorado, 93 pages.

Maddaus, W.O. 1987. *The Effectiveness of Residential Water Conservation Measures*, Journal AWWA, 79(_):52-58.

Maidment, D.R., S.P. Miaou, D.N. Nvule and S.G. Buchberger. 1985. *Analysis of Daily Water Use in Nine Cities*, Center for Research in Water Resources, CRWR Report 201, University of Texas, 67 pages.

Mayer, P.W., W.B. DeOreo, E.M. Opitz, J.C. Kiefer, W.Y.Davis, B.Dziegielewski, and J.O. Nelson, 1999. Residential End Uses of Water, American Water Works Association, Denver, CO, 310 pages.

Morgan, W.D. and J.C. Smolen. 1976. Climatic indicators in the estimation of municipal water demand, *Water Resources Bulletin*, 12(3):511-518.

National Oceanic and Atmospheric Administration. 1974. *Climates of the States, Volume 1-Eastern States*, Water Information Center, Port Washington, New York.

Newell, G.F. 1982. *Applications of Queueing Theory*, Chapman and Hall, 303 pages.

Orndorff, J.R. 1966. Domestic Water Use Differences in Individual Well and Public Water Supplies, *Report III on Phase 2 of the Residential Water Use Research Project*, Johns Hopkins University, 98 pages.

Parzen, Emanuel. 1962 *Stochastic Processes*. Holden-Day, Inc. San Francisco, 324 pages.

Perry, P.F. 1981. Demand forecasting in water supply networks, *ASCE Journal of Hydraulic Engineering*, 107(9):1077-1087.

Ross, S.M. 1985. *Introduction to Probability Models*, Academic Press, Orlando, 502 pages.

Rothstein, E. 1992. Water demand monitoring in Austin, Texas, *Journal of the American Water Works Association*, 84(10)52-58.

Russac, D.A.V., K.R. Rushton and R.J. Simpson. 1991. Insights into domestic demand from a metering trial, *Journal of the Institution of Water Engineering and Management*, 5(6):342-351.

Schade, T. G. 1996. *Water Demands and Travel Times in a Residential Dead-End Loop*, MS Thesis, University of Cincinnati, 64 pages.

Schlumberger Inc. (1992) Neptune T-10 Water Meter Specifications. 4 pages.

Shvartser, L., U. Shamir and M. Feldman. 1993. Forecasting hourly water demands by pattern recognition approach, *ASCE Journal of Water Resources Planning and Management*, 119(6):611-627.

Small, M.J. and J.R. Mular. 1987. Long-term pollution degradation in the unsaturated zone with stochastic rainfall infiltration, *Water Resources Research*, 23(12):2246-2256.

Solley, W.B., E.B. Chase and W.B. Mann. 1983. *Estimated Use of Water in the United States in 1980*, U.S. Geological Survey Circular 1001, Alexandria, VA. 56 pp.

Sullivan, J.P. and E.M. Speranza. 1992. Proper meter sizing for increased accountability and revenue, *Journal of the American Water Works Association*, 84(7):53-60.

Swamee, P.K., V. Kumar and P. Khanna. 1973. Optimization of dead end water distribution systems, *ASCE Journal of Environmental Engineering*, 99(2):123-134.

Walski, T.M. 2001. The wrong paradigm–Why water distribution system optimization doesn't work, *ASCE Journal of Water Resources Planning and Management*, 127(4):203-205.

Wells, G. 1994. *Statistical characteristics of residential water demands*, MS Thesis, University of Cincinnati. 92 pages.

Wu, L. 1996. *Stochastic model of flow in the periphery of a water distribution system*, PhD Dissertation, University of Cincinnati. 191 pages.

BIBLIOGRAPHY on WATER QUALITY

Allen, M.J. 1980. The occurrence of microorganisms in water main encrustations. *Journal of the American Water Works Association*, 72(11):614.

Antoun, E.N., J.E. Dykseen and D.J. Hiltebrand 1999. Unidirectional flushing: a powerful tool. *Journal of the American Water Works Association*, 91(7):62-71.

APHA, AWWA, WEF (American Public Health Association, American Water Works Association, and Water Environment Federation). 1994. *Standard Methods for the Examination of Water and Wastewater*. 19th ed. Denver, Colo.: APHA.

AWWARF (American Water Works Association Research Foundation). 1996. *Tracer Studies in Water Treatment Facilities: A Protocol and Case Studies*. Denver, Colo.: AWWARF.

Axworthy, D.H. and B.W. Karney. 1996. Modeling low velocity/high dispersion in water distribution systems, *ASCE Journal of Water Resources Planning and Management*, 122(3):218-221.

Bezanson, G., S. Burbridge, D. Haldane, and T. Marrie. 1992. In situ colonization of polyvinyl chloride, brass, and copper by Legionella pneumophila, *Canadian Journal of Microbiology*, 38:328-330.

Bhamidimarri, S. M. Rao and T. T. See. 1992. Shear Loss Characteristics of an Aerobic Biofilm, *Water Science and Technology*, 26(3-4):595-600.

Biolog 1993. *GN MicroPlate Instructions for Use* Biolog, Inc. pp. 10.

Bird, R.B., W.E. Stewart and E.N Lightfoot. 1960. *Transport Phenomena*, John Wiley and Sons, Inc. New York

Biswas, P., C. Lu, and R.M. Clark. 1993. A model for chlorine concentration decay in pipes, *Water Research*, 27(12):1715-1724.

Block, J. C. 1992. Biofilms in Drinking Water Distribution Systems, *In: Biofilms-Science and Technology*, Melo, L. F., T. R. Bott, M. Fletcher, and B. Capdeville, Kluwer Academic Publishers, Netherlands, 469-485.

Boulos, P. F., Altman T., Jarrige, P. A., and Collevati, F. 1995. Discrete simulation approach for network-water-quality models, *ASCE Journal of Water Resources Planning and Management*, ASCE, 121(1), 49-60.

Breed, R. S., E. G. D. Murray, N. R. Smith. 1957. *Bergey's Manual of Determinative Bacteriology*, The Williams & Wilkins Co., Baltimore, MD, 1094 pages.

Campos-Herrero, M. I., A. Bordes, H. Rodriguez, A. Perera, B. Gonzalez, and A. Conde 1997. *Pseudomonas stutzeri* Community-Acquired Pneumonia Associated with Empyema: Case Report and Review, *Clinical Infectious Diseases*, 25:325-326.

Carter, J.T., E. W. Rice, S. G. Buchberger, and Y. Lee. 2000. Relationships Between Levels of Heterotrophic Bacteria and Water Quality Parameters in a Drinking Water Distribution System. *Water Research* 34(5), 1495-1502.

Characklis, W. G. and K. C. Marshall 1990. *Biofilms*, John Wiley & Sons, Inc., New York, 795 pages.

Clark, R.M. 1998. Chlorine demand and TTHM formation kinetics: a second-order model, *ASCE Journal of Environmental Engineering*, 124(1):16-24.

Clark, R.M., F. Abdesaken, P.F. Boulos and R.E.Mau. 1996. Mixing in distribution system storage tanks: its effect on water quality, *ASCE Journal of Environmental Engineering*, 122(9):814-823.

Clark, R.M., L.A. Rossman, and L.J. Wymer. 1995. Modeling distribution system water quality: regulatory implications, *ASCE Journal of Water Resources Planning and Management*, 121(6), 423-428.

Clark, R.M., W.M. Grayman, J.A. Goodrich, Deininger R.A., and K. Skov. 1994. Measuring and modeling chlorine propagation in water distribution systems, *ASCE Journal of Water Resources Planning and Management*, 120(6), 871-887.

Clark, R.M., W.M. Grayman, R.M. Males and A.F. Hess, 1993. Modeling contaminate propagation in drinking water distribution systems, *ASCE Journal of Environmental Engineering*, 119(2):349-364.

Clark, R.M., and J.A. Coyle. 1990. Measuring and modeling variations in distribution system water quality, *Journal AWWA*, 82(8):46-53.

Clark, R.M., W.M. Grayman, and R.M. Males. 1988. Contaminant propagation in distribution systems, *ASCE Journal of Environmental Engineering*, 114(4):454-466.

Costas, M., B. Holmes, S. L. W. On, and D. E. Stead. 1992. Identification of Medically Important Pseudomonas Species Using Computerized Methods, *In: Identification Methods in Applied and Environmental Microbiology*, Blackwell Scientific Publications, pp. 1-19.

Costerton, J. W., Z. Lewandowski, D. E. Caldwell, D. R. Korber, and H. M. Lappin-Scott. 1995. Microbial Biofilms, *Annual Review in Microbiology*, 49:711-745.

Costerton, J. W. and R. T. Irvin. 1981. The Bacterial Glycocalyx in Nature and Disease, *Annual Review of Microbiology*, 35:299-324.

Crosby, G.B., A.M. Kramer and A. Shanker. 1994. Using a computer model to assess water quality in a water distribution system — a case history, in *Proceedings AWWA Engineering Design Conference*, Cincinnati, Ohio, pp 179-191.

Deininger, R.A., R.M. Clark, A.F. Hess, and E.V. Bernstam. 1992. Animation and visualization of water quality in distribution systems, *Journal AWWA*, 84(7):48-52.

De Zuane, J. 1990. *Handbook of Drinking Water Quality: Standards and Control*. New York.: Van Nostrand Reinhold.

Donlan, R. M. and W. O. Pipes. 1988. Selected Drinking Water Characteristics and Attached Microbial Population Density, *AWWA Journal of Research and Technology*, 80:70-76.

Dukan, S., Y. L. P. Piriou, F. Guyon, and P. Villon. 1996. Dynamic Modeling of Bacterial Growth in Drinking Water Networks, *Water Research*, 30(9):1991-2002.

Elton, A. L.F. Brammer and N.S. Tansley. 1995. Water quality modeling in distribution networks, *Journal AWWA*, 87(7):44-52.

Federal Register, Thursday June 29, 1989, Vol 54, No 124, pg 27567.

Geldreich, E. E. 1996. *Microbial Quality of Water Supply in Distribution Systems*, Lewis Publishers CRC Press, Inc., Boca Raton, Florida, 504 pages.

Gibbs, R. A., J. E. Scutt, and B. T. Croll. 1993. Assimilable Organic Carbon Concentrations and Bacterial Numbers in a Water Distribution System, *Water Science and Technology*, 27(3-4):159-166.

Grayman, W.M., R.M. Clark and R.M. Males. 1988. Modeling distribution system water quality: dynamic approach, *ASCE Journal of Water Resources Planning and Management*, 114(3):295-312.

Haas, C. N. and J. B. Rose. 1996. Distribution of Cryptosporidium Oocysts in a Water Supply, *Water Research*, 30(10):2251-2254.

Hargesheimer, E., O. Conio and J. Popovicova (editors). 2002. Online Monitoring for Drinking Water Utilities - A Cooperative Research Report, AWWA Research Foundation and CRS Proaqua, Denver, Colorado. 425 pages.

Hermanowicz, S. W. and F. L. Filho. 1992. Disinfection and Attachment of Bacterial Cells, *Water Science and Technology*, 26(3-4):655-664.

Herson, D. S. and K. H. Baker. 1982. New Approaches to Measuring Total Bacterial Populations in Water Supplies, *Journal American Water Works Association*, 74:537-539.

Holder, G. A. 1992. Analyses of Rate Data for Biofilm Studies, *Water Science and Technology*, 26(3-4):713-722.

Holt, J.G., N.R. Krieg, P.A. Sneath, J.T. Staley and S.T. Williams. 1994. *Bergey's Manual of Determinative Bacteriology 9th ed.*, Williams & Wilkins, Baltimore, 787 pages

Islam, M.R., M.H. Chaudhry and R.M. Clark. 1997. Inverse modeling of chlorine concentration in pipe networks under dynamic conditions, *ASCE Journal of Environmental Engineering*, 123(10):1033-1040.

Jain, D. K., S. Stroes-Gascoyne, M. Providenti, C. Tanner, and I. Cord. 1997. Characterization of Microbial Communities in Deep Groundwater from Granitic Rock, *Canadian Journal of Microbiology*, 43:272-283.

James M. Montgomery, Consulting Engineers, Inc. 1985. *Water Treatment Principles and Design*. : John Wiley & Sons, Inc.

Kapoor, V., C.T. Jafvert and D.A. Lyn 1998.Experimental study of a bimolecular reaction in Poiseuille flow, *Water Resources Research*, 34(8):1997-2004.

Keevil, C.W., J. Rogers, and J.T. Walker. 1995. Microorganisms at Large: Potable-Water Biofilms, *Microbiology Europe*, 3(6):10-15.

Keevil, C. W., C. W. Mackerness, and J. S. Colbourne. 1990. Biocide Treatment of Biofilms, *International Biodeterioration*, 26:169-179.

Kerneïs, A., F. Nakache, A. Deguin, and M. Feinberg. 1995. The Effect of Water Residence Time on the Biological Quality in a Distribution Network, *Water Research*, 29(7):1719-1727.

Kersters, I., L.Van Vooren, L. Verschuere, L. Vauterin, A. Wouters, J. Mergaert, J. Swings and W. Verstraete. 1997. Utility of the biolog system for characterization of heterotrophic microbial communities. *System. Applied Microbiology*. 20:439-447.

Kim, Y.H. 1993. Evaluation of Redox Potential and Chlorine Residual as a Measure of Water Disinfection. *Proceeding to the 5th International Water Conference*. 108-113.

Kuch, A. and I. Wagner. 1983. A mass transfer model to describe lead concentrations in drinking water. *Water Research*. 17(10):1303-1307.

Lahti, K. 1993. Microbial quality of drinking water in some Finnish distribution systems. Water Science and Technology, 27(3-4):151-154.

LeChevallier, M. W., N. Welsh, and D. B. Smith. 1996. Full-Scale Studies of Factors related to Coliform Regrowth in Drinking Water, *Applied and Environmental Microbiology*, 62(7):2201-2211.

LeChevallier, M.W., Lowry, C.D., Lee, R.G., and Gibbon, D.L. 1993. Examining the Relationship Between Iron Corrosion and the Disinfection of Biofilm Bacteria. *Jour. AWWA*, 85(7):111-123.

LeChevallier, M. W., W. Schulz, and R. G. Lee. 1991. Bacterial Nutrients in Drinking Water, *Applied and Environmental Microbiology*, 57(3):857-862.

LeChevallier, M. W., C. D. Lowry, and R. G. Lee. 1990. Disinfecting Biofilms in a Model Distribution System, *AWWA Journal of Research and Technology*, 82(7):87-99.

LeChevallier, M. W. 1990. Coliform Regrowth in Drinking Water: A Review. *AWWA Journal of Research and Technology*, 82(11):74-86.

LeChevallier, M. W., T. M. Babcock, and R. G. Lee. 1987. Examination and Characterization of Distribution System Biofilms, *Applied and Environmental Microbiology*, 53(12):2714-2724.

Lee, S.H., J.T. O'Conner, and S.K. Banerji. 1980. Biologically mediated corrosion and its effects on water quality in distribution systems, Journal of the American Water Works Association, 72(11):636-645.

Lieu, N.I.,Wolfe, R.L., and Means, E.G. 1993. *Optimizing Chloramine Disinfection for the Control of Nitrification*. Jour. AWWA, 85(2):84-90.

Liou, C.P. and J.R. Kroon. 1987. Modeling the propagation of waterborne substances in distribution networks, *Journal AWWA*, 79(11):54-58.

Lytle, D.A. and M.R. Schock. 1996. *Stagnation time, composition, pH and orthophosphate effects on metal leaching from brass*, EPA/600/R-96/103, USEPA, Cincinnati, Ohio, 171 pages.

MacLeod, R. A., P. R. MacLeod, and M. Berthelet. 1997. On the Origin of Variants of the Marine Bacterium *Deleya aesta* 134 Able to Grow at Low Na^+ Concentration, *Canadian Journal of Microbiology*, 43:868-878.

Males, R.M., W.M. Grayman and R.M. Clark. 1988. Modeling water quality in distribution systems, *ASCE Journal of Water Resources Planning and Management*, 114(2):197-209.

Manz, W., U. Szewzyk, P. Ericsson, R. Amann, K.-H. Schleifer, and T.-A. Stenstrom. 1993. In Situ Identification of Bacteria in Drinking Water and Adjoining Biofilms by Hybridization with 16S and 23 S rRna-Directed Flourescent Oligonucleotide Probes, *Applied and Environmental Microbiology*, 59(7):2293-2298.

Marshall, K. C. 1988. Adhesion and Growth of Bacteria at Surface at Surfaces in Oligotrophic Habitats, *Canadian Journal of Microbiology*, 34:503-506.

Means, E. G., L. Hanami, H. F. Ridgway, and B. Olson. 1981. Evaluating mediums and plating techniques for enumerating bacteria in water distribution systems, *Journal of the American Water Works Association*, 73:585-590.

Mercurio, J. W. 1997. *Delineation of the Wellhead Protection Area and Characterization of the Background Water Quality for the City of Milford, Ohio*, M.S. Thesis, University of Cincinnati.

Mikels, M.S. 1992. Characterizing the Influence of Surface Water on Water Produced by Collector Wells. *Jour. AWWA*, 84(9):77-84.

Montgomery Consulting Engineers. 1985. *Water Treatment Principles and Design*, John-Wiley and Sons, Inc, New York, NY, 696 pp.

Mueller, R. F. 1996. Bacterial Transport and Colonization in Low Nutrient Environments, *Water Research*, 30(11):2681-2690.

Ohio Administrative Code (2002), Primary Standards for Ohio Public Water Supplies, Chapter 3745-83, revised March 1, 2002.

Owens, J.H., G. Berg, R.S. Summers, E.W. Rice, D.J. Reasoner, and R.M. Clark, 1994. Survival of a fraction of heterotrophic bacteria indigenous to the Ohio River after prolonged exposure to free chlorine. *In Proceedings AWWA Annual Conference*, June 19-23, 1994, New York, pp 357-366.

Pedersen, K. 1990. Biofilm development on stainless steel and PVC surfaces in drinking water. *Water Research.* 24(2):239-243

Prévost, M., A. Rompré, H. Baribeau, J. Coallier, and P. Lafrance. 1997. Service Lines: their effect on microbiological quality, *Journal AWWA*, 89(7):78-91.

Rajeshwari, R., J. Yashitola, A. P. K. Reddy, and R. V. Sonti. 1997. Characteristics of Stationary-Phase Variation Affecting Virulence in *Xanthomonas oryzae pv. oryzae*, *Canadian Journal of Microbiology*, 43:862-867.

Reasoner, D. J. and E. E. Geldreich. 1985. A New Medium for the Enumeration and Subculture of Bacteria from Potable Water, *Applied and Environmental Microbiology*, 49(1):1-7.

Reasoner, D.J. and E.W. Rice (1993) Bacterial Nutrients and Distribution Drinking Water Quality, in Strategies and Technologies for Meeting SDWA Requirements, edited by Clark, R.M. and R.S. Summers, Technomic Publishing Co., Inc, pp. 305-317.

Rice, E.W., Scarpino, P.V., Reasoner, D.J., Logsdon, G.S., and Wild, D.K. 1991. Correlation of Coliform Growth Response With Other Water Quality Parameters. *Jour. AWWA*, 83(7):98-102.

Rossman, L.A., P.F. Boulos and T. Altman. 1993. Discrete volume-element methods for network water-quality models, *ASCE Journal of Water Resources Planning and Management*, 119(5):505-517.

Rossman, L.A., R.M. Clark and W.M. Grayman. 1994. Modeling chlorine residuals in drinking water distribution systems, *ASCE Journal of Environmental Engineering*, 120(4):803-820.

Rossman, L.A., and P.F. Boulos. 1996. Numerical methods for modeling water quality in distribution systems: a comparison, *ASCE Journal of Environmental Engineering*, 122(2):137-146.

Rusin, P. A., J. B. Rose, and C. P. Gerba. 1997. Health Significance of Pigmented Bacteria in Drinking Water, *Water Science and Technology*, 35(11-12):21-27.

Schock, M.R., D.A. Lytle and J.A. Clement. 1995. Effect of pH, DIC, orthophosphate and sulfate on drinking water cuprosolvency, EPA/600/R-95/085, USEPA, Cincinnati, Ohio, 95 pages.

Servais, P. A. Anzil, and C. Ventresque. 1989. Simple method for determination of biodegradable dissolved organic carbon in water. *Applied and Environmental Microbiology*, 55(10):2732-2734.

Shamir, U. and C.D.D. Howard. 1991. Topics in modeling water quality in distribution systems, *Proc. Water Quality Modeling in Distribution Systems*, AWWA Research Foundation, Feb. 3-5, Cincinnati, OH, 311-321.

Sharp, W.W., J. Pfeffer, and M. Morgan. 1991. Insitu chlorine decay rate testing, *Proc. Water Quality Modeling in Distribution Systems*, AWWA Research Foundation, Feb. 3-5, Cincinnati, OH, 311-321.

Sheskin, D.J. 1997. *Handbook of Parametric and Nonparametric Statistical Procedures*, CRC Press, 719 pages.

Sibille,I., L. Mathieu, J.L. Paquin, D. Gatel, J.C. Block. 1997. Microbial characteristics of a distribution system fed with nanofiltered drinking water, *Water Research*, 31(9):2318-2326.

Skadsen, J. 1993. Nitrification in a Distribution System. *Jour. AWWA*, 85(7):99-103.

Skerman, V. B. D. 1967. *A Guide to the Identification of the Genera of Bacteria, 2nd ed.,* The Williams & Wilkins Co., Baltimore, MD, pp. 303.

Spino, D. F. 1985. Characterization of Dysgonic, Heterotrophic Bacteria from Drinking Water, *Applied and Environmental Microbiology*, 50(5):1213-1218.

Smith, D.B., A.F. Hess and D. Opheim. 1989. Control of distribution system coliform regrowth, pp 1009-1029, *Proceedings Water Quality Technology Conference*, Phil, PA.

Smith, D.B., A.F. Hess and S.A. Hubbs. 1990. Survey of distribution system coliform occurrence in the United States, pp 1103-1116, *Proceedings Water Quality Technology Conference*, San Diego, California.

Snoeyink, V. L. and D. Jenkins. 1980 *Water Chemistry*. J. Wiley & Sons, New York City, New York.

Stumm, W. and Morgan, J.J. 1981. *Aquatic Chemistry: An Introduction Emphasizing Chemical Equilibria in Natural Waters*. 2nd ed. :John Wiley & Sons, Inc.

Tchobanoglous, G. and Schroeder, E.D. 1985. *Water Quality: Characteristics, Modeling and Modification*. : Addison Wesley.

The Prokaryotes A Handbook on the Biology of Bacteria: Ecophysiology, Isolation, Identification, Applications (1992) Springer-Verlag, New York, Inc. Volumes. I-IV.

Turakhia, M. H., K. E. Cooksey, and W. G. Characklis. 1983. Influence of a Calcium-Specific Chelant on Biofilm Removal, *Applied and Environmental Microbiology*, 46(5): 1236-1238.

United States Environmental Protection Agency. 1997. *Water on Tap: A consumer's guide to the nation's drinking water*, EPA 815-K-97-002.

van der Kooij, D., H. R. Veenendaal, C. Baars-Lorist, D. W. van der Klift, and Y. C. Drost. 1995. Biofilm Formation on Surfaces of Glass and Teflon Exposed to Treated Water, *Water Research*, 29(7):1655-1662.

van der Kooij, D., A. Visser, and W. A. M. Hijnen. 1982. Determining the Concentration of Easily Assimilable Organic Carbon in Drinking Water, *Journal of the American Water Works Association*, 74:540-545.

van der Wende, E., W. G. Characklis, and D. B. Smith. 1989. Biofilms and Bacterial Drinking Water Quality, *Water Research* 23(10):1313-1322.

van der Wende, E., W. G. Characklis, and J. Grochowski. 1988. Bacterial Growth in Water Distribution Systems, *Water Science and Technology*, 20(11-12):521-524.

Vasconcelos, J.J. et al., 1997. Kinetics of chlorine decay, *Jour AWWA*, 89(7):54-65.

Wolfe, R.L., Means, E.G., Davis M.K., and Barrett, S.E. 1988. Biological Nitrification in Covered Reservoirs Containing Chloraminated Water. *Jour. AWWA*, 80(9):109-114.

Yagupsky, P. and R. Dagan. 1997. *Kingella kingae*: An Emerging Cause of Invasive Infections in Young Children, *Clinical Infectious Diseases*, 24:860-866.

ABBREVIATIONS

ANOVA	analysis of variance
AOC	assimilable organic carbon
APHA	American Public Health Association
ATI	Analytical Technologies, Inc.
Awwa	American Water Works Association
AwwaRF	Awwa Research Foundation
BA	high nutrient blood agar
BDOC	biodegradable dissolved organic carbon
BGP	biofilm growth potential
BPS	booster pump station
°C	degrees Celsius
$C_j(t)$	number of customers arriving in residence j during interval (0,t)
CFU	colony forming units
Cl_2	chlorine
cm	centimeter
CMU	continuous monitoring unit
Cov[X,Y]	covariance between X and Y
CPVC	chlorinated polyvinyl chloride
CSTR	continuously stirred tank reactor
d	duration of water use pulse
D_j	diameter of pipe in link j
DC	direct count of bacteria
DLA	deadend loop apex
DMA	demand monitoring area
DO	dissolved oxygen
DOC	dissolved organic carbon
E[.]	expected value
e.g.	for example
F*	F-test statistic
°F	degrees Fahrenheit
$F_j(q,t)$	cdf of flows through link j at time t
FCV	flow control valve
gpm	gallons per minute
gpcd	gallons per capita-day
GA-1	global average one, (simple arithmetic average)
GA-2	global average two, (weighted average)

ABBREVIATIONS (cont'd)

G -	Gram negative
$G_j(q,k)$	cdf of flows into residence j with k busy servers
GIS	geographic information system
H*	Kruskall-Wallis one-way ANOVA test statistic
HPC	heterotrophic plate count
ISE	ion specific electrode
i.e.	that is
j	index for homes in block, $j=1,2,\ldots,n$
JTU	Jackson turbidity unit
k	index for busy servers, $k=0,1,\ldots$
$K_j(t)$	number of busy servers in residence j at time t
L	liter
Lpm	liters per minute
L_N	volume of water in pipe link N
MA	moving average
mg	milligram
MG	million gallons
MRD	modified Robbins device
mV	millivolt
N	number of homes in block
N	nitrogen
NO_3	nitrate
NOM	natural organic matter
NPRP	nonhomogeneous Poisson rectangular pulse
NSF	National Science Foundation
NTU	nephelometric turbidity unit
ORP	oxidation reduction potential
p	significance level
P_0	Probability of stagnation (flow is zero)
$P_j(k,t)$	probability of k busy servers in residence j at time t
$p_j(t)$	probability that any server in residence j is busy at time t
PAC	Project Advisory Committee
PCA	plate count agar
pdf	probability density function

ABBREVIATIONS (cont'd)

PFR	plug flow reactor
PoE	point of entry (service line into a residence)
PoU	point of use (appliance/server inside a residence)
PoN	point of no return (entry to a dead-end main)
ppcd	pulses per capita-day
ppm	parts per million
PRP	Poisson rectangular pulse
PVC	polyvinyl chloride
$Q_j(t)$	flow into residence j at time t
q	flow rate
R*	Spearman's Rho test statistic
Re	Reynolds number
REUW	residential end uses of water
R2A	low nutrient agar
SCADA	supervisory control and data acquisition
SDWA	Safe Drinking Water Act
SERP	single equivalent rectangular pulse
SORB	string of random blocks
S.U.	standard units (pH)
t	time
t_0	start time for a water demand
t_n	end time for a water demand
T	pulse duration
T*	Student's t-test statistic
TC	total coliform analysis
TOC	total organic carbon
TSA-SB	tryptic soy agar with 5% sheep's blood
U*	Mann-Whitney test statistic
USEPA	U.S. Environmental Protection Agency
V	pulse volume
$V_j(t)$	pulse volume at servers in residence j
Var[.]	variance
WST	Wallace Street water storage tank
WTP	water treatment plant
Y	pulse intensity
$\bar{Y}_k$	kth average intensity in a smoothed water use

ABBREVIATIONS (cont'd)

η	amplification factor for water demand process, $\eta = \rho/[1-\exp(-\rho)]$
ϕ	average volume of water pulse
ψ^2	variance of water pulse volume
α	mean of water pulse intensity at a busy server
β^2	variance of water pulse intensity at a busy server
τ	average duration of water pulse
ω^2	variance of water pulse duration
$\Lambda_j(t)$	mean of Poisson arrival process over interval (0,t) at residence j
λ	arrival rate for customer in a residence
$\lambda_j(t)$	time dependent Poisson arrival rate for customers at residence j
μ	mean service rate in a residence
ρ	utilization factor for residence, $\rho = \lambda/\mu = \lambda\tau$
$\rho(t)$	time dependent utilization factor
*	denotes a quantity summed over a block of n homes.

See also Table 6.1 in the main report.

IWA Publishing's authorised EU representative for General Product Safety Regulations is Diane D'Arras, 15 rue Duret, 75116 Paris, France, e-mail: safety@iwap.co.uk.

Printed and bound by CPI Group (UK) Ltd, Croydon, CR0 4YY
17/09/2026
02218799-0007